Rethinking Science Education

Philosophical Perspectives

A volume in
Science & Engineering Education Sources

Series Editor
Calvin S. Kalman, *Concordia University*

Science & Engineering Education Sources

Calvin S. Kalman, Series Editors

College Teaching and the Development of Reasoning (2009)
edited by Robert G. Fuller, Thomas C. Campbell,
Dewey I. Dykstra, Jr., and Scott M. Stevens

Using and Developing Measurement Instruments in Science Education: A Rasch Modeling Approach (2010)
by Xiufeng Liu

Rethinking Science Education: Philosophical Perspectives (2014)
by Roland M. Schulz

Dedication

This work is dedicated to

my wife,

Elke

who has been an unfailing source of love, support and encouragement throughout the many years in which this work came into being

and to my two sons,

Armin and Kai

I thank them for constantly reminding me of the nature of the educative process through their sense of wonder of the world around them, their questions and desire to know, but especially their joy of exploration and play, and flights of imaginative fancy. May you always be fascinated by questions and endure the quest for understanding.

Rethinking Science Education

Philosophical Perspectives

by

Roland M. Schulz, PhD
Imaginative Education Research Group (IERG) at Simon Fraser University in Vancouver

INFORMATION AGE PUBLISHING, INC.
Charlotte, NC • www.infoagepub.com

Library of Congress Cataloging-in-Publication Data

A CIP record for this book is available from the Library of Congress
http://www.loc.gov

ISBN: 978-1-62396-714-7 (Paperback)
978-1-62396-715-4 (Hardcover)
978-1-62396-716-1 (ebook)

Printed in the United States of America

CONTENTS

FOREWORD

This book presents a philosophy of science education as a research field as well as for curriculum and teacher pedagogy. It seeks to re-think science education as an educational endeavor by examining why past reform efforts have been only partially successful, including why the fundamental goal of achieving scientific literacy after several "reform waves" has proven to be so elusive. The identity of such a philosophy is first defined in relation to the fields of philosophy, philosophy of science, and philosophy of education. Several arguments are provided why and how a philosophy of science education can support science teachers in classrooms. The perspective to be taken on board is that to teach science is to have a philosophical frame of mind—about the subject, about education, about one's personal teacher identity.

Considering science education as a research discipline, it is emphasized that a new field should be broached with the express purpose of developing a discipline-specific philosophy of science education (largely neglected since Dewey). A conceptual shift towards the philosophy of education is needed, thereto, on developing and demarcating true educational theories, which could in addition serve to reinforce science education's growing sense of academic autonomy and independence from socio-economic demands. Two educational metatheories are contrasted, those of Kieran Egan and the Northern European *Bildung* tradition, to illustrate the task of such a philosophy. Egan's cultural-linguistic metatheory is presented for two primary purposes; it is offered as a possible solution to the deadlock of the

Rethinking Science Education: Philosophical Perspectives, pp. xi–xii

science literacy conceptions within the discipline, and regarding practice, examples are provided how it can better guide the instructional practice of teachers, specifically how it reinforces the work of other researchers in the history and philosophy of science (HPS) reform movement. Those in the HPS movement value historical, philosophical, and sociological studies of science, as well as integrating narrative and story-telling into learning science.

Considering curriculum and instruction, a philosophy of science education is conceptualized as a "second order" reflective capacity of the teacher. This notion is also aligned with Shulman's original idea of pedagogical content knowledge (PCK) for science teachers' developing pedagogy and professionalism, as a transformative model. It is argued that for educators the nature of effective science learning must be informed by an analytical examination of several key pedagogical factors where philosophy and its sub-disciplines play indispensable and decisive roles. Philosophical acumen allows for critical self-reflection, but equally for improved inspection of curriculum, policy, and epistemological and popular media issues as they arise. Philosophy of education enhances educators' ability to scrutinize and clarify educational aims and rationales of courses and curriculum, to help unearth buried educational ideologies or those suggested by social interest groups insisting on reforms, while recognizing the vital role of educational metatheory. Philosophy of science allows for insight into personal and textbook epistemologies, to provide for an authentic *nature of science* discussion and integration. Lastly, philosophy of language reveals how the *nature of language* fundamentally influences curriculum, teacher-talk and student comprehension of subject content, often in a concealed manner. Philosophy of science education seeks to encompass and illuminate all such aspects, and to exemplify its duty, two specific case studies linked to the latter two are offered: an examination of the realism/instrumentalism debate as discussed in history and philosophy of science scholarship, and the scrutiny of Dewey's language views from a Gadamerian hermeneutic perspective, contributing to recent research on language and science literacy in classrooms.

PREFACE

While writing this Preface to Roland Schulz's *Rethinking Science Education: Philosophical Perspectives*, I received an e-mail from a student I had taught in 1976, nearly 40 years ago. The email said:

> Over the years, I have often thought about the impact your philosophy of education course had on me. It was at the time very thought provoking and stimulating and stood me in good stead over my many years as a teacher and librarian.
>
> Anyway .... I just thought I would acknowledge after all these years (and all the current edu babble) that I still reflect fondly on a Dip Ed year that I had anticipated would be rather "ho hum", but instead turned out to be something very different. Thank you!

This is a nice email for any teacher to receive. For current purposes it identifies two themes in the book: Namely that philosophy of education should be part of the curriculum of all teacher training programmes, and that development of philosophical awareness and competence is something that has practical benefits in a teacher's professional life. The "back story" of the email is also germane to the subject matter of this book: When the above student was taught there were seven philosophers of education in the UNSW School of Education, subsequently the total number of staff in the school has increased but the number of philosophers has decreased

Rethinking Science Education: Philosophical Perspectives, pp. xiii–xvii

to one. This dismal local picture is writ large across the world; at least the world of Anglo-American schools, colleges and faculties of education where philosophy is at best "on the ropes." Schools of education and teacher training programmes are increasingly given over to applied, classroom-management courses and practice teaching in schools. Learning theory and some form of "cultural studies" or "social studies" occupies what little theory-space there is available in the programmes.

This problem is a very old one. Charles Silberman writing 40 years ago on the then crisis in U.S. education (is there ever a crisis-free period?) said that the way out of the crisis was first to recognise that:

> The central task of teacher education, therefore, is to provide teachers with a sense of purpose, or, if you will, with a philosophy of education. This means developing teachers' ability and desire to think seriously, deeply, and continuously about the purposes and consequences of what they do -about the ways in which their curriculum and teaching methods, classroom and school organization, testing and grading procedures, affect purpose and are affected by it. (Silberman 1970, p. 472)

Schulz agrees with this; and his book can be seen as spelling out the position for the case of science education. Science teachers, as with all teachers, do need to think "seriously and deeply" about the subject they teach; this is another way of saying that they need to think about the history and philosophy of science (HPS). Teachers, need an educational compass. But there are many available all pointing in different directions. Philosophy is necessary to craft an educational compass that points to educational north, and that can contribute to a teacher's identity as a genuine educator, their social standing and be a guide in their myriad professional decision making.

A noteworthy, if depressing, feature of Anglo-American philosophy of education has been its neglect of science education, and the specific philosophical considerations that arise in the teaching of science. The huge corpus of Analytic Philosophy of Education which dominated the professional field in the final decades of the last century is almost devoid of questions about science education, or any analysis informed by the history and philosophy of science.

There is, for instance, no mention of these topics in the influential books of Richard Peters, books that largely defined the field (Peters, 1966, 1967) or in collections such as the major three volume anthology *Education and the Development of Reason* (Dearden, Hirst, & Peters, 1972) that was the flag-ship of analytic philosophy of education. These books and associated literatures were the staple of philosophy in teacher education programs including science teacher education. The neglect of science and HPS in this philosophy of education (PE) literature is the more puzzling as across the world, science curricula were being overhauled in the post-Sputnik era, and at the same time HPS had its own "Sputnik-moment" with the publication of Kuhn's (1962/1970) *The Structure of Scientific Revolutions* and the consequent tsunami of scholarly studies that washed over *all* university faculties. It could have been expected that formal PE would have engaged with both the science curricular and the scholarly HPS enterprises. But with some exceptions this did not happen. Schulz's book, pleasingly, addresses this neglect of science by formal, disciplinary, philosophy of education.

Against the contemporary "applied" tide, Schulz presents "muscular" arguments for why PE needs to be utilized in science education, and also why such philosophy needs to be conjoined with philosophy of science (PS) (and presumably by the history of science although this dimension is not elaborated).

Importantly Schulz is not just contributing to the fairly well-recognized and established project of showing how philosophy can be utilized in solving theoretical, curricular and pedagogical problems in science education; the project of philosophy *for* science education that has been developed by a number of philosophers including Joseph Schwab, Gerald Holton, Michael Martin and others. Rather he is engaged in a more singular and ground-breaking project of forging a philosophy *of* science education (PSE). He argues that there are philosophies *of* physics, biology, chemistry, mathematics, economics, and so on in part because each of these areas constitute an academic discipline. He then shows that science education has over the past 50 years become a discipline and so the pursuit of a comparable philosophy of science education is warranted, and such PSE can, as Richard Peters maintained, shape and direct the discipline; it does not have mere "observer" status.

Schulz sets himself a particularly difficult task in formulating his PSE because he wants PSE to be above partisan philosophical disputes. Just as there are practical, theoretical and metaphysical disputes in all the above mentioned disciplines, so also of course there are in science education. For Schulz, a serious PSE will need to be cognizant of these differences and strive for a universality that accommodates them. Partisan PSE is not particularly novel or difficult. There have been Catholic, Islamic, Stalinist, Hindu, Christian Fundamentalist, New Age, and other such philosophies *of* science education that come to science education from the outside and lay down metaphysical, epistemological and ethical guidelines or constraints for the teaching and learning of science. In recent decades Constructivism has done PSE duty, as can be seen in widely made comments such as:

> To become a constructivist is to use constructivism as a referent for thoughts and actions. That is to say when thinking or acting, beliefs associated with

> constructivism assume a higher value than other beliefs. For a variety of reasons the process is not easy. (Tobin 1991, p. 1)

Many have pointed out philosophical and pedagogical problems with constructivism, but these problems are not Schulz's concern; they are just internal disputes for PS. For Schulz Constructivism fails as a PSE because it fails his non-partisan test for an adequate PSE (as would Catholicism, Islam, Stalinism and so on). He is formulating a kind of transcendental PSE; a PSE that is required by the very meaning of "science," "education," "learning," "understanding" and "teaching." It should be obvious that this is a more difficult task than just formulating any externally derived or generated PSE. What is less obvious is whether, with this self-imposed constraint, Schulz succeeds in his task. His project is akin to formulating aims of education without dependence on any particular 'ism' or ideology or worldview. For this latter task, the modest goal is to provide some common core of aims that can be derived from the very idea of education that hopefully all specific "isms" can acknowledge, and agree upon, leaving a disputed penumbra of aims where it is not expected that consensus can be achieved. The hope of liberal educational theorists is that the common core can be expanded.

As Schulz says:

> This book is such an attempt to *think science education anew, and to offer constructive advice from philosophy of education* by utilizing "conceptual schemes" based on the metatheory of [Kieran] Egan as well as advice from *philosophy of language* by utilizing the philosophical hermeneutics of [Hans-Georg] Gadamer. (p. 25, this volume, emphasis in original)

It is a very clearly argued and exceedingly well documented book. How much he is able to substantively (as distinct from linguistically) move from philosophy *and* science education to philosophy *of* science education will be one of the major interests of readers; but even opening up the latter domain for discussion and argument is itself a positive contribution of the book.

Michael R. Matthews
School of Education
University of New South Wales Australia

REFERENCES

Dearden, R. F., Hirst, P. H., & Peters, R. S. (Eds.). (1972). *Education and the development of reason* (3 vols.). London, England: Routledge & Kegan Paul.

Kuhn, T. S. (1970). *The structure of scientific revolutions* (2nd ed.). Chicago, IL: Chicago University Press. (Original work published 1962)

Peters, R. S. (1966). *Ethics and education*. London, England: George Allen and Unwin.

Peters, R. S. (Ed.). (1967). *The concept of education*. London, England: Routledge & Kegan

Silberman, C. E. (1970) *Crisis in the classroom: The remaking of American education*. New York, NY: Random House.

Tobin, K. (1991). *Constructivist perspectives on research in science education*. Paper presented at the annual meeting of the National Association for Research in Science Teaching, Lake Geneva, Wisconsin.

ACKNOWLEDGMENTS

This book is the culmination of years of work, research, course reflections, discussions, and various other kinds of input from many different people. I would especially like to acknowledge the support of Dr. Charles Bingham of Simon Fraser University (SFU) in Vancouver, my senior PhD supervisor. He has been very supportive throughout the time of the development of my thesis work, from which this book emerged. Our many discussions concerning philosophy and language have been very enjoyable and much appreciated. Dr. Mark Fettes of SFU I likewise wish to thank for his many critical comments and analysis of the ideas. In general I would like to recognize the support of the community of professional teacher educators in Professional Programs at the university.

There have also been a few other individuals whom I wish to explicitly mention who have made significant contributions to my own intellectual development and who have left their mark in one way or another on this book. Dr. Kieran Egan of SFU has been very helpful with encouragement, insight, and various critical comments, which have contributed to the eventual publication of earlier versions of some chapters. Dr. Michael Matthews of Australia (University of New South Wales) has been very supportive, and I wish to acknowledge the stimulating intellectual environment of the International History, Philosophy, and Science Teaching (IHPST) group. As a former elected graduate student representative serving on the executive council, I had the privilege of observing the organization become officially established. The valuable presentations and newfound friend-

Rethinking Science Education: Philosophical Perspectives, pp. xix–xx

ships at the conferences have enriched my beliefs and views of science and how science education could be enhanced and positively shaped. I want to also acknowledge Dr. Calvin Kalman, physicist and consummate physics educator (Concordia University), whose friendship, suggestions, and encouragement have enabled this book to be seen into print.

I am also grateful to Dr. Mark Battersby (Capilano University)—mentor, fellow philosopher and friend—for many fruitful discussions on science and philosophy, for his encouragement and critical comments on some chapters. Finally, thanks are owed to my lifelong friend, Dr. Don Lint, scientist and physician, for our countless conversations on life, science, philosophy, and religion with many a fine cigar and cognac.

I would like to acknowledge the rights holders for their kind permission to re-publish previous work, either my own or from another author:

Cengage Learning Inc. for permission to reproduce Figure 4.1, taken from Ronald N. Giere's Understanding Scientific Reasoning, 3E. © 1991 www.cengage.com/permissions

Springer Science Business Media for their kind permission to reproduce aspects of my previous publications. They have been reproduced in whole or in part throughout this book as follows:

Significant sections of Chapter 2 and Chapter 3 have appeared separately in two papers in the journal *Science & Education:*

(2009a). Reforming science education: The search for a philosophy of science education (Part I). *Science & Education, 18,* 225–249.

(2009b). Reforming science education:Utilizing Kieran Egan's educational metatheory (Part II). *Science & Education, 18,* 251–273.

Sections 1.1 and 1.4 from Chapter 1, section 4.1.1 from Chapter 4, section 5.1.2 from Chapter 5 have appeared previously in a recently published international handbook:

(2014). Philosophy of education and science education: A vital but underdeveloped relationship. In M. R. Matthews (Ed.), *International handbook of research in history, philosophy and science teaching* (Vol. II, Ch. 39, pp. 1239–1315). Dordrecht: Springer.

Most of Chapters 1 and 5 are unpublished, and along with most of Chapter 4 appear here for the first time.

INTRODUCTION

Philosophical Perspectives on Science Education

INTRODUCTION

> *It was through the feeling of wonder that men now and at first began to philosophize ... but he who asks and wonders expresses his ignorance ... thus in order to gain knowledge they turned to philosophy.*
>
> —Aristotle (*Metaphysics*, 1998, p. 9)

What sort of thing is a "philosophy of science education?" As implied by the included subjects, is it more concerned with philosophy, or with science, or with education? Then again, is it intended to focus more on problems in philosophy of science, or philosophy of education as related to learning *science*? Moreover, can it be undertaken, this "philosophy"—one that purports to be a philosophizing *of science education*? What would comprise its scope and tasks? And who would undertake it?

Science education has roots going back over a century, although as a discipline and research field it is relatively new, only having come into its own in the U.S. since the 1920s (National Association of Research on Science

Rethinking Science Education: Philosophical Perspectives, pp. 1–9

Teaching, NARST, was created in 1928), arguably perhaps only since the "Sputnik shock" of 1957 and the professionalism that followed. In physics, Arnold Arons began discussions with Robert Karplus, in 1968, and then moved to the University of Washington. There he began a collaboration with Lillian C. McDermott. This collaboration led to the formation of the Physics Education Group at the University of Washington. This was the formal beginning of a new field of scholarly inquiry for physicists: physics education research. In chemistry, the ChemLinks project was initiated by Brock Spencer of Beloit College (principal investigator and project director for the ChemLinks Coalition, 1994–2002) and developed with members of the Midstates Science and Mathematics Consortium. Discipline-based educational research in mathematics began around 1988. Dubinsky, at Georgia State University, began his research by extending Piaget's work. In biology there is the BioQUEST Curriculum Consortium (Beloit College). This project was founded in 1986 by John Jungck, editor of The BioQuest Library. BioQUEST is a group of educators and researchers committed to providing students with biology research and research-like experiences. The consortium began with an initiative of the Commission on Undergraduate Education in the Biological Sciences, established by liberal arts college biologists in the 1960s. These research fields, as very new fields of scholarly inquiry, are still finding their way among their respective academic departments at some universities (for their true worth is not yet universally acknowledged). Yet despite its longer history some would argue that science education is still developing an *identity* (see Fensham, 2004). What would comprise the role of a "philosophy of science education" (PSE) *for* science education, especially in light of establishing its identity and further contributing to its professionalization? What would its value be—assuming it had one—and what could it contribute?

This book seeks primarily to examine the relationships among the three fields of philosophy, philosophy of education (PE), and philosophy of science (PS) to inquire how philosophy could better contribute to improving science curriculum, teaching and learning, and, above all, science teacher education. The value of philosophy *for* science education in general remains underappreciated at both pedagogical levels, whether the research field or classroom practice. While it can be admitted that philosophy has been an area of limited and scattered interest for researchers for some time, it can be considered a truism that modern science teacher education has tended to bypass philosophy and philosophy of education for studies in psychology and cognitive science, especially their theories of learning and development (which continue to dominate the research field; Lee, Wu, & Tsai, 2009). A major turn encompassing philosophy would thus represent an *alternative approach* (Roberts & Russell, 1975).

Science education is known to have borrowed ideas from pedagogues and philosophers in the past (e.g., from Rousseau, Pestalozzi, Herbart, and Dewey; DeBoer, 1991), however the sub-field of *philosophy of education* has been little canvassed and remains on the whole an underdeveloped area. At first glance such a state of affairs may not seem all too surprising since science education is mainly concerned with educating students about particular science subjects or disciplines. But this necessarily implies a tight link between content and education. Hence, if education is to mean more than mere instructional techniques with associated texts to encompass broader aims including ideals about what constitutes an educated citizen (i.e., defining "scientific literacy") or foundational questions about the nature of education, learning, knowledge, or science, then philosophy *must* come into view (Nola & Irzik, 2005). As is known, an *education in* science can be, and has been, associated with narrow technical training, or with wider liberal education, or with social relevance (STSE reforms), or lately with "science for engineers" (U.S. STEM reforms), an updated version of the older vocational interest.[1] Yet all these diverse curricular directions imply or assume a particular educational philosophy, which is rarely clearly articulated (Matthews, 1994; Roberts, 1988).

At second glance then, and viewing science education in a broader light, being principally at home in education unavoidably implies an excursion into philosophy of education. In fact, it avoids this sub-field of philosophy at its own peril, as argued elsewhere (Matthews, 1997, 2002). Equally, there are lessons to be learned from its own past, yet most science teachers and too many researchers seem little aware of, or even concerned to know about, the rich educational philo-historical background of science education as it has developed to the present, whether in North America, Europe, or elsewhere (some examples are Mach, Dewey, Westaway, and Schwab; DeBoer, 1991; Gilead, 2011; Matthews, 1991, 2014). In fact, recent critical reviews insist that educators must acknowledge and respond to how past historical developments have molded science education while continuing to adversely shape the current institutionalization of school science (Hudson, 2994; Jenkins, 2007; Rudolph, 2002). A central concern of this book is to emphasize the value of philosophy in general and philosophy of education in particular. It will be claimed that an awareness of the worth of these fields can have positive results for further defining the *identity* of both the science teacher as professional (Van Driel & Abell, 2010; Clough & Olson, 2008) and science education as a research field (Fensham, 2004). The perspective to be taken on board is that to teach science is to have a philosophical frame of mind—about the subject, about education, about one's identity.

OVERVIEW OF THE BOOK AND ITS ARGUMENT

This brings us to the present investigation and the organization of the book. My personal route of discovering the need of a PSE arose from my own experiences working both in science classrooms and with pre-service science teachers, and linking these experiences to the widely-recognized issue of the problematic (or unsuccessful) nature of implementing effective reforms in science education. What is true of my experience—that one's own philosophizing usually emerges out from the problems of the socio-educational environment one is immersed in and concerned about—corresponds with Dewey's (1916/1944) insight "that philosophical problems arise because of wide-spread and widely felt difficulties in social practice" (p. 328). Or perhaps, put simply, my discovery resulted from "a sense of wonder" and disorientation, as Aristotle had identified the origin of philosophy (as quoted at the beginning of this introduction). It was the inability of my secondary students to significantly improve their grades in their junior and senior classes, regardless of my pains taken at direct instruction, tutoring, modified group and inquiry activities, that brought me to re-examine and question my entire pedagogical approach.

My junior science students were interested but not overly enthused with the predominant curricular academic diet and showed little improved comprehension at year end, time and again. I was stunned, for I believed my experience with over 10 years of classroom teaching (including international experience), my enthusiasm, the concern I showed for their learning, my continued professional interest in reading science literature and following and incorporating the latest scientific findings, would have borne better fruit for my students. It was not that my senior students did not perform well on state-mandated, standardized final exams; on the contrary, many did exceptionally well for the most part. This discovery was also a puzzlement. Moreover, what I continued to observe was a lack of deep understanding, with little comprehension of what science or physics was about and meant to them personally, how it had developed or could be used, and generally, for most of them, little interest to pursue it further.

At about this time of discouragement and bafflement, I became more critical of the academic science curriculum and sought out what I believed to be more accurate historical portrayals of scientific development than had been commonly presented in the official textbooks I had been using. I also occasionally incorporated stories and philosophical aspects into my science teaching, especially in my senior physics classes, with special focus on the Copernican revolution. This became an extension to the regular dry and difficult material on gravity, Kepler's and Newton's laws. (I slowly discovered glaring historical errors and embarrassing simplifications of theories, method, and scientific progress—for example, how atomism and key experiments were discussed in chemistry textbooks. See Niaz, 2000, 2002.) But even while some students became genuinely enthused, others questioned the relevance of history and philosophy of science (HPS) for their science courses with their pre-set exams directly tied to college placement. Most students in senior classes, not surprisingly, had developed a very instrumental attitude to learning and the content of their courses. Yet through these experiences, I was gradually led to science education research, the large body of literature on misconceptions, constructivism, and the HPS reform movement. This trajectory, looked at in hindsight, was probably inevitable.

So, it can rightly be said that I did not so much "discover" philosophy of science education as it discovered me. PSE is needed because a "wider dimension" of science education *reform* has until now been missing. It is then not surprising that Chapter 2 is concerned with *why* the solutions offered have so far been proven inadequate. Why have the problems not been satisfactorily resolved? How can one overcome the deadlock in the historically identified essential goals of science education, which seem to be at cross-purposes (Bybee & DeBoer, 1994)? Is there a successful resolution that the community could accept concerning the contentious and divergent meanings of science literacy (Roberts, 2007)? Thereto, why is learning of science concepts so difficult, why is understanding so elusive, and do there exist significantly better conceptions of education and ways of learning heretofore not fully considered despite several decades of research (Egan, 1997)? To that end, the following research question was adopted:

Research Question

> *What can a philosophy of science education contribute towards improving conceptual understanding of science and science education as a discipline?*

This broad question will be broken down into specific questions that are addressed in the respective chapters that follow:

1. How can a philosophy of science education contribute to the reform of science education goals and the conception of literacy? (Ch. 2)
2. What does it mean to be educated in science? (Ch. 2)
3. What does conceptual understanding entail? (Ch. 2/3)
4. How does one best come to learn science? (Ch. 2/3)
5. What is a "metatheory" of education? (Ch. 2)
6. Why does science education require a metatheory? (Ch. 2)
7. What is Kieran Egan's metatheory about and how can it improve science education? (Ch. 3)

8. What is the contribution of a philosophy of science education to the *nature of science* (NoS) discussion? (Ch. 4)
9. What is the contribution of a philosophy of science education to the *nature of language* (NoL) discussion in science education? (Ch. 5)

From this list of questions one clearly notices that this book hopes to play its part in analyzing and contributing solutions to those problem sets listed later under the "Scope" for PSE inquiry (Section 1.5). It is further to be understood that any solutions suggested and ideas proposed are tentative and meant to contribute to continued scholarly discussion and research.

THE ARGUMENTATION OF THE BOOK

The argumentation of the book is to rethink science education as an educational endeavor by examining it from several philosophical perspectives: the need to shift its theoretical research emphasis towards philosophy of education, educational theory, and philosophical theorizing in general; an analysis of key learning issues in the physical sciences; the re-examination of its fundamental goals—especially *science literacy*—and the relevance and yet problematic status of epistemology and *nature of science* (NoS) for curriculum and teachers; the value of considering Egan's educational *metatheory*; and finally the need to scrutinize the role and *nature of language* (NoL) in Dewey and science education from a Gadamerian-hermeneutic perspective. It is argued that for educators the *nature of science learning* (NoSL) must be informed by a critical consideration of educational theory, NoS, and NoL. Some of these themes are themselves not new, of course, but what is original is the way they are treated and brought together from the standpoint of philosophy and its sub-disciplines.

The *first chapter* provides the framework for a "philosophy of science education" (abbreviated PSE throughout the book). It takes the position that it represents a *synthesis* or integration of (at least) three fields, being philosophy, philosophy of science (PS), and philosophy of education (PE). Each field is examined carefully in turn, and the chapter provides several arguments why philosophy, and philosophy of education in particular, is of value, especially for science teachers and their developing identity, epistemology, professionalism, and pedagogical content knowledge.

The *second chapter* has several aspects structured around the evidence of the widely recognized issue of the problematic (or unsuccessful) nature of implementing effective reforms in science education. The book takes the perspective that the problem is to be located at the nexus of at least three key factors: (1) competing school-based paradigms, with their own largely exclusivist conceptions of scientific literacy that are tied to competing interest groups; (2) the prevalent disregard of philosophy of education, in conjunction with (3) the ongoing neglect of generating discipline-specific educational theories (not restricted to learning theories, and which could help clarify curriculum and goals). Common "crisis-talk" is seen to mask competing paradigms and interests. Hence it is suggested that a turn towards more philosophical-based reflection and educational theory in the discipline is required, with the proposal that a metatheory may offer an answer to the dilemmas we face. The nature of metatheories is then discussed and clarified. Finally, more reasons are provided why science education requires a metatheory of education, being linked with the options left to the community in the literacy debate.

The *third chapter* then concentrates on presenting and describing Egan's metatheory and providing some examples how it can contribute in concrete ways to improving science education. As a preliminary to this examination, a comparison is made between the educational systems of the Anglo-American "curriculum" tradition and the German-Norse *Bildung* tradition, since the latter is characterized by use of a renowned metatheory. Other parallels are also drawn, in particular the role of a "philosophy of science education" as linked to European *Didaktik* analysis and Shulman's pedagogical content knowledge (PCK). Once Egan's metatheory has been explicated in some detail, some links are drawn to other researchers who have emphasized the importance of using narrative and the history of science for improving instruction and learning. On the other hand, a metatheory of education only addresses the issue of the nature of science learning (NoSL) and leaves the question of the nature of subject content largely aside.

The *fourth chapter* turns to the key issue of how to address the need to problematize the content knowledge of both school science and science disciplinary subjects, especially the problem of the image of science as it is found in epistemologies of curriculum design and textbooks, which a general metatheory of education (with its educational philosophy) must draw upon and is intended to educate students about. The educational significance of science subject content knowledge is analyzed and critiqued as belonging to three possible content-structured categories: epistemic, contextual, and socialization. The history and philosophy of science (HPS) reform movement is referenced, as well as how it is attempting to address these epistemological issues (and several myths associated with current pedagogy), including the issue of improved NoS understanding and integration to better reshape the content knowledge base. Problems associated with NoS are considered too. Finally, a detailed case study examining the long-standing realism/instrumentalism problem in recent PS debates is offered. Some lessons are then drawn for science teachers. It is recognized that the problem of the NoL cannot ultimately be ignored and that a hermeneutic perspective on language and a move "beyond epistemology" may also contribute to improved learning and expanding teachers' PCK.

The *fifth* chapter turns to the *nature of language* (NoL). It briefly summarizes the literature on language as a relatively recent research field in science education, and discusses the conjectured "interpretative turn" in philosophy of science and language. It acknowledges that the so-called turn "beyond epistemology" and towards hermeneutics is contentious. Nonetheless, one thesis of the book suggests that a focus on Gadamer's *philosophical hermeneutics* and his conception of language can offer profound insights into the nature of learning that has been largely ignored in science education research. To that effect, a significant portion of the chapter is dedicated to a NoL case study involving a critical examination of Gadamer's and Dewey's notion of language. It is argued that Dewey's "symbol and tool" view continues to dominate science education to its detriment. The shift from epistemology to ontology is discussed, and shortcomings also in a Vygotskian "tool view" of language are identified. Some shortcomings in Egan's metatheory are also discussed. These shortcomings need to be considered when looking at student learning problems and suggest areas for continuing research.

OTHER FINAL CONSIDERATIONS

A *philosophy* of *science education* (PSE) must of necessity address and include dimensions of *three* fields of inquiry, which themselves embrace their own venerable historical traditions and specific contemporary research programs: philosophy, science, and education. It is understood that an attempt to consider each to the degree of depth and completeness they require would be an undertaking not achievable in the kind of work here suggested. This much must be acknowledged, and hence one must at the outset set limits as to the direction and scope of the inquiry under investigation. Regarding *philosophy*, the scope will be narrowed to aspects of *philosophy of science* (especially epistemology) and *philosophy of education* (especially educational theory) as they bear on the arguments and intention of the book. Here one should quickly add, however, neither will it be possible to fully explore the significance of these two sub-disciplines of philosophy—albeit not being as old and well-established as the aforementioned three—which certainly has substantial bearing on the present project. Thereto, some dalliance with respect to language theory and philosophical hermeneutics will be included. Regarding *science,* the focus will be primarily on physics (including its historical development), to a lesser extent on chemistry, but also with some reference to geology. Finally, regarding *education,* the primary concern will be the literature and research programs pertaining to science education (including referencing physics education research [PER] and chemistry education research [CER]), together with a discussion of metatheories, of curriculum, learning theories and conceptual change literature, and the relatively recent social-cultural research tradition. As well, there will be a concern to address what sort of ramification a "philosophy of science education" has for *science teacher education*. This will include examining Shulman's original conception of PCK and the German-Norse educational tradition of *Bildung*. The book is clearly a work of exploratory and interdisciplinary scope, enriched and encumbered by the character such work entails, which will judiciously draw on the fields of inquiry mentioned in order to address not only some pressing contemporary problems of science education reform but equally to make the wider case for the creation and establishment of a new sub-field of inquiry science education.

NOTE

1. The prominent U.S. National Science Teachers Association (NSTA) has made STEM a central reform emphasis (see www.nsta.org/stem). References for the other more common science classroom curricular emphases are Aikenhead (1997b, 2002b, 2007); Carson (1998); DeBoer (1991); Donnelly (2001, 2004, 2006); Pedretti and Nazir (2011); Roberts (1982, 1988); Schwab (1978); Witz (2000); and Yager (1996).

CHAPTER 1

MAKING THE CASE FOR AND DEFINING THE IDENTITY OF *PHILOSOPHY OF SCIENCE EDUCATION*

Surveying the Terrain

1.1. PHILOSOPHY OF SCIENCE EDUCATION FRAMEWORK

To be clear from the start, there is no attempt made here at formulating a particular philosophical position thought appropriate for science education, in contrast to such discussions having taken place in *mathematics education* for some time. In that field several educators have articulated and debated the notion of a "philosophy of mathematics education," for example, Platonism and foundationalism versus social constructivism and fallibilism (Ernest, 1991; Rowlands, Graham, & Berry, 2011). On the other hand, it will be stressed that the development of a "philosophy *of* science education," that is, an "in-house philosophy" for the field, could be significant for reforming science education. It can be acknowledged that math educators have been in the forefront of attempting to establish

Rethinking Science Education: Philosophical Perspectives, pp. 11–62

a "philosophy of" for their educational discipline while science educators in the main have not yet come to consider or value such an overt evolution in their field. Such an endeavor urges exploration of new intellectual territory. The sign of the times seems ripe for such an investigation ever since the science educational field became staked out by opposing, even irreconcilable positions "from positivism to postmodernism" (Loving, 1997).

In the past constructivism was once seen by many educators as a kind of "philosophy" (though not expressed as such) that was to serve the role as a "new paradigm" of science education. Today many consider this view to be mistaken (Matthews, 2002; Phillips, 1997, 2000; Suchting, 1992).[1] This judgment has come about largely because many supporters at the time did not reflect seriously enough about the philosophical underpinning of its various forms—cognitive, metaphysical, epistemological.[2] Constructivism remains a dominant and controversial topic in education, but one lesson to be had from the heated debate of the past three decades is that absence of philosophical training among science educators became apparent (Matthews, 2009b; Nola & Irzik, 2005). Another lesson learned is the absence of any explicit discussion regarding educational philosophy, even though constructivism in some corners was brashly substituted for one. In hindsight it is surprising that constructivism—which after all still finds its principal value as a learning theory (and perhaps teaching method)—could be mistaken as a dominant kind of "philosophy of" science education at the neglect of broader aims and concerns relevant to educational philosophy, as to what it *means* to educate someone in the sciences. And science education once again showed unawareness of its own history, since Dewey (1916/1944) and Schwab (1978) had previously addressed such concerns. At minimum the case of constructivism had illustrated—although not widely recognized—how interwoven, if not dependent, science education in the academy had become with certain psychological ideas and philosophy of science (notably its Kuhnian version; Matthews, 2003a).

In light of this background, we will develop the discipline of a philosophy of science education (PSE). With this project in mind, one can draw attention to two useful aspects pertaining to philosophy in general that can come to our aid and contribute to improving science education and developing such a philosophical perspective: the ability of philosophy to provide a *synthesis* of ideas taken from associated disciplines with their major educational implications, and providing what can be called "philosophies of." In this way it will be shown how philosophical thought can be brought to bear directly on educational ideas and practice.

1.1.1. The Synoptic Framework

The role and value that philosophy itself and its two important subdisciplines of *philosophy of science* (PS) and *philosophy of education* (PE) can have is illustrated by the representation below. Note that "philosophy of science education" (PSE) is then understood as the *intersection* or *synthesis* of (at least) three academic fields. For each respective field of study some individual points are stressed that comprise core topics of interest to science education pertinent to each, but it is meant to be illustrative, not exhaustive:

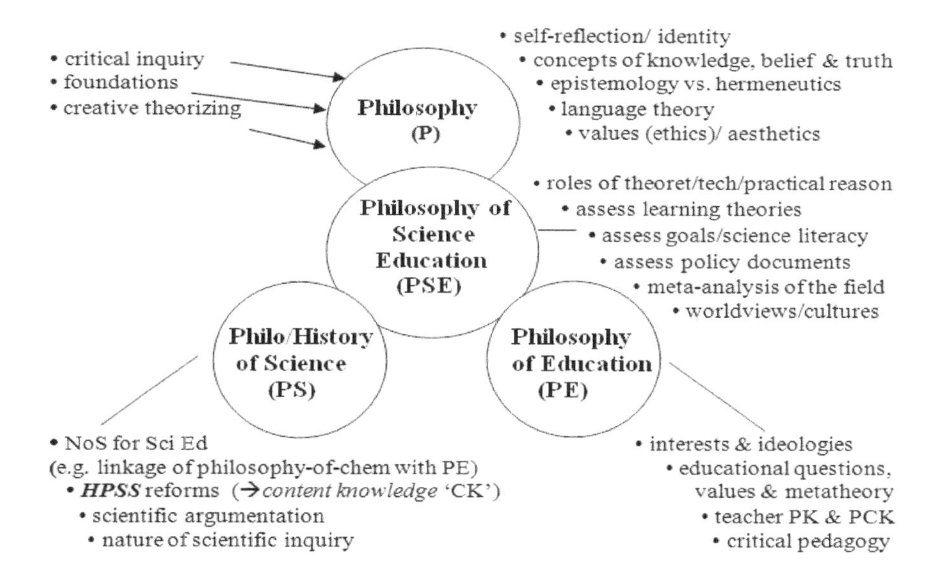

Figure 1.1. Philosophy of Science Education Framework.

The framework in itself assumes neither prior philosophical positions (e.g., metaphysical realism or epistemological relativism) nor pedagogical approaches (e.g., constructivism, multiculturalism, socio-political activism, etc.). As a graphic organizer it does provide science teachers and researchers with a holistic framework to undertake analysis of individual topics and perhaps help clarify their own thinking, bias, and positioning with respect to different approaches and ideas. The main point is to show that any particular PSE as it develops for the teacher or researcher should take into consideration, and deliberate upon, the discourses pertinent to the three other academic fields when they impinge upon key topics in science education. At minimum it should contribute to helping develop a philosophic mindset.

In sum (as Figure 1.1 shows), any philosophy of science education (PSE) is foremost a *philosophy* and as such receives its merit from whatever value is assigned to philosophy as a discipline of critical inquiry. (This value may not appear at all obvious to science educators.) Furthermore, such a philosophy would need to consider issues and developments in the philosophy, history, and sociology of science[3] and analyze them for their appropriateness for improving learning *of* and *about* science. Finally, such a philosophy would need to consider issues and developments in the philosophy of education and curriculum theory and analyze them for their appropriateness for education in science, as to what that can *mean*, how it could be conceived and best achieved. A fully developed or "mature" PSE can be understood as an integration of all three fields. It ultimately aims at improving science education as a research field as well as assisting teachers in broadening their theoretical frameworks and enhancing their practice.

1.1.2. Providing "Philosophies of"

Philosophy today has evolved into several specialized sub-disciplines. These include philosophy of science, of education, of mathematics, of technology, of history, of religion, and others, which can collectively be called "philosophies of." It is especially the first two that are of immediate concern to us when developing one for ourselves, as Figure 1.1 illustrates. And yet this conceptualization is not as new as it may appear. Over 40 years ago the philosopher Israel Scheffler summarized the value of these "philosophies of" for science educators:

> I have outlined four main efforts through which philosophies-of might contribute to education: (1) the analytical description of forms of thought represented by teaching subjects; (2) the evaluation and criticism of such forms of thought; (3) the analysis of specific materials so as to systematize and exhibit them as exemplifications of forms of thought; and (4) the interpretation of particular exemplifications in terms accessible to the novice. (Scheffler, 1970, p. 392)

He understood these "philosophies of" would provide invaluable components to a science teacher's identity and preparation, in addition to the common three of subject-matter competence, practice in teaching, and educational methodology. Especially the inclusion of philosophy of science (PS) topics he considered vital to allow teachers to be "challenged to reflect deeply on the foundations" of their subjects and "to relate their reflections to the task of teaching" (Scheffler, 1970, p. 388).

Matthews (1994a, 1994b, 1997) helped popularize Scheffler's earlier vision, whose call for inclusion of PS has been broadly acknowledged today though unfortunately little implemented worldwide in teacher education programs.[4] He has expanded upon Scheffler's line of reasoning to include additional pedagogical and professional arguments. An improved pedagogy, for example, should include several aspects: wisely evaluating constructivism and the educational aims of curricular documents, integrating history, philosophy, and sociology of the sciences (HPSS) topics, developing critical thinking; allowing science courses to show a "human face," and at minimum making science more interesting and understandable. Enhancing professionalism requires teachers to develop a wider perspective of their subject and its role in education, including becoming versed with topics and questions associated with science and society concerns. These would include religion and science, "multicultural science," feminism, techno-science, environmental ethics, animal rights, and others.

In short, philosophical questions concerning both education and science are at the heart of the science education profession, many of which have kept, and continue to keep, teachers, researchers and curriculum developers engaged. Broadly speaking, they encompass essential concerns immediately identifiable with the two fields of philosophy of education (PE) and philosophy of science (PS):

> educational ones about the place of science in the curriculum, and how learning science contributes to the ideal of an educated citizen and the promotion of a modern and mature society. The questions also cover the subject matter of science itself. What is the nature of science? What is the status of its knowledge claims? Does it presuppose any particular worldview? The first category of questions constitutes standard philosophy of education (PE); the second category constitutes philosophy of science (PS) or history and philosophy of science (HPS). (Matthews, 2002, p. 342)

The teacher's professional role today has in some cases also come to include co-creating, advising on, and assessing so-called national science "standards" documents. Since the 1990s several countries around the world have sought to define curriculum "standards," which harbor considerable agreement on nature of science policy statements (McComas & Olson, 1998).

> Clearly all these curricular exhortations depend on teachers having philosophical acumen and knowledge in order to understand, appraise, and enact them. This requires a mixture of philosophy of science (to understand the substantial claims), and philosophy of education (to interpret and embrace the objectives of the curricula). (Matthews, 2000, p. 343)

In the sections below, the intention is to further elaborate on the worth of philosophy as a subject in general, but especially philosophy of education since this topic is usually overlooked. Philosophy of science for educators will only be glossed (above all its newer sub-specialties), as this topic has been an active area of research. An in-depth case study involving philosophy of science will be presented later in Chapter 4.

1.2. THE RELATION TO *PHILOSOPHY*

1.2.1. Preliminary Considerations: Philosophy Quo vadis?

As the title implies, the topic and theme of this book is one that concerns providing philosophical perspectives on science education. This theme as presented and elucidated herein is original and exploratory; it seeks to bring teachers and researchers along on a *search*, a *looking for*, a route to a destination, through uncharted science education territory. It is quite common to distinguish the path from the goal itself. When the goal is known ahead of time, for instance climbing to a known craggy summit, one often has various options of choosing which path might prove the most expedient, or the most beneficial (or perhaps both), to arrive at the desired destination. Depending upon one's own predispositions, one may be willing to make that extra effort if the path is known to be arduous but the view from the summit known to be spectacular. But what if the destination is a new one, the path barely trodden, the goal shadowy, and the summit shrouded in mist? The route to be taken becomes much more difficult to make out, and the effort questioned. Pioneers are then ridiculed as fools or later lauded as visionaries depending upon their respective failure or success. In either case there are risks involved. This book is such a risk, as it implies such an endeavor. Why? Because the position taken here is indeed *new*: that science education both as academic inquiry and as a field of practice requires the development of an in-house *philosophy* in order to execute effective reforms and progress as a research discipline and as a profession.

The risk is that, as with all things new and especially with educational matters, one is at first startled and curious, but soon one begins to question the value and relevance if no immediate practical applications can be discerned. Indeed the harsh *utility* criterion has come to serve as a predominant tribunal of judgment for new reform ideas and efforts in science education, of which, it can be sadly admitted, there have been too many in the last century, several in the last 50 years alone. Some teachers and researchers may view this as a good thing, since it functions as a rule-of-thumb (certainly from the perspective of practitioners) of filtering well meaning but abstract academic (or "top down") initiatives that appear for most classroom purposes unproductive. That said, one must concede that the serious problem for science education today (and education in general) is the ongoing gap between university-based research and classroom applications—not just pertaining to "pure" academic subjects like philosophy—arising for different reasons, which continues to occupy the concern of the research community (Boersma, Goedhart, De Jong, & Eijkelhof, 2005; Millar, Leach, & Osborne, 2000). Yet even if one *has* begun to recognize the value of philosophy, or perhaps better put as appreciating critical thinking and philosophical analysis for curriculum development *and* for one's own practice, which it seems to me many *experienced* teachers eventually come to admit, there remains the issue whether one thinks science education itself requires such an encompassing philosophy as a constituent *goal,* as something required of its self-conception and identity, to help it "mature" and better able to develop as an autonomous discipline. And even *if* this much is admitted, one may still question the *route* to get there, and hence the feasibility of ever accomplishing anything so grand.

Philosophy has always entailed certain risks, as the lives of Socrates and Aristotle in antiquity, Giordano Bruno, Spinoza, Rousseau and Fichte at the start of the modern age, or much more recently Hannah Arendt and Karl Jaspers (because of the Third Reich) exemplify. With the eventual establishment of academic philosophy as a profession at the universities in the 19th century, most risks have been thankfully mitigated, but risks remain today, albeit of a less ominous kind. It has also been historically associated since at least the time of Nietzsche (1890s) with the metaphor of climbing heights if for no other reason than attaining clarity, a better overview of the metaphysical landscape (or scientific or political, etc.), and possibly of attaining a new vision. *The premise of the book is that there is the necessity of both for improving science education.* In other words, the need to re-examine and re-think the common problems of the science educational terrain—associated with goals (especially *science literacy*), curriculum, learning issues, nature of science and language—from fresh philosophical perspectives and with the intent of providing solutions offering new ideas, new research pathways, and possibly a new vision is quite relevant and necessary.

Yet philosophy as a discipline has certainly not been without its critics. One should not overlook, as skeptics will be keen to remind us, that philosophers over the ages have tended to promise more than they have delivered, and history bears ample witness to the collection of constructed and cast off philosophical *systems* (from Plato to Aquinas, from Descartes to Kant, to German idealism, Hegel, and Marx). Nietzsche was among the first (along with Kierkegaard) to be rightly critical of all-encompassing philosophical "systems" and argued against their attempted project. In our time, in the wake of the collapse of epistemic *foundationalism* generally, including the linguistic-based logical positivism of the English-speaking world, Wittgenstein, Rorty, Taylor, and others have accepted this attitude and arrangement as the new philosophical status quo. Continental philosophers such as Heidegger, Gadamer, Derrida, Habermas, and Foucault, arriving at similar conclusions in appraising the Western tradition, have afterwards gone in different directions, but some in parallel with American pragmatism. The tremendous wake has left behind many troubling questions among philosophers, now widely known, such as what the contemporary role of epistemology should be (with related ramifications for philosophy of science). And above all, if philosophy itself—understood as philosophy with a capital "P," as Rorty puts it, the age-old conflict between "Platonists" and "positivists" to achieve the definitive appraisal of knowledge (epistemology), reality (ontology), and virtue (ethics)—has possibly come to an "end" (Baynes, Bohman, & McCarthy, 1987; Heidegger, 1977; Rorty, 1982).

When examining the recent quarrels among professional philosophers as to whether or not philosophy as a discipline has now finally managed to find its "proper object" and "goal," including the right way (or "systematic methodology") to go about solving philosophical problems, one discovers soon enough a sharp contrast of views among adherents within the Anglo-analytic tradition, for instance spotlighting such familiar figures as Michael Dummett, who has answered in the positive, and Richard Rorty, who remains skeptical and has answered in the negative. While Dummett would front the significance of Frege in finally having helped philosophy establish its proper domain and tasks (all the while recognizing the many previous failed historical attempts to ascertain this), Rorty would recognize instead the importance of three—namely, Dewey, Wittgenstein and Heidegger—"precisely because they have helped overcome the very conception of philosophy that Dummett and so many professional philosophers accept" (Bernstein, 1983, p. 6).[5] Bernstein would see the contrasts between Dummett and Rorty (also between Popper and Feyerabend in the philosophy of science) as indicative of a wider cultural debate spanning other disciplines, especially the social sciences, which he identifies as the antithesis of objectivism and relativism in Western thought. In line with Rorty, Gadamer, and others, he too would see the *epistemological task* of traditional philosophy coming to a close: "we are coming to the end—the playing out—of an intellectual tradition (Rorty calls it the 'Cartesian-Lockean-Kantian tradition')" (Bernstein, 1983, p. 7). He observes a "new conversation" emerging about human rationality going "beyond objectivism and relativism," moving from epistemology to *hermeneutics* and *praxis*, or what elsewhere has been called "the interpretative turn" (Hiley, Bohman, & Shusterman, 1991).

A philosophy of science education (PSE) must of course be cognizant of these quarrels and developments both in philosophy and philosophy of science: first of all to help it examine its own self-understanding as a "philosophy," regarding its own tasks and what it can possibly accomplish (which will be discussed shortly), and secondly, to partake of this current conversation—the position taken here is that science learning can be improved if the "interpretative turn" is also recognized and valued (some implications will be further examined in Chapter 5).

But to avoid misunderstanding, what is suggested here is nothing even remotely comparable to "system building" or attempts to place philosophy back upon its former pinnacle position, as arbiter over the sciences—or education—as a sort of ultimate epistemological tribunal in judgment of other worldviews and knowledge claims, as it had claimed for itself since the Enlightenment era when it stole the crown from theology. It remains rooted instead in the firm ground of philosophy's original Socratic calling, the worth of examining one's *way of life* (one's academic discipline and pedagogy), entailing conceptual clarity, critical questioning, as well as cautious speculation and theorizing. This is a kind of living project that Wittgenstein, Rorty, and Gadamer would subscribe to, and indeed many philosophers and teachers would go along with (granted, what is ignored in this version is the proper role of reason).[6] Such a humble yet noble perspective of philosophy hardly commands an end; if anything it demands a *continuing*, perhaps in some sense and with respect to the topic to be examined, a new beginning.

The point here is to show the relevance of philosophy for the *inquiring mind*, even when simultaneously, and on the other hand, the *irrelevance* card is often raised because of frustration over the apparent *lack of progress* on questions towards solutions by great thinkers in areas of vital human concern, including education, science, and schooling. The typical objections to philosophy as a subject preoccupied with obtuse speculation, arcane technical jargon, and unresolved disputes all remote from everyday matters are familiar and to be lamented. (In this sense it has faithfully followed in the footsteps of a scholastic theology it supplanted at the universities). Dewey, though, would expect from us more circumspection when philosophy wrestles with the complexities and dilemmas of those problems embedded in life (or educational) issues:

> If there are genuine uncertainties in life, philosophies must reflect that uncertainty. If there are different diagnoses of the cause of a difficulty, and different proposals for dealing with it; if, that is, the conflict of interests is more or less embodied in different sets of persons, there must be divergent competing philosophies.... But with reference to what is wise to do in a complicated situation, discussion is inevitable precisely because the thing itself is still indeterminate. (Dewey, 1916/1944, p. 327)

Such differences and disputes, moreover, Dewey sees arising from thinking that is embedded within actual social experiences and practices. This also appears to be the case concerning the central conflicts within the science education community, where researchers and practitioners are inevitably preoccupied with problems associated with different phenomena and segregated facets of the discipline. This book will follow his insights on "conflicting proposals and interests" when examining the disputes over science literacy (Chapter 2) understood as competing paradigms of what science education should be about. In addition, the requirement of philosophy of science education (PSE) for the science education community is advanced exactly for the reason he mentions: in my interpretation, *the need for some sort of philosophical platform of "discussion" to elucidate the problem "thing" of science education reform issues*.

Is there actually any time for teachers preoccupied with overcrowded secondary curricula, stressed by high-stakes testing, and worried about their students learning in diverse school cultural environments, to stop, think, and ask the *essential educational questions* again: "what does it *mean* to be educated?" (expressed very broadly)—and for our science education profession, "what does it mean to be educated *in science*?" (or in physics, in geology, etc.?). Thereto, "how is an education in science contributing to human flourishing?" These crucial questions should be kept distinct from—though they are all too often *subsumed* by—the question harboring a *social utilitarian* criterion so prevalent among the English-speaking nations: "what do we educate people in science *for*?" Are there in fact answers to these kinds of questions that can be "cashed in" for teachers and their usual "instrumental" stance for "what works," but in noteworthy ways that can help transform their thinking and pedagogy? This book will show that there is. Yet what is of utmost importance is to recognize that to entertain such questions and concerns is to presume the invaluable and obligatory nature of a philosophy of education for instruction, but especially when taking the broader view, for *fundamental science education reform* intending to invigorate a *discipline* concerned with curriculum development and research, teacher education, and meaningful student learning of science.

There are those skeptical thinkers like Wittgenstein (1958), of course, who once famously stated that philosophy "leaves everything as it is" (*Philosophical Investigations*, #124; in a passage where he specifically dismisses *foundationalism*), and would instead have its value reside in things other than any sort of policy or institutional "reform" or restructuring (for example, "therapy," "to shew [sic] the fly out of the fly-bottle"; *PI*, #309). Then again, Marx had astutely stated a good century before Wittgenstein that philosophy had only *interpreted* the world, but the point was now to *change* it—and this as a critique of the reigning academic and system-heavy philosophy of Hegelianism of his day. And who would dare deny Marxism has not contributed, no matter how controversial, to radical intellectual and socio-political change? Or deny the horrific *adverse* transformations (both personal and communal) wrought by fascist and communist philosophies? History witnesses that philosophy is at its most transformative when it develops as a *worldview*—one thinks of Aristotelianism Neo-Platonism, or Locke, or Rousseau, or atomism, or evolution; it need not be expressly political, although it could have those ramifications—think Nietzsche, Marxism, or social Darwinism. In 1919 the young Heidegger (1919/2008) took the position that philosophy had a dual purpose with respect to worldview: the *immanent task* of philosophy is to develop one, and, alternatively, it sets the *limit* of philosophy, which must stand against it as a "critical science" (p. 9).

The question of the degree to which science itself harbors philosophical worldviews, either of necessity as belonging to its epistemological or socio-cultural make-up or rather as consequence resulting from implications of its theories, is an important debatable subject—though one science teachers or the curriculum rarely entertain. Yet there are clear historical precedents (e.g., Copernican revolution, materialism, reductionism, scientism, eugenics). The attention this topic deserves in science education has only recently been addressed (Matthews, 2009a).

In the field of education, one cannot deny the consequences of Dewey's philosophy, especially in the development of science education (DeBoer, 1991), although too many today are not familiar with it. For science teachers the reality of philosophy as implicit worldview equally cannot be denied, for it pervades their subject and curriculum in diverse guises, such as objectivism, reductionism, and positivism (among others), though regrettably many will not recognize it. The argument that science teachers in particular need to become more philosophically circumspect regarding the dimensions of their own teaching (e.g., nature of authority) and about science epistemology and worldview as presented in curricula (the image of knowledge or nature of science being projected) was first raised by Robinson (1969) and later Roberts and Russell (1975) some time ago, yet it remains more acute than ever today (Hodson, 2008; Roberts & Oestman, 1998). For instance, there are presently those among critical pedagogues, postmodernists, multi-cultural science educators, and others (e.g., Aikenhead, 1997a, 2002a; Foucault, 1980, 1989; Loving, 1997; Hodson, 1993a; Lyotard, 1979/1984) who would go much further, making sinister accusations against the Western science curriculum (if not against so-called "Western science" *per se*). They charge that its unspoken epistemology is tainted with an ideology labeled *scientism*, which is indisputably exclusivist and oppressive while it innocently parades as a vital knowledge tool for globalizing interests, socio-economic progress, and enlightenment.

Here we have *two unavoidable critical claims* about science teaching and curriculum (the "hidden" philosophy behind curriculum, and that it is supposedly oppressive) for arguing the necessity of developing a PSE for teachers, their profession, and the academic research field in general. These reasons are certainly grave enough, especially if the claims have merit, and yet if such a philosophy could in addition incorporate *educational* discussions and meta-theoretical dimensions that could show improvements in science conceptual understanding alone, that of itself it seems (one would hope) would be enough to convince most science educators and researchers of the worth of the undertaking and establishment of this new inquiry field.

Let us nonetheless return to Wittgenstein's observation and examine further the common accusation of philosophy's ineffectual worth, its lack of progress, its inability to penetrate the veil of existence and hence its lost status, in short, it being rather "useless," taken to mean of little practical value. For he had uttered a skeptical statement of the task and value of philosophy that corresponds with a widely-held notion today—although he still saw its worth rather differently, to help dispel nonsense problems: "the philosopher's treatment of a question is like the treatment of an illness" (*PI* #255). Yet the problem of its *practical relevance* is as old as Aristotle and the thorny relation between *theoria* and *phronésis* (between "theoretical" and "practical" reason) as discussed in his books. For right at the beginning of his *Metaphysics* (1998), he deliberately raises this issue as to what sort of thing "philosophy" is. He contrasts the knowledge gained through experience of particular objects or events with theoretical knowledge gained *from* these experiences that allow us to derive general propositions or "universals" that can go beyond them. He admits that to know particulars is invaluable, since from them comes practical action—for example, in the case of medicine to cure the real person in the concrete situation (not some general category "man")—and conversely having a mere "theoretical account" without experience often leads to error. This is the advantage of practice, and it is the view many teachers, especially pre-service teachers, would recognize and immediately share. But the truly knowledgeable, so Aristotle continues, develop expertise in both modes of doing (*techné,* as an art or skill) *and* theorizing (*theoria*), and it is these that are generally accorded more prestige. Such status he associates with the ability of *techné* and *theoria* (today more commonly termed "technical" and "theoretical" reason) to having identified underlying causes and principles that usually evade the mere practitioner. In so doing he consciously asserts two value judgments, one according greater worth to reasoning and contemplation over practice, and the other defining the worth of philosophy as residing in uncovering those causes and principles as an end-unto-itself—or as I would interpret it, the *search for understanding* broadly conceived. So viewed, it retains sole *intrinsic worth*: "So it is clear that we seek it for no other use but rather, as we say, as a free man is for himself and not for another ... for it alone exists for its own sake" (Aristotle, 1998, p. 9). As the translator phrased this viewpoint, "philosophy is supremely useless and supremely elevating" (p. 4).

Now such a statement may sound quite absurd (and help reinforce the opinion of the irrelevance of philosophy) to those in science education whose ear resonates with the loud tones of utilitarian intentions and avowed moral projects, especially among those reformers preoccupied with singing the goals of science education (and "literacy" conceptions) to a drumbeat of socio-utilitarian or socio-political tunes (e.g., the STS movement or "science as socio-political action"; Hodson, 2003; Roth & Désautels, 2002; Yager, 1996). Not to be misunderstood, no disparagement of such declared ventures is intended here. My point is a different one: what usually remains unacknowledged is that such sentiments and arguments assume a predetermined position for "philosophy" implicitly. To be sure, it presumes the appropriateness of a socio-political philosophy of education. One implied standpoint taken by such movements, or so it appears, is that *if* philosophy is to have any merit at all (as a useful PSE), it must first of all (if not exclusively) perform a duty and render service—in other words, have *instrumental worth*. That may indeed be *one* of its functions, as is currently taken on board by the "critical-emancipatory" research tradition in science education (Kyle, Abell, Roth, & Gallagher, 1992). But this tradition relies heavily (and self-acknowledged) on a Marxist-Freire and Habermasian influenced philosophy of education and worldview, also termed "critical pedagogy" (Apple, 1990; Blake & Masschelein, 2003; Freire, 1970)—some weaknesses of which will be discussed later in Chapter 3. But whether this should be the *sole role* of a PSE or even its *best* one—indeed, whether such a philosophy of education should be presumed—is exactly what needs to be first established, and that means analyzed, argued, and not assumed. This situation exhibits another crucial task of PSE, that such *hidden* philosophies of education (PE) *must be brought to the surface and revealed as they impinge upon the purposes, values, and identity of science education.*

The view taken here is, first of all, to step back and align with the original intention of Aristotle, being the need for a search for understanding, and hence *a PSE stands on its own intrinsic merits*. As Dewey (1916/1944) wrote: "On the side of the attitude of the philosopher and of those who accept his conclusions, there is the endeavour to attain as unified, as consistent, and complete an outlook upon experience as is possible. This aspect is expressed in the word *philosophy*—'love of wisdom'" (p. 324, italics in original). But this is not meant to imply, secondly (as Dewey certainly would have objected to such a conceived split between theory and practice), that it could not perform a dual role, to indeed serve some very useful functions (inclusive of the aforementioned question of duty), such as either helping better demarcate science education from psychology, language, or science studies—with critical appraisal when utilizing theories from outside disciplines—or analyzing and clarifying science education goals (including "literacy"), among others, all of which could contribute to helping ground science education as an independent, mature discipline. This broader perspective for a developed or "mature" PSE, however, must conceive of the valuation "useful" rather differently, namely, that *understanding* in actual fact performs quite an important practical role in and of itself, for it inevitably, and sometimes immediately, forces change in thinking and doing, by blending contemplation and action. In short, the search for understanding is never "useless" (as the "process" philosopher Whitehead [1929], himself saw), only that the ways in which *theoria* is interwoven with *techné* or *phronésis* may not be obvious or generally foreseen.

With that admission I wish to avoid not only a misunderstanding but a danger inherent to certain perspectives of the task of any philosophy as well as the assumed *role* that "theory" itself should take—both issues being identified as problems belonging to, and long exhibited in, the history of philosophy and philosophy of education, including everyday thought. My view bears much in common with the well-known positions of both pragmatism (e.g., Dewey, Rorty) and philosophical hermeneutics (e.g., Gadamer), which strongly deny any strict separation between theory and practice in reality, and which would characterize such dichotomies as artificial because *mind* can never be truly divorced from the cultural-communal nature of language and social activity. Rather such assumed positions as the isolation of philosophy and the divorce of "pure theory" from life or lived practice result instead, according to Dewey's (1916/1944) perspective, from two sources: from *social strictures* and from perceived *dualisms* due to the nature of thinking and experience. The latter has occupied philosophical thinking for centuries and are easily identified as deep problems having preoccupied the field, as some still do today (intellect and emotions, particulars and universals, knowing and doing, mind and body, individual and society, authority and freedom, science and arts, culture and vocation). By the former is meant those social conditions and values inherent to a society that privileges passivity and the contemplative life, elevating abstract reasoning and theorizing above practical reasoning and agency (also placing the worth of academic life above vocations).

What is here proposed, in other words, for PSE is, in the *first instance,* that it be taken as *theoria* as Aristotle and Kant understood it, namely in the sense of *critical* reasoning but with practical import. Although admittedly, the former philosopher also meant from work withdrawn, a contemplative and consummate life as the ultimate human achievement, while the latter meant the supremacy of epistemology because of its detached sovereignty capable of objectively assessing the worth of all knowledge, beliefs, and actions. Both of these evaluations have certainly come to typify modern academic as well as everyday notions of philosophy and "theory" (though it should be stressed the Aristotelian notion was communal, whereas the modern one tends to be individual-subjectivist). But one need not hold to the implied wider conceptions of these two well-known philosophers while still maintaining that *theoria* understood in the widest sense could fruitfully imply an analysis and critique of all *three* kinds of rationality: namely, *theoretical*, *technical* (or instrumental), and *practical*—themselves first identified by Aristotle (1998, p. 154).

Such a view does not automatically implicate one in any sort of prioritizing among them, and the temptation to do so must be resisted. For it is common nowadays, especially with the devaluation (if not dismissal) of "theory" that has come along with postmodernism (Lyotard, 1979/1984) and with the rediscovery of "practical reason" in educational studies (see Dunne & Pendlebury, 2003), to not only contrast technical reason (*techné*, or production) with practical reason (*phronésis*)—which can have its merits—but to associate (if not mistakenly to subsume under or outright equate) *scientific theory* with the former, at the same time as elevating the latter at the denigration of the first. One proceeds by creating the new hierarchy: practice over theory and production (together), to supplant the antiquated Greek hierarchy of theory-practice-production. Heidegger certainly inverted this order initially, as did Dewey, but whereas the later Heidegger appears to have returned to the classical Greek schema, Dewey did not (Hickman, 2001). So the implications of the notion of *theoria* casts a wide conceptual and historical net, it is acknowledged, but no such prior mentioned commitments or disputes are meant to be either implied or ignored by my use and understanding of the term *theoria,* or the view here taken on philosophy.

Having said this, what is proposed for a PSE in the *second* instance is that its task is neither merely *passive* nor just *critical*, but also *constructive*:

> Philosophy is not [passive] in the sense that it involves merely making explicit what was implicit in the forms of thought that have developed [contra Hegel]. Neither is it merely critical in the sense that it involves merely challenging such systems of thought. It can pass over into attempts to reconstruct conceptual schemes and think out anew the basic categories necessary for describing the world [of science education]. (Peters, 1966, p. 61)

This book is such an attempt to *think science education anew, and to offer constructive advice from philosophy of education* by utilizing "conceptual schemes" based on the metatheory of Egan (Chapter 3), as well as advice from *philosophy of language* by utilizing the philosophical hermeneutics of Gadamer (Chapter 5).

In the next section, the merit of philosophy will be taken up in earnest by presenting several arguments for its unquestionable efficacy, in particular for science teacher education.

1.2.2. The Value of Philosophy

Philosophy is an academic discipline that seeks to establish a systematic reflection on reality, however it may be construed. Its analytic function, often termed rational inquiry, involves critical appraisal of different topics, beliefs, and schools of thought.[7] Because of the complexity of the world around us (both natural and artificial), philosophy has been traditionally divided into separate major fields of study (first accredited to Aristotle) such as metaphysics, epistemology, logic, ethics, aesthetics and politics. These fields individually have either major or lesser bearing on science education directly. The *first two* have played a significant historical role pertaining to our understanding of the nature of reality, of knowledge and of science.[8]

- *Ontology*: the branch of philosophy (metaphysics) that concerns itself with the most general questions of the nature or structure of reality: what "is" or "what is *being*?" and existence. It examines natural and supernatural claims and asks about the feature of essences (e.g., are natural kinds, like species, universal or nominal?). Questions regarding *scientific* ontology are concerned with ascertaining the status (or validity) of the products of human creativity or discovery; included are scientific models and theoretical entities (e.g., gene, field, black hole, tectonic plates, etc.), evaluated as to their truth (realism) or merely useful (fictive) construct to solve problems and fit experimental data (empirical adequacy).
- *Epistemology*: the branch of philosophy that studies the nature of knowledge, its scope, foundations, and validity; it deals with theories of knowledge, distinctions between believing and knowing, and justification. *Scientific* epistemology is concerned with describing and ascertaining the nature of both the body of known scientific facts and theories (degree of certainty) and the production of new knowledge (i.e., scientific inquiry). *Personal* epistemologies are commonly taken to include individual beliefs, views, and attitudes about a particular subject; hence they can be considered a "personal knowledge framework" (i.e., "what do you know about 'X', and how do you know (it)?"). Two competing views of epistemic justification are *foundationalism* and *coherentism*.[9]

As mentioned, the significance of *philosophy of science* for science education is generally recognized today—though more so among researchers than science teachers themselves (Duschl, 1994; Hodson, 2008; Matthews, 1994)—while philosophy per se is accorded much lesser importance, notwithstanding the limited forays by some researchers into its sub-fields, which can be acknowledged (e.g., language studies, post-structuralism, hermeneutics, scientific argumentation, "critical theory"). Why this situation has arisen and persists is an open question and would require its own socio-empirical research and hence is not of immediate concern of this review. But it remains an important question that should be pursued, as it could reveal much about our community, about how science education is perceived and undertaken. In other words, it aims at the core of the self-understanding of science education as profession and identity (Fensham, 2004).

A familiar question posed by pre-service and science teachers alike is: "What does philosophy have to do with science?" Or, more succinctly and less pejoratively, "how can any sort of 'philosophy' contribute to helping my students better understand difficult *scientific* concepts?" Such questions implicitly assume of course a deep divide between science and philosophy, certainly between science education and philosophy.[10] While science teachers need not be openly hostile to philosophy, they certainly appear indifferent. Much responsibility can be laid at the door of the academy, its structure, culture, and teacher training. Their attitudes and preparation effectively expose much about how teacher identity is formed,[11] about preconceptions of knowledge, but also about the nature of university science education and scientific specialization, including the nature and influence of science textbooks.[12]

Classroom teachers tend to be more concerned with valuable but mundane matters of decision-making regarding immediate instruction, learning, and assessment. For them as pertains their professional duties and identity, these concerns have little if anything to do with philosophy—or so it would seem. A consequence of this disregard makes providing educational rationales of their thinking and practice a challenge: "When planning lessons, teachers often struggle when asked to express how they decide what science content within a discipline is worth teaching. Rationales are post-hoc and rarely reflect deep thinking about the structure of the discipline, or how students learn" (Clough, Berg, & Olson, 2009, p. 833).[13] Their struggles become quite apparent when further asked to give an explicit account of their "philosophy of teaching" or "philosophy of learning." And this counts not just for content teaching and conceptions of learning but equally for providing truly *educational* objectives for either their individual courses or overall science education.[14] Seldom are the contextual aspects of teaching the subject matter made explicit even though *seven* competing "curriculum emphases" have been identified in science educational history (Roberts, 1988). In effect, particular curricular emphases bear witness to buried educational philosophies. The teaching profession itself is mired in a scenario of what Roberts (2007) has astutely identified as two substantial conflicting "visions" of science education.[15] These facts alone warrant developing philosophical acumen for teachers.

If this picture as sketched is indicative of teacher training and science education culture, then emphasizing the significance of philosophy, especially philosophy of education (PE), would require a paradigm shift in thinking. Exactly this sort of thing had been recommended by Jenkins (2000) for effective reform of that culture, although the present proposal would encompass a wider scope than was initially suggested.[16]

Philosophy in truth cannot be avoided, and not just for analyzing national "standards" documents, providing coherent rationales, or detecting curricular ideologies. Science teachers inadvertently find themselves in its territory when confronted by diverse events, such as: (1) explaining common scientific *terms* (like "law," "theory," "proof," "explanation," "observation," etc); or (2) student-driven *quandaries* ("how do we know X?"; "do models reflect reality?"; "why are we studying this?" etc.); or (3) when teacher and pupil together come across science-related public *controversies* (e.g., climate change, nuclear weapons, evolution versus intelligent design, etc.)—never mind popular beliefs and media reports (e.g., astrology or alien abductions). Such occurrences usually illustrate that "philosophy is not far below the surface" in any classroom (Matthews, 1994, p. 87). Moreover, the scientific tradition (as an integral part of Enlightenment culture) based on rationality, objectivity, and skepticism that teachers have inherited is equally challenged by strands of pseudoscience, irrationality, and credulity of the times (Hodson, 2009; Slezak & Good, 2011). How can teachers illustrate these differences, especially the distinction between valid and reliable knowledge claims from invalid ones (or natural from supernatural claims) without philosophical preparation? Yet it's not just the classroom, contemporary media discourse, or pop culture that is infused with questions, beliefs, claims, and counter claims of philosophical significance, but likewise the evolution of science itself.

When science is seen historically, its development has always been interwoven with philosophical interests and debates, whether concerning epistemology, logic, metaphysics, or ethics (the major sub-fields of philosophy proper). A quick survey makes this evident: from debates on the nature of matter or motion in Ancient Greece, to questions of logic, method, and truth with Galileo and Kepler during the Copernican revolution (or Descartes and Newton in the Enlightenment); also Lyell and Darwin concerning the age of the Earth or origin of species, respectively, in the 19th century (which saw the realist controversy about atoms in chemistry revived). Although such controversies have ceased among professional scientists, right down to our present age philosophical controversies exist, whether concerning the onto-epistemological debates in quantum mechanics or reduction in chemistry.[17]

The history of science, furthermore, is not simply a survey of fantastic discoveries, ideas, and theories as too many textbooks would imply, but is equally littered with discarded concepts and discredited theories (e.g., ether, epicycles, humors, electric fluids, vital force, quantum orbits, phlogiston, phrenology, caloric, Lamarck, Lysenkoism, etc.). Can teachers distinguish between quasi-histories and pseudo-histories? Or unmask how subject matter is organized to reflect the typical linear, non-controversial and progressive accumulation of scientific knowledge, imitating the myth of "convergent realism?" (Kuhn, 1970, 2000; Laudan, 1998a). The textbook's and one's personal view of scientific knowledge and its development both presume prior philosophical commitments (e.g., positivism? empiricism? naïve realism? critical realism? social constructivism?).

Regarding ethics, one should not forget that Socrates was condemned on moral and religious grounds—as were Bruno, Galileo, Spinoza, Rousseau, and Darwin (though not exclusively). Eugenics, once the scientific "hard core" of the social Darwinism movement, was considered a legitimate topic of scientific research less than a century ago. Even modern physics cannot escape this subject, ever since Oppenheimer made the self-incriminating remark that physicists "had known sin" by developing the atomic bomb. The American philosopher C. S. Pierce had stated: "Find a scientific man who proposes to get along without any metaphysics ... and you have found one whose doctrines are thoroughly vitiated by the crude and uncriticised metaphysics with which they are packed" (as cited in Matthews, 1994, p. 84). Studies in history, philosophy, and sociology of the sciences (HPSS) have made this claim abundantly apparent. These fields cannot be either ignored or glossed over during science teacher education, but require time and attention for the emergence of an adequate PSE.

We have already noted the worth of philosophy (along with key aspects mentioned above) to lie in providing teachers with both (1) the perspective for synthesis of their educational enterprise by developing a PSE framework, and (2) making available to them in-depth studies termed "philosophies of." Linked to the latter, coming again to PS (appearing as the PS corner of the Figure 1.1 triangle), teachers need to be made aware that in the past 20 years new avenues of scholarship have been developing *within* the sub-field itself to help them expand their foundational understanding of their specialty (e.g., philosophy of chemistry, philosophy of biology).[18] Here questions concerning major issues in subject matter content that bear directly on senior courses are being discussed. For example, there is dissention whether laws and explanations in biology and chemistry are of the same order and function as those in physics—normally taken for granted in PS literature.

Such "cutting edge" philosophical research has acute ramifications for secondary and post-secondary education, expressly *subject epistemology*, including nature of science discourse (Irzik & Nola, 2011; Jenkins, 2009; Matthews, 1998a, 2014).[19]

In addition to the above mentioned reasons, the worth of philosophy plainly lies in *self-reflection*. This means nothing less than to reassess one's own practice, educational ideas, and aims, even going so far as to re-evaluate one's own constructed socio-cultural science teacher *identity*. Along with suggesting "philosophies of," Scheffler also argued that science teachers require philosophy as a "second order" reflective capacity into the nature of their work, their understanding of science, and their educational endeavors. He considered this capacity analogous to the role PS plays when examining science:

> The teacher requires ... a general conceptual grasp of science and a capacity to formulate and explain its workings to the outsider.... No matter what additional resources the teacher may draw on, he needs at least to assume the standpoint of philosophy in performing his work.... Unlike the researcher [or the academic] he cannot isolate himself within the protective walls of some scientific specialty; he functions willy-nilly as a philosopher in critical aspects of his role. (Scheffler, 1970/1992, p. 389)[20]

These proposals of Scheffler can equally be associated today with requirements to enhance teachers' "pedagogical content knowledge" (PCK; Abell, 2007; Kind, 2009; Shulman, 1987; Van Driel, Verloop, & de Vos, 1998), which not only means developing *their epistemology* of science (Matthews, 1994, 1997) but in addition their familiarity with *philosophy of education topics* (Matthews, 2002; Waks, 2008). (Though it needs mentioning that often PCK research has ignored epistemological and especially PE components.) Again, Figure 1.1 displaying the PSE framework identifies these important aspects and illustrates how they are related to, and embedded within, the three corresponding dimensions of philosophy, PS, and PE.

Philosophy in a nutshell then corresponds to the ancient Socratic dictum to examine oneself, and that "the unexamined life is not worth living." Transposing this motto, "the unexamined pedagogy is not worth doing"; in fact, it's unsuccessful (as conceptual change research has uncovered), if not harmful (i.e., indoctrination into scientism[21]). Such an examination aligns with Kant's famous definition of Enlightenment as the emergence from one's self-imposed immaturity (due to reliance upon outside authority), the ability to freely make use of one's own faculty of reason, to "have courage to use your own understanding!" (Kant, 1784/1974). This ambition is inherent of course to the *liberal education tradition* (Carson, 1998; Matthews, 1994; Stinner, 1989), the objective sought after when teachers desire students to "think for themselves"—easily an identifiable historical goal of science education (DeBoer, 1991, 2000; Schwab, 1978). This is inclusive of the newer critical thinking movement (Bailin, 2002; Siegel, 1988, 1989; Smith & Siegel, 2004). The primary focus here, however, is upon the further development of teachers' critical thinking and competence, their own capacity to judge not only curricular and policy documents, but above all their pedagogy, epistemological assumptions, and educational beliefs. (Whether implied by their textbooks: e.g., naïve realism, inductivism, pseudo-history—or proposed by science educational literature: e.g., STEM, STSE, constructivism, postmodernism, science-for-social-action, etc.). The topic of *critical thinking* is well-trodden ground in PE, although researchers seldom avail themselves of this literature (Bailin & Siegel, 2003; Siegel, 2003).

Finally, as Wittgenstein (1958) stated, philosophy can even be *therapeutic*. Implied for our theme this means it can alert science teachers to implicit *images* of science and philosophies of education they may hold unaware. Perhaps they have internalized these through practice or originally picked them up through teacher training from university professors promoting their own pet educational ideas and theories. Indeed, the teacher may have developed strong opinions about HPSS or social justice topics, "but the point of education is to develop students' minds, which means giving students the knowledge and wherewithal to develop informed opinions" (Matthews, 1997, p. 171). In any case, translating Pierce's statement above with science educators in mind, one can write: "Find a science educator who proposes to get along without any philosophy of education ... and you have found one whose goals, perceptions, and methods are thoroughly vitiated by a crude and uncriticised one with which they are packed."[22] While the textbook epistemology is often concealed, a teacher's epistemology and educational theory is usually pieced together during his or her career and rarely made explicit.

In summary, philosophy cannot be gone around, for as a discipline of critical inquiry it allows analysis into curriculum, textbooks, learning, best practice, and identity. Relooking at our previous PSE triangle (Figure 1.1), this includes: (1) offering conceptual clarity; (2) unmasking ideologies (social, political, educational); (3) sorting out foundational aims, values, and teacher identities; (4) providing perspectives and theoretical frameworks, as well as synoptic and integrative approaches; and (5) possibly even utilizing *creative* theorizing as solutions to pressing problems (discussed below in Section 1.4 on educational theory).

1.3. THE RELATION TO *PHILOSOPHY OF SCIENCE*

How would a philosophy of science education (PSE) be related to the field of philosophy of science (PS)? This question brings us to the bottom left corner of Figure 1.1. Thus far we have attempted to ascertain the status and worth of PSE as a "philosophy" and established its location within historical philosophical developments. We have presented several arguments to reinforce the value of philosophy in general for developing a teacher's PCK and emerging educational critical mindset. Moreover, we have already briefly identified the significance of the newer PS subdisciplines for science teacher education and professionalism. It probably goes without saying that any such philosophy *of science education* that has as its main concern the education of persons into the disciplines of science (narrowly construed)—or better phrased, an education of the scientific enterprise (widely construed)—in addition must consider what studies into the epistemology and *nature of science* (NoS) have also revealed.

Here, though, we are on well-trodden ground, no matter how difficult the trek in the past has been through muddy disputed soil and foggy epistemological terrain in the academic field and associated scholarship. For there have been cases not only of historical overlap in the development between the two fields of science education and PS in the past, as examples, Mach in Europe and Phillip Frank and Robinson in America (Frank, 1957; Matthews, 1991; Robinson, 1969), but also cases where the two fields have diverged and not taken notice of each other. It is now known among researchers that the well-funded and large-scale curricular reforms enacted predominately by government ministries and scientists in a "top-down" fashion in the 1950s/1960s were strongly informed by an inductivist perspective on scientific methodology and science learning, which itself reflected (though not solely) the then-reigning ahistorical positivist views in the philosophy of science (the "received view"; Suppe, 1977). The terrain of problems associated in particular with defining *scientific methodology and authentic inquiry* appropriate for science classrooms and laboratory work has been recognized for some time now as especially thorny (DeBoer, 1991; Hodson, 1996; Schwartz, Lederman, & Crawford, 2004).

Yet even at the time of the early 1960s when the philosophy sub-discipline was undergoing a major upheaval and shifted into a "post-positivist" phase with the works of Hanson, Toulmin, Kuhn, Feyerabend, and others (Giere, 1999; Suppe, 1977), science education remained largely wedded to the older philosophy. By 1985, Richard Duschl could complain there had been "25 years of mutually exclusive developments" between the two, with science education (e.g., textbook and teacher epistemology) imbued as it were with inductivism and positivism, lagging decades behind. In other words, the influence of some sort of PS was always present to some degree within science education and its changing "curricular emphases" (Roberts, 1982) in the last hundred years or so of its advancement—whether acknowledged by teachers or not—just that on some occasions in certain eras the influence of that sub-discipline was more pronounced and explicit than at other times. Moreover, it is clear that changes in one field take decades before they impact the other.

Today, at least on the academic front, the situation has changed considerably. The revival of the *history and philosophy of science* (HPS) reform movement (its origins go back almost a century as well) in the 1990s has now allowed for the "rapprochement" in earnest of the once separate academic disciplines of history, philosophy, and sociology of the sciences with science education. The founding in 1992 of the new journal *Science & Education,* dedicated exclusively to HPS reforms, has allowed a common platform for academic cooperation among the diverse scholarly fields to share insights for the express purpose of advancing science education research. But its potential has not yet been fully realized for science teacher education or science classrooms (Clough & Olson, 2008; Nashon, Nielsen, & Petrina, 2008; Matthews, 1994 2014; Turner & Sullenger, 1999).[23]

For science teachers themselves, most have some sort of rudimentary "philosophy" of teaching their subject matter (more implied than expressed), say, keeping close to the "structure of the discipline" or, more commonly today, "teaching by inquiry." Unfortunately, as already some earlier studies had shown, their beliefs (or as I would prefer, personal "philosophies" or epistemologies) quite often do not match their practice in classrooms (DeBoer, 1991; Hodson, 1993b; Welch, Klopfer, Aikenhead, & Robinson, 1981). Regardless, studies have shown that some *image of science* is either implied (if not fully recognized) or overtly taught in the way teachers use *language* in instruction and arrange activities to justify the nature of lab or fieldwork (Lemke, 1990; Tasker & Osborne, 1985; Wellington & Osborne, 2001). Students receive some lasting impression of what science is and how it is done from what is articulated and displayed in class—termed "companion meanings" by Roberts and Oestman (1998). This is further influenced by what textbooks deliberately convey about science *content knowledge* (CK) and "processes" (or "method"), or other *nature of science* (NoS) terms (e.g., law, hypothesis, theory, etc.), including the kind of distorted historical portrayal of the subject content (Allchin, 2003), often referred to in general as the "epistemology of school science" (Cawthron & Rowell, 1978).

Overall, considering what is conveyed about the nature and history of science, this is done quite poorly, which is now widely recognized and been the subject of very many research studies (Lederman, 2007; McComas, Almazroa, & Clough, 1998). In brief, a false sense of inquiry as a universal and singular "scientific method" is disseminated (Bauer,

1992)—one of about 15 myths that have been identified to infuse school science (McComas, 1998). Moreover, in secondary and tertiary classrooms, a narrow curricular focus and assessment on the "what" instead of the "how" of science—that is, a preoccupation on formal, decontextualized CK organized into topic structures of specialized disciplines instead of how knowledge is obtained and confirmed along with the actual process of theory change and discovery—too often result in an image of science in static "final form" (Duschl, 1990; a problem already identified by Mach and Dewey many decades earlier as a major obstacle to science learning). This pertinent observation echoes Schwab's (1962) prior complaint that such instruction implies science is to be understood as a "rhetoric of conclusions," while such structures at the same time implicitly assume and project a so-called "positivist epistemology." It is then not all too surprising that science teachers themselves usually associate their identity closely with the science of their subject specialty, with preconceived views about knowledge and instruction (their "epistemological beliefs") as the direct transmission of ahistorical "objective facts" as conveyed in their technical textbooks.

Sutton (1996) has insightfully described how the *language* of a scientific idea changes from its initial discovery through to research paper and textbook formulation, especially how textbooks distort and mask the *historicity* of scientific language and theory (key educational issues first pointed out by Thomas Kuhn himself). He has argued that students' *beliefs* about science are shaped by their beliefs and use of language (following Lemke, 1990), which are intimately shaped by textbook "talk" and classroom conversation, reinforcing Duschl's "final form" indictment. (The problem of the nature of language and its key role in science learning and curricula is examined further in Chapter 5.) All these factors are said to contribute to the low levels of science literacy and the ongoing poor understanding of science among the public (Bauer, 1992; Miller, 1998; Yager, 1996).

The ability to change this particular debilitating quality of the character of science textbooks appears to be extremely limited for the foreseeable future (although exceptions exist, for example, the physics textbook of Holton & Brush, 2001), hence chances for improvement of HPS awareness and inclusion seem better suited to be met in the areas of teacher in-service and pre-service teacher training.[24] Matthews (1994) writes with respect to teacher epistemology:

> A teacher's epistemology or theory of science influences the understanding of science that students retain after they have forgotten the details of what has been learnt in their science classes ... this ought to be as sophisticated and realistic as is possible in the circumstances. [Unfortunately] a teacher's epistemology is ... largely picked up during his or her science education; it is seldom consciously examined or refined. (pp. 204–205)

Interest in studying the epistemology of school science had already produced several significant prior studies (Cawthron & Rowell, 1978; Elkana, 1970; Hodson, 1985; Smolicz & Nunan, 1975), and research continues into both teacher and student epistemologies (as examples, Abd-El-Khalick & Lederman, 2000; Désautels & Larochelle, 1998; Driver, Leach, Millar, & Scott, 1996; Meichstry, 1993; Ryan & Aikenhead, 1992). Such studies have too often shown, sadly, that an improved alignment between teacher and nature of science epistemologies that Matthews identified as of central importance is still far from being realized. Nonetheless, the importance of *nature of science* (NoS) understanding for improved student comprehension of what science is all about has been an avowed objective of science education for almost a century, and has recently been reemphasized on a global scale (as mentioned). But how effective these policy documents are remains arguable (Lederman, 1998, 2007), and for the most part teachers remain indecisive, if not poorly prepared, to successfully implement HPS aspects into their instruction. This situation has been investigated in a paper by Höttecke and Silva (2011), who spotlight four major obstacles with regard to physics education that continue to hinder effective implementation.[25]

The situation has only marginally improved since the time of Gallagher's (1991) paper, where he had identified the inadequate philosophy and history of science preparation that students receive from both their specialist undergraduate science courses in the academy and during teacher training in education faculties. Since then there have certainly been efforts undertaken in some countries (Argentina, Australia, Canada, Denmark, Finland, France, Germany, Greece, Italy, S. Korea, some U.S. states) to improve the HPS education of aspiring science teachers, although the success remains largely uneven and highly dependent upon individual researchers and teacher educators at given universities and faculties.[26] One suggestion of this book is that a PSE, when shaping a teacher's *pedagogical content knowledge*—following Shulman's (1987) perspective—would seek to *influence a science teacher's specialist CK through HPS awareness and education* (examined in Chapter 4). From a *hermeneutic perspective*, this would mean to allow the horizon of a teacher's "epistemology" (or understanding and beliefs of subject content and teaching) to expand through awareness of the need of nature of science insight and HPS integration as essential components of his or her PCK (examined further in Chapter 5). With a developed personal PSE, as argued here, teachers' own tightly held attitudes and beliefs, which so often form obstacles to HPS implementation, would be changed and conducive to different modes of seeing and instructing, because a personal philosophy of education forms part of the core of our emotional center and identity as teachers.

Any discussion involving the *nature of science* (NoS), it will be admitted, is itself contested terrain among respective academic disciplines (to be examined in the next section). This has certainly complicated the issues surrounding HPS inclusion and integration (Kelly, Carlsen, & Cunningham, 1993; Leonard, 2003; Matthews, 1998a, 1998b, 2003a, 2003b; Rudolph, 2000). While a PSE as articulated here would continue to support previous reform efforts towards HPS inclusion (e.g., Hodson, 2008; Monk & Osborne, 1997; Stinner, Mcmillan, Metz, Jilek, & Klassen, 2003), it has the additional task of canvassing issues as debated in the philosophy, history, and sociology of science disciplines in order to better ascertain the appropriate understanding of these issues for use in science learning, curriculum policy, and HPS integration—showing both strengths and limitations for the classroom. This book will contribute to this discussion *by attempting to uncover aspects of the current debate in the philosophy of science concerning the nature of science, especially the realism/instrumentalism controversy as relevant for science education* (in Chapter 4).

There still remains the question about the much-debated so-called "disunity" of the sciences (Galison & Stump, 1996), also possibly indicated by the current fragmentation of philosophy of science into various new subdiscipline specialties, as mentioned earlier (e.g., philosophy of biology, philosophy of chemistry, etc.). If one accepts the claim that there cannot exist any one "philosophy of science" (although there certainly exists an academic journal by that name) any more than a common language of science can be said to exist, but rather a multiple, then similarly the question arises if not a multiplicity of philosoph*ies* of science education (PSEs) may need to be considered and separately developed. Without wanting to address the complexity of the disunity debate, one can be sympathetic to the gist of the argument, and that developing diverse "philosophies of science education" may indeed be a possible (or even probable) future evolution of the general idea that has been here initially suggested.

On the other hand, a "philosophy of science education" may be only marginally affected by sub-specializations in the philosophy of science. As an example, while a teacher's CK in chemistry may need to be better informed by research in the philosophy of chemistry (one crucial component of PSE would involve stressing this factor), nonetheless a PSE is more concerned with how such CK can be made to fit with the requirements of an educational metatheory and its concern with the cognitive-emotive *developmental stage* of the learner, with respect to this subject matter. In other words, a teacher's CK and the curriculum are not at the forefront for learning science (although they are invaluable dimensions), as is commonly done. Rather, they are evaluated in light of philosophy of education (PE) and the learner's age-developmental mind-frame as befits what it means to *educate* a person in the sciences. This emphasis necessarily shifts the focus to the substance of a teacher's PCK, educational *aims,* and educational *philosophy*.

1.4. THE RELATION TO *PHILOSOPHY OF EDUCATION*

Thus far in the discussion we have examined the relation of PSE to *philosophy* and *philosophy of science* (PS). We now consider the subject of philosophy *of education* (PE)—the third and final corner of the Figure 1.1 triangle. This subject, one could argue, is probably the most essential since teachers' PE emerges from the core of their being and identity as professionals; it provides them with rationales for their teaching, it contributes to selecting curricular materials, scope, and sequence, and influences their attitudes and beliefs in substantial ways.

One is amazed to discover, however, that this academic field has tended to be little referenced, if not almost entirely ignored, not only by science teachers in particular, but the science education research field in general. Here exists an unknown landscape that has been barely explored. A possible exception is the original significance attached to this field by some authors within the revived HPS reform movement in the early 1990s (Matthews, 1997, 2002; Siegel, 1989), although that movement too, appears lately (last 15 years) to have lost sight of its merits (Matthews, personal communication, 2009). Other than the seminal authority of John Dewey (above all in North America), whose recurring influence has ebbed and flowed in time from the 1920s onwards, it can readily be admitted that no other philosopher has since managed to significantly shape the science educational endeavor (DeBoer, 1991). One major feature of this book will be *an appraisal of Dewey's philosophy of education, chiefly his conception of language in light of the "interpretative turn"* (as discussed in Chapter 5).

Philosophy of education (PE), as mentioned, is a branch of philosophy. It seeks to address questions relating to the aims, nature, and problems of education. As a discipline it is "Janus-faced, looking both inward to the parent discipline of philosophy and outward to educational practice.... This dual focus requires it to work on both sides of the traditional divide between theory and practice, taking as its subject matter both basic philosophical issues (e.g., the nature of knowledge) and more specific issues arising from educational practice (e.g., the desirability of standardized testing)" (Siegel, 2007).

Thoughtful consideration of educational practice and assessing science curriculum is normally considered part of a teacher's professional competence, hence some sort of philosophical thinking can be justifiably attributed to educators and researchers. What is of issue is the view that science educators can be encouraged to philosophize on a broader and systematic scale, and they can profit from PE studies (using their in-depth deliberations on theory and practice).

1.4.1. The Neglect of Philosophy of Education

If as Aristotle had intimated (by the opening quote) philosophy begins when one is filled with wonder—a state of being that can arise when confronted with some dilemma (hence one's *lack* of knowledge)—then the neglect to articulate a *systematic philosophy of* (PSE) for one's own science pedagogy (let alone the research field) causes one to ponder why so little effort and time has been invested into the subject. The consequences have not been a minor matter—confusion over educational *aims* including the "science literacy" debate, its meaning, and competing "visions"[27]; science education's dependence on socio-utilitarian ideologies and competing group interests; science teachers' confusion about their identity and purpose, including the divide between belief and practice, and so on (refer to Aikenhead, 1997a, 1997b, 2007; Bybee & DeBoer, 1994; Donnelly, 2004; Donnelly & Jenkins, 2001; Pedretti, Bencze, Hewitt, Romkey, & Jivraj, 2008; Shamos, 1995; Witz & Lee, 2009; Yager, 1996).

Jenkins (2001) has rightfully complained that the research field is too narrowly construed and suffers from "an over-technical and instrumental approach" at the expense of other perspectives, such as neglecting historical studies. Although some recent research work can be taken as mitigating this charge (Gilead, 2011; Jenkins, 2007; Olesko, 2006; Simon, 2013), even his perceptive critique had failed to mention the worth of philosophical studies. The inertia of traditionalism at the upper levels had prompted Jenkins' surprising call for a paradigm shift, as mentioned—but this is serious talk, nothing less than a plea for somber philosophical contemplation and reorientation. Even at the post-secondary level the need to reform introductory science classes has received increased attention, especially with some new findings in physics education research (PER) indicating that the dominant textbook- and lecture-based instruction in large classrooms is unwittingly producing an anti-scientific mind.[28] The appearance in time of three identified public "crises" regarding school science education (1957, early 1980s, late 1990s, as elaborated in Chapter 2 [Schulz, 2009a]) and the apparent inability of different "reform waves" to provide for major, long-lasting changes could in turn suggest that a shift towards a more concentrated educational-philosophical examination of the problems lies at hand.[29] It can be argued that the general lack of consideration of educational philosophy and theory, that is, a *philo-educational failure,* could help account for why curricular reforms are particularly vulnerable to the political whims (or "ideologies") of various stakeholder groups, an enduring situation several researchers have taken notice of.[30] It could, for example, better inform policy deliberations when diverse stakeholders are at odds over what should count as science education (Fensham, 2002; Roberts, 1988).[31]

Fensham (2004) argues in his important book *Defining an Identity* that science education is still searching for ways to characterize its own "identity" as a discipline. (His comprehensive survey canvasses the views and backgrounds of 76 prominent researchers in 16 countries, active from the 1960s to the present.) One would like to suppose that helping to define such an identity would include philosophy, especially a PSE. And it is not only the identity of the *discipline* that is of issue here, but as referred to in the previous section, that of the classroom professional as well. Hence, it might appear the time has come for science education to return to some philosophical ground work, to come to value PE, and in turn, for the research field to inaugurate and develop a new *fourth area* of inquiry—philosophic-historical. This one added next to the common three of quantitative, qualitative, and emancipatory, in support of arguments made previously by others for its development as a "mature discipline" (Good, Herron, Lawson, & Renner, 1985; Kyle et al., 1992; Yager, 1984).

But Fensham's book, with the sole entry of PE on one page alone (where the significance of Dewey is also cited), bears ample evidence of the disregard of this subject topic for researchers and science teachers alike.[32] One can infer from the evidence to date that the worth of any sort of meta-analysis of their discipline and pedagogy seems to hold little value for the majority, thereto the need to bring systematic educational-philosophical reflection to bear on research, curriculum, and teaching.

This claim is further evidenced by a simple perusal of four research *handbooks* published thus far, where the subject of PE (including topics "philosophy," "educational theory," "curriculum theory") is missing entirely (Abell & Lederman, 2007; Fraser & Tobin, 1998; Fraser, Tobin, & McRobbie, 2012; Gabel, 1994). This absence is likewise attested by recent publications of European handbooks of research in the field (Boersma et al., 2005; Psillos et al., 2003). Crossing over the other way, most handbooks or "guides" of PE exhibit the same paucity by avoiding science education, though art education, moral education, knowledge, feminism, postmodernism, critical thinking, and critical pedagogy as subjects remain prevalent.[33] Two exceptions exist: Curren (2003) and Siegel (2009). Comparing both fields, the claim is reinforced by an inspection of the respective leading research journals in both PE and science education for the past 30 years, which exhibit an almost complete disregard of the opposing field (barring exceptions). What one finds is that only a handful of philosophers write for the science education journals, and even fewer science educators publish in philosophy of education.[34]

If an examination of the preparation of science education researchers is any indication of the kind of academic preparation science teachers themselves receive (before they become researchers), then another look at Fensham's *Identity* book as commented on by Matthews (2009b, p. 23) is revealing. He notes that "the interviews reveal that the overwhelming educational pattern for current researchers is: first an undergraduate science degree, followed by school teaching, then a doctoral degree in science education" (citing Fensham, 2004, p. 164). As Matthews observes, unfortunately "most have no rigorous undergraduate training in psychology, sociology, history or philosophy." Fensham (2004) himself comments that at best, "as part of their preparation for the development tasks, these teachers had opportunities to read and reflect on materials for science teaching in schools and education systems that were different from their own limited experience of science teaching" (p. 164).[35] Matthews concludes that Fensham's survey reveals an overall "uncritical adoption of idealist and relativist positions" among researchers, and that poor academic preparation is a reason why "shallow philosophy is so evident in the field."[36] It certainly appears as if the inadequate science teacher preparation in PE is mirrored by the widely recognized fact of the inadequate preparation with respect to philosophy, history, and sociology of science.

1.4.2. Historical Background of Philosophy of Education and Science Education

With an eye fixed solely on the mutual historical developments of both fields, this neglect is rather difficult to explain, especially because science education is after all about *education*, with natural focus on the science specialty. But philosophy and education have roots that are intertwined in history long past, convincingly traceable back to Plato (*Meno*; *Republic*). Every major philosopher in the Western tradition, from Plato (in Ancient Greece) to Kant (European Enlightenment) to Dewey (modern industrial America), have proposed educational projects of some kind (Frankena, 1965; Rorty, 1998). As Amelie Rorty (1998) correctly points out: "Philosophers have always intended to transform the way we think and see, act and interact; they have always taken themselves to be the ultimate educators of mankind" (p. 1). Understood in this way, Dewey (1916/1944) was on the mark when he famously phrased the view that the *definition* of philosophy is "the theory of education in its most general phases" (p. 331)—although most professional philosophers today would probably not construe it as such.

It was the Enlightenment's "project of modernity" (Habermas, 1987b)—first begun in the 17th century—that was expressly formulated as an *educational project*, and that saw in the new science of the day an instrument for personal and socio-political liberation (Gay, 1969/1996; Matthews, 1989). It is of course in full awareness of this intellectual and cultural heritage that postmodernists like Lyotard (1979/1984) would outright dismiss the "grand narrative" of this project with its associated role and *image* of science as an emancipatory and positive force, including those science educators convinced by his critique (Loving, 1997; Nola & Irzik, 2005; Rorty, 1984; Schulz, 2007).[37] In fact the popularity of strands of post-structuralist and postmodernist thinking among some researchers bears witness to the recent discovery of the value of philosophy for the field (Zembylas, 2000, 2006).

Looking much further back in time (again at the *Metaphysics*), Aristotle identifies the man of knowledge—one who has attained expertise either via *techné* or *theoria* (instrumental or theoretical reason)—as the one who is plainly able to teach what he has learned, and as such draws one distinguishing feature of the philosopher. To be a philosopher was to be a teacher. Conversely, to be a teacher implies one must do philosophy (of one form or other). Science educators, seen in this light, are inescapably located within a venerable philosophical tradition *along with* the newer scientific one that they usually and exclusively tend to associate themselves with—though, here too, not fully aware of the latter's cultural roots and significance.

The first mention of philosophy of education as a distinct field of study was in Paul Monroe's *Cyclopedia of Education*, published 1911–1913 (Chambliss, 1996). Philosophy of education, depending upon the given nation and its educational traditions, can be viewed as a relatively new discipline or not. As Hirst (2003, p. xv) points out, "philosophical inquiry into educational questions" was more established in the U.S., Germany, and Scandinavia, whereas in the UK philosophy of education as a discipline first came into its own in the 1960s. It was dominated by analytic philosophy and accounts of schooling, although in ethics Kantianism was the major influence. In the United States, the American Philosophy of Education Society had already been founded earlier in 1941, along with the Deweyan journal *Educational Theory* in 1951. It was the pragmatist philosopher and educationalist John Dewey in his influential work *Democracy and Education* (1916/1944) who had conceived of PE to be a study worked out on an experiential basis—in other words, that educational ideas were to be applied and tested in practice. He also considered that theory and practice were interdependent in a kind of feedback loop mutually learning from and reinforcing each other. This stood in contrast to the earlier views of the Englishman Herbert Spencer who instead conceived of education as an inductive science and where PE would serve as a kind of scientific method.

Alternatively, on the Continent in Northern Europe, very different views about education had been developing. The ideas of Kant, Schiller, Herder, Herbart, Humboldt, and others had contributed to create the influential *Bildung* paradigm in the 19th century.[38] It has become established as the *Bildung/Didaktik* tradition, whose conception of education dominates the German-speaking world and the Nordic countries.[39] Today this paradigm is not without its detractors, for by the 1960s this tradition had itself begun to clash with the "critical theory" of the Frankfurt school (Blake, Smeyers, Smith, & Standish, 2003; Blake & Maschelein, 2003; Smeyers, 1994). It continues to engender much debate among educational thinkers and philosophers alike, both in Europe and English-speaking countries. Thereto, advocates of both traditions—Anglo-American "curriculum" and *Bildung/Didaktik*—came together in the 1990s to open dialogue comparing the relative benefits of each (Gundem & Hopmann, 1998; Jung, 2012; Vásquez-Levy, 2002).

The *Bildung* paradigm itself actually represents an *educational metatheory* (Aldridge, Kuby, & Strevy, 1992), a type of "grand theory" in education, very few of which have been constructed in modern times (inclusive of Dewey and Egan; Polito, 2005). It immediately raises the question of the worth and relation of educational theory to practice, whose merits are currently being contested in philosophy of education (Carr, 2010).

The direct link between *Bildung* and science education[40] has been drawn only recently, notably in Fensham's *Identity* book (2004) and by Witz (2000).[41] Sjöström (2013) argues for the significance of *Bildung* for improving interest in, and learning of, chemistry, moving beyond "pure content" to knowledge *about* chemistry and society. Fensham provides a highly informative discussion, explaining the concept and significance of *Bildung* when contrasting the Norse/German tradition with the content knowledge-driven Anglo-American tradition. He contends that a serious shortcoming of the so-called "curriculum tradition" of the English-speaking world is its consistent disregard of metatheory (discussed further in Chapter 2).[42] *He advises that science education should acquire one*.

Another interesting aspect about the *Bildung* paradigm can be noted: it exercised an indirect influence via Herbart's ideas on the philosopher-scientist Ernst Mach. While Mach's impact on Einstein's thinking is generally recognized, his educational ideas are hardly known in the English-speaking world. Already back in the late 19th century he had been politically active for educational reforms, including improving teacher education, and is credited with founding and co-editing the very first science education journal in 1887, *Journal of Instruction in Physics and Chemistry* (Matthews, 1991, 1994). Siemsen and Siemsen (2009) argue his rediscovery at present could provide significant contributions to current European reform efforts.

On a final note, for the English-speaking nations, the U.S. was in the forefront of the establishment of both disciplines (science education and philosophy of education) that have developed in tandem—simultaneously but separately in the early 20th century. One would think that because of this pedigree, and in some cases of clearly overlapping interests (as exhibited in the important case of Dewey), that science education would be more cognizant, and science teacher training more reflective, of their common roots. Unfortunately, on this matter science education seems to suffer amnesia on both counts, for if it can be admitted that "philosophy of education is sometimes, and justly, accused of proceeding as if it had little or no past" (Blake et al., 2003, p. 1), then this certainly rings true of science education.[43]

The call for a PSE is not only to raise awareness of this forgotten earlier period, but *to identify the need to create a sub-discipline within educational studies* that, although new, nonetheless has substantial historical roots going back into the science-educational, but especially the philosophical-educational past.

Why science educators do not associate themselves just as intimately with PE is a fascinating question, one that cannot be pursued here. It almost certainly has a lot to do with several factors (such as: the prestige of science in society, how disciplinary knowledge is structured, how their own university science education proceeded, and, not least, how they were trained as educational professionals).[44] What are called "foundations in education" courses, which usually include studies in the history and philosophy of education, are often optional for pre-service science teachers, depending upon the prerequisites of their attending institutions.[45]

1.4.3. Philosophy of Education Today

Coming at last to the present historical culmination, PE has today progressed to become a respectable, established sub-discipline in philosophy. It comprises evolving research fields, a sizeable literature, professional organizations, professorial chairs, and several leading journals.[46]

There now exist two *handbooks* (Bailey, Barrow, Carr, & McCarthy, 2010; Siegel, 2009), but also a *guide* (Blake et al., 2003), *companion* (Curren, 2003), and dictionary of key concepts (Winch & Gingell, 1999). An *encyclopedia* of PE is also on hand (Chambliss, 1996). These can be sought out by science educators to familiarize themselves with the current discussion, inclusive of disputes regarding different topics of individual interest to them. Several newer and older *introduction* texts are also available (e.g., Barrow & Woods, 2006; Tibble, 1966), including Carr (2003) and Noddings (2011). For educators seeking immediate information, several encyclopedia articles exist providing succinct, comprehensive overviews of PE (accessible online: Phillips, 2008; Siegel, 2007).

1.4.4. The Value of Philosophy of Education

Philosophical questions bearing on the different facets of science curriculum, teaching and learning, must be addressed and inspected by the thoughtful educator.[47] Questions whether pertaining to: (1) chief educational goals and course objectives, or (2) assessing learning theories, or (3) bearing on nature of science and techno-science related issues—thereto, the character of scientific research, knowledge, and societal applications as related to curriculum or policy reforms. Questions also pertaining to who enacts and benefits from such reforms with respect to interests and ideologies. And all this often in spite of, not because of, state-mandated and pre-packaged "content knowledge" curricula:

> What are the aims and purposes of science education? What should be the content and focus of science curricula? How do we balance the competing demands of professional training versus everyday scientific and technological competences versus the past and present interactions of science with society, culture, religion and worldviews? What is the structure of science as a discipline and what is the status of its knowledge claims? What are the ethical constraints on scientific research and what are the cognitive virtues or intellectual dispositions required for the conduct of science? What is the meaning of key scientific concepts such a theory, law, explanation, and cause? (Matthews, 2002, p. 342)

If it is indeed true, for example, that pre-college and first year college level science courses are primarily about "technical pre-professional training," then vital questions need to be asked about what differences should exist between training and education in science. It raises cultural, epistemological and political questions about the nature of school science: whether, for instance, it's truly reflective of the NoS (in some form) or more reflective instead about courses performing a "gate-keeping" function by limiting access to higher education (a socio-political role)—this in turn reflecting norms of school culture and assimilation (as critical pedagogy perspectives contend).[48] Does a hidden cultural bias exist (as "cultural studies" perspectives contend)? Should the worth of school physics and chemistry education, say, be mainly determined by "political/instrumental value" (prerequisites to college entrance courses; Aikenhead, 2006)? If so, this would raise more disturbing questions about the nature of, or links among, socialization, training, and perhaps indoctrination (into scientism). There can be little doubt that in such cases a given "vision" of what constitutes "science education" is in place (with hidden "companion meanings"; Roberts & Oestman, 1998).

At minimum it should raise questions about subject epistemology, or the preeminent *value* placed upon a certain kind (Gaskell, 2002). Such topics, though, have been a staple of PE disputes for quite some time—inclusive of deliberating the difference between hidden aims and genuine educational aims of curriculum and schooling (Apple, 1990; Posner, 1998), or the differences between education and indoctrination (Snook, 1972). Not to forget, previous science education reforms have too often been associated with several past "crises" (as cited) that were themselves linked with wider socio-economic problems in society: were these just pseudo-crises manipulated by science education stakeholders and their interest groups? What educational values/views inform such groups and their policies?[49] Again, similar questions are addressed in PE.

Philosophy of Education and the Nature of Science Debate

If we focus on one fundamental topic, the *nature of science* (NoS) debate, one can glean insight into how PE can illuminate a common problem in science education (a general elaboration of NoS and epistemological issues will be addressed later in Chapter 4). Because NoS is an involved topic, we will instead in this sub-section zero in on only one aspect of this debate, the key question: "who defines science for science educators?" The scientific experts within isolated academic disciplines (as is common)? Philosophers of science? Historians? Sociologists? Or those within cultural and women's studies? Postmodernist-type thinkers and critics? Or possibly students and teachers themselves, according to some versions of social constructivist theory?

The NoS topic alone has been recognized as one chief aim of science education for over 50 years, yet to this day there exists a poor record of achievement worldwide (Lederman, 2007). This fact is due to several interrelated causes, not least of which is the entrenchment of traditionalism (conventional discipline-based paradigm)—but more so the reality that NoS is itself a contested field in HPSS studies. The "science wars" (initially launched by the *Sokal hoax*; 1996a, 1996b) and their aftermath have made the issue public, and science teachers are inadvertently involved in a contest that is being fought in the academy.[50] Researchers can certainly be found on either side, running the gauntlet from "positivism to postmodernism" (Loving, 1997; Turner & Sullenger, 1999).[51]

These polarized camps have made the business of science education a messy and complicated affair—it has become increasingly difficult to navigate a pedagogical course between competing views "from diehard realism to radical constructivism" (Rudolph, 2000, p. 404). At best consensus can be found that several common classroom myths must be exposed, including talk of "scientific method" (Bauer, 1992; Feyerabend, 1988; Hodson, 1998; Jenkins, 2007). Teachers clearly require substantial philosophical background to familiarize themselves with the issues, but even if consensus could be achieved (which seems unlikely), the question cannot be solely confined and determined on HPSS grounds. This decision would leave entirely untouched the related *pedagogical question* how that (would-be) conception of science plays a role in the education of the student, as to what educational aim(s) school science is ultimately expected to achieve.[52] In other words, for the educational setting the question "what counts as science?" must be allied with "what counts as science education?"[53] The historian may have something to say (e.g., correcting pseudo-history in textbooks), at other times the philosopher of science (e.g., correcting misleading epistemology inherent to textbooks), other times the sociologist, and so on, each depending upon the context of instruction and in coordination with desired educational objectives and policy deliberations of stakeholders.

If, say, NoS knowledge is taken to be an *end* (an aim in itself), then an implicit "philosophy" would be "academic rationalism" (Eisner, 1992)—whose objective could be associated with "knowledge for knowledge's sake," building "mind" (possibly even critical thinking), and likewise similar sounding ideals coupled to a typical knowledge-driven educational metatheory (Egan identifies it with Plato's historic project).[54] This *can* equally be squared with science teaching within the conventional academic paradigm, though providing subject content with *context* (Roberts' "Vision I"); on the other hand, NoS combined with "critical thinking" as *means* to create critical-minded citizenry to strengthen democracy in society would couple NoS teaching with Deweyan-type educational metatheory (Egan identifies this educational tendency with a form of socialization; Roberts "Vision II"). There are tensions here, which may not be reconcilable[55]—tensions also inherent to liberal education (e.g., aims for the individual and society can clash considerably); they are certainly topics of concerned debate in PE. Not to be forgotten, there are those who wish to teach NoS because it stands alone—the *intrinsic* worth to learn about authentic science (or science as a cultural force); others, however, see it subservient to other ends—for advancing critical thinking (itself), or chiefly addressing science-societal issues (Zeidler, Sadler, Simmons, & Howes, 2005), or yet again, for emancipation (critical pedagogy) and socio-political action (Hadzigeorgiou, 2008; Hodson, 2003, 2009; Jenkins, 1994).[56]

What is really of issue here, though hardly recognized, is how (and which) *epistemic aims* of science education (e.g., knowledge, truth, justification)[57] can or should be met, either apart from, or linked with, or perhaps subordinated to, other identified *moral* and *political* aims of education (e.g., autonomy, human flourishing, citizenship, social justice).[58] A common and depressing feature of several reform programs (e.g., STSE, SSI, socio-political activism) is the notable confused state of their several suggested educational aims. Moreover, it can be asserted that such avowed and increasingly popular projects for science education as identified presuppose educational metatheory of some kind, whose existence is either assumed or overlooked.[59] Engagement with philosophy of education debates about, and analyses of, *indoctrination* can be an antidote to such political-activism programs simply replacing unthinking science lessons with uncritical acceptance of whatever causes teachers or researchers might be energized about. As Erickson has stated (2007), the science education community "needs to develop pedagogical models that make explicit the normative premises about aims" (p. 33) in its discourse on scientific literacy.[60] Whenever the topic of educational aims arises, the neglect and need of PE becomes only too evident.[61] The time has come for the community to strive for clarity and prioritization concerning which fundamental aims the field can and should achieve (Bybee & DeBoer, 1994).

In any event, the NoS topic clearly raises foundational *philo-educational* questions. It presupposes one already has a conception and answer to the prior question "what is the ultimate aim of science education?" (or, e.g., of physics education?). Hence, "which aspect of NoS should be selected to achieve one's overall educational aim?" But this question is itself dependent upon clarifying and answering other prior questions, such as "what does it *mean* to be educated in science?" and "how is such an education related to human flourishing?" These should ideally all be addressed before the subsidiary question "what do we educate people in science *for*?"—often the common starting point of curriculum thinking and policy decision making, which begins first with the *instrumental* value position, and presuming the priority of *social utility*. The former should not be approached as "mere academic questions" during teacher preparation, for they aim at the heart of what the profession and teacher identity is all about. Yet it should be clear that they cannot be answered without reference to educational philosophy and theory—while the utility rationale, alternatively, already presupposes a particular one. In other words, it requires of the science educator a *philosophical valuation* of subject content and aims, and an awareness of the broader educational purpose of the science educational field, including some personal positioning among available educational/ curriculum theories (Scott, 2008).

Educational Theory and Science Education

It is clear that working without a clearly thought-out educational theory will lead to pedagogical confusion, a lack of curricular cohesion and a mix-up of key educational instructional aims, which the teacher is seeking to inculcate. But what exactly is meant by "educational theory?" Or as further testified as an expansion upon that notion, what is meant by educational "metatheory?" (A brief overview of the former will be discussed below, while the latter will be examined in the next chapter).

To talk of "educational theory" in general is first of all to recognize that it has undergone shifts in meaning ever since Western philosophy began contemplating educational matters in Ancient Greece. For the sake of brevity (and hazarding oversimplification), one charts a course from there to the current age by noting how its worth and purpose have undergone several changes, not only when specifying what *aims* to target, but *who* should carry the prime duty, namely, either philosophers, educationalists, or, in our age, empirical scientists (Carr, 2010; Phillips, 2010).

The priority in antiquity (Plato, Aristotle, Cicero) was to establish the grounds for knowledge to improve moral virtue (the "Good"), but conceived more along *a priori* philosophical lines—hence the emphasis on reason and rationality. This tendency took "an empirical turn" with Rousseau, progressivism, and the rise of the scientific Enlightenment. This science-inspired propensity has continued right down to the primacy of developmental psychology in our age, "the view that the study of human cognition, emotional and social growth and learning ought to be scientifically grounded" (Carr, 2010, p. 38). Largely lost sight of along the way was the previous prominence of moral virtue required to remodel society—reclaimed later in different guises by Deweyan theory (of social adaptation or reconstruction), critical theory/pedagogy, and *Bildung*. The post-war positivistic, language-based "analytic revolution" in philosophy (or "linguistic turn" as Rorty opined) that arose in the U.S. and England facilitated the "new" PE in the 1960s (e.g., Scheffler and R. S. Peters, respectively).

The "analytic school" in education had sought to improve teacher professionalism by augmenting the usual study of the "doctrines of the great educators" with added philosophical analytical skills to help sort out educational language and thinking (which they had diagnosed as incredibly confused). They also sought to combine their reform effort with guidance sought from research in the social sciences. It allowed for neat separation between the roles of philosopher and scientist, a dualism between theory and practice, and essentially pictured *educational theory as applied science* (a view Piaget held into the 1970s). Needless to say, the "post-analytic revolt," which came afterwards, challenged and rejected many of the previous guiding views and assumptions, including its dualism, its epistemological objectivism and deficient language theory, and its philosophy of science (the so-called "received view"; Suppe, 1977).

In its wake diverse, contemporary "schools of thought" (Barrow, 2010) have championed various anti-theory, anti-foundationalist and assorted postmodernist, constructivist and socio-political views. These in turn certainly suffer problems of their own (not to be appraised here), suffice to note others have recently come to reprieve the status of theory.[62] Its proponents not only take issue with anti-theory and postmodern-type arguments, but equally with previous analytic inspired views, and dismiss the secondary reliance of educational theory on the social sciences, or worse, its reduction to a mere branch of the field (Carr, 2010; Egan, 1983, 2002, 2005a).[63] They have reasserted the worth of philosophy to deliberate upon educational theory independent from constraints they see placed upon it, especially from scientific psychology.[64] They advocate in spirit that PE should once again claim its own unique, rightful place, neither accepting subordinate status nor intending to displace the social sciences, rather seeking complimentary standing. An educational metatheory then, when developed independently and critically from philosophical perspectives, and with sole regard to pedagogical and curricular substance, could very well insist on the difference between psychological and educational development. In sum: *the essential merit of metatheory lies in creating curricular coherence, properly transposing subject content knowledge for the learner, and defining and steering educational aims.*

Overview of Philosophy of Education Subjects and Questions

It is the view of the present author that teachers as well as researchers when becoming more conversant with the ideas and disputes as argued by philosophers of education will help them (at minimum) gain insight and perhaps (at maximum) resolve problems related to issues of *common interest* (the nature and kinds of aims; the nature of language and learning, knowledge and truth, educational theory; feminism, multiculturalism; education for citizenship; critical thinking; ideology, interests and curriculum; indoctrination, etc.). The field of PE is a veritable mine of ideas, posed problems and suggested solutions. This holds true whether the *approach* to PE is simply to:

1. study prominent philosophers and their views on education (e.g., Plato, Aquinas, Rousseau, Kant, Whitehead, Scheffler, Foucault)[65];
2. study educational thinkers and their philosophical positions (e.g., Schiller, Herbart, Dewey, Peters, Freire, Hirst, Egan, Noddings);
3. study sub-branches of philosophy and their relevance to education (e.g., PS, moral and political philosophy, or aesthetics);
4. study "schools of thought" in education (e.g., idealism, realism, Thomism, Marxism, existentialism, critical theory, postmodernism)[66]; or
5. study the philosophical questions of ultimate concern (e.g., the nature of being, of knowledge and cognition, the ideal of an educated person, autonomy).

There is intellectual insight and pedagogical profit to be had in any of these approaches (Barrow, 2010). For the more practical-minded science educator though, the approach to PE could imply instead a focus on specific, contemporary educational questions. Here Amélie Rorty's (1998) list of essential PE questions serves to illustrate the "down-to-earth" PE approach, when *transposed* onto science education:

> What are the directions and limits of public [science] education in a liberal pluralist society?... Should the quality of [science] education be supervised by national standards and tests? Should public [science education] undertake moral education?[67]... What are the proper aims of [science education]? (Preserving the harmony of civic life? Individual salvation? Artistic creativity? Scientific progress? Empowering individuals to choose wisely? Preparing citizens to enter a productive labor force?) Who should bear the primary responsibility for formulating [science] educational policy? (Philosophers, ... rulers, a scientific elite, psychologists, parents, or local councils?).[68] Who should be educated [in science]?[69] How does the structure of [scientific] knowledge affect the structure and sequence of learning?... What interests should guide the choice of [science] curriculum? (pp. 1–2)

It's quite clear that common questions and concerns exist and one would have expected more cross-disciplinary discourse than has heretofore existed.

On the other hand, it is not here being suggested that a consensus is to be found among philosophers of education on such questions. In fact there are important disagreements and even diversity of interest and approaches to the solutions, as different PE "schools of thought" display (analytic, existential, phenomenological, postmodern, critical theory, etc). Indeed, philosophy more often divides than it unites, and as one contemporary education philosopher admits: "missing in the present world of diversity of interests is the classic sense of a quest for philosophic unity" (Chambliss, 1996). As Scheffler stressed, "philosophies of" are not forged by some harmony of agreed-upon, sealed discourses. Instead they

> do not provide the educator with firmly established views ... on the contrary, they present him with an array of controversial positions. But this array, although it does not fix his direction, liberates him from the dogmatisms of ignorance, gives him realistic apprehension of alternatives, and outlines relevant considerations that have been elaborated in the history of the problem. (Scheffler, 1970/1992, p. 391)

The point is not that some sort of philosophical unity should be either expected or had among philosophers or science educators, although of course consensus on common fundamental issues is to be desired. Rather, the nature of the discourse and sophistication of the debate can help illuminate those problems and issues by which science educators are confronted and continue to struggle with, or have misconstrued, have overlooked, or for too long avoided.

The constructive purpose of PSE becomes immediately obvious, *to ascertain the* legitimacy *or* scope *of either borrowed ideas or transferred theories from "outside" disciplines into the educational field*. In light of these important developments, this book asks if a *partial shift in research direction towards philosophy of education and a reorientation of thinking is required for both science teacher training and science education as a field of research* (Chapter 2).

It is suggested here that another important role for PSE is the need to not only *locate* hidden philosophies with their educational theories (or "meta"theories, e.g., Plato; Rousseau; Whitehead; Dewey, critical pedagogy) but to *actively create* such theories—or another possibility, to *show the relevance* of those educational metatheories that *exist* as proposed in PE or curriculum studies, to illustrate their worth for improving science teaching and learning (Egan, 1997; Scott, 2008; Waks, 2008; Walker, 2003). To that end this book will *describe and examine Kieran Egan's cultural-linguistic metatheory,* including a brief comparison with the *Bildung-centered Didaktik tradition of Central and Northern Europe*. The purpose will be to expound upon its relevance for solving common problems not only engendered by student learning of science but equally engendered by unresolved internal disputes with respect to science education goals, especially scientific literacy (Chapter 2).

1.5. THE *NATURE OF PHILOSOPHY OF SCIENCE EDUCATION* (PSE): SCOPE AND POSSIBILITIES

With the previous discussion in mind and the various aforementioned italicized suggestions, we now come to the topic of summarizing the scope and possibilities of a PSE as well as the direction and organization of this book.

In this book I develop the framework for a "philosophy" *of and for* science education *itself*. The creation of any new discipline will only be acknowledged insofar as a group of scholars working on certain issues and problems find that their research interests can no longer be accommodated and foresee the need to move in new directions of inquiry. My view concurs with Scerri and McIntyre (1997), who defended their new discipline of "philosophy of chemistry" with the following words:

> Of course, there are always some who wish to drag their feet; to deny the legitimacy of the interests of those who first make their way into a new discipline. Yet ultimately, if the discipline has enough people who are interested in it, and who are willing to chip away at the parochialism of those who would deny it, a new field of inquiry gradually emerges. (p. 211)

One envisions that PSE could serve as an academic forum where fundamental goals, criteria for content selection, and critiques of epistemologies, research methods, learning theories and instructional strategies can be debated, along with the focus on developing in-house educational metatheories—especially how such theories can inform practice—can be nurtured. Also it should help clarify the relationship between educational theories and philosophy itself, which has its own concerns, as Hirst (1966) first noted.

The *scope* of its field of inquiry and analysis of central issues can be organized around at least *nine* topics or problem sets:

1. goals of science education (and science literacy);
2. development/analysis of educational metatheories for science education;
3. nature of science (NoS) as suitable for science education;
4. nature of science learning (and critique of learning theories);
5. nature of science teaching and assessment;
6. nature of ideology and interest in curriculum; criteria for content selection;
7. nature of language (NoL) in curriculum, instruction, and learning;
8. relation of science literacy and goals to socio-technology issues education; and
9. relation of science and science education to *worldviews* and cultures.

This is not meant to serve as an exhaustive list, and it is to recognize that some topics have already found an audience for discussion,[70] especially, items 1, 2, 3, 4, 6, and 7. Related areas of research and inquiry are suggested in Figure 1.1. The topic regarding item 7 is a relatively new field of science education research (Fensham, 2004) but is found wanting for present lack of comprehensive philosophical treatment. Chapter 5, which discusses the *nature of language* by contrasting Dewey's and Gadamer's conceptions, is meant as a contribution in this direction, linked to the previous input of Eger (1992; philosophical hermeneutics) and Lemke (1990; semiotics). Of the nine research questions mentioned earlier in the Introduction, which seek to address some of the above named problem sets, a few nonetheless address new topics, while others venture onto well-trodden ground (question #8) or familiar territory (question #9), but from a PSE perspective.

Given the above list, it is recognized, however, that no such single academic platform currently exists to bring the topics into better focus, for enhanced dialogue and more comprehensive analysis, although the listed topics and debates have appeared sporadically over time and for different reasons, and some are scattered across different research journals and handbooks. And the items on the list aim to acknowledge the internal conflicts within the field. Hence, I do not mean to imply that others have not already been engaged in aspects of what is here described as a *philosophy* of science education. But because of the ongoing sub-specialization within science education, I fear the consideration requisite to such concerns and the need for a broader, integrated inquiry field is, and continues to be, lost sight of. Fensham (2000, 2004) and Jenkins (2000, 2001) have dropped tantalizing hints in this direction. Certainly what Kyle et al. (1992) have suggested would move within the orbit of such a sub-discipline (to delineate the field of research inquiry itself), as would an analysis of various "research programs" of learning theories (item 4; Anderson, 2007; Erickson, 2000).

The *possibilities* for reform and what a reorientation might imply: There are at least *three areas* for potential reform:

1. science education as a field of research (Jenkins, 2001; Kyle et al., 1992);
2. science teacher education and curriculum (Abell, 2007; Fensham, 2004); and
3. science education as a discipline, defining its *identity* (Fensham, 2004).

With respect to Area 1: As mentioned before, Jenkins (2001) had justifiably complained that the field of science education research has been too narrowly construed (in part as a critique directed at the two earlier *Handbooks* of Fraser and Tobin, 1998, and Gabel, 1994), being overly preoccupied with teaching and learning as practice, and thereto, suffering from "an over-technical and -instrumental approach" (p. 11), at the expense of other perspectives, such as "the long-standing neglect of historical studies of science education, and more widely, of research not related directly to the teaching and/or learning of science" (p. 12). This should certainly encompass explicit educational philosophical reflection on *educational* theory (which he fails to mention). Along with the *three* predominant kinds of research in science education today—quantitative, qualitative-interpretive, and critical-emancipatory (Fensham, 2002; Kyle et al., 1992)—*I would argue for the pressing need of a fourth, philosophical (or philosophic-historical).*

With respect to Area 2: it remains to be clarified how a PSE would impact the understanding of teacher education programs, especially its relation to the ongoing research on PCK and its elaboration (Abell, 2007; Kind, 2009), which itself seems to have a hidden or elusive nature. Nevertheless, Fensham in his important *Identity* book (2004) has himself recently argued for an alignment of the *Bildung* tradition and PCK. He sees in the *Bildung* tradition a critical element for content selection and curriculum analysis that is missing in the typical aphilosophical CK-focused science curricula of English-speaking nations. What he fails to recognize, however, is that the *Bildung* tradition itself represents an example of an educational metatheory, on par with Egan's (1997) cultural-linguistic metatheory. An important task of PSE would be a critical comparison of these metatheories, their relation to PCK, and their impact on science teacher education (see above "Scope," item 2). (A brief discussion is undertaken at the start of Chapter 3; see also Figure 3.1).

With respect to Area 3: My suggestion is that an educational metatheory could ideally contribute to grounding science education as an autonomous discipline by helping distance it from its historical dependence on psychological metatheories and economic or socializing ideologies.[71] It would contribute to further defining its *identity*, in line with Fensham's arguments (2004), as he has raised the issue. Although this claim is not addressed by Fensham and the researchers he had canvassed, it is mentioned at points in Chapter 2.

It stands to reason one could accept points 1 and 2 above as belonging to the proper inquiry field of PSE, and hence argue for its inauguration, and not be committed to point 3. That is, one need not hold that science education is either a discipline nor that it requires metatheories characteristic to it—although I will argue the opposite case. Then again, as I understand what belongs to the essence of philosophy as mentioned above, it means a project that involves not only analysis but also the creation of new ways of seeing and doing education. In other words, *the active development of educational philosophies with their concomitant metatheories appropriate for the sub-discipline of science education.*

NOTES

1. "Regrettably, much of the constructivist literature relating to education has lacked precision in the use of language and thereby too readily confused theories of knowledge with ideas about how students learn and should be taught" (Jenkins, 2009, p. 75).
2. The literature on constructivism is vast. Critiques are found in Davson-Galle (1999), Geelan (1997), Grandy (2009), Kelly (1997), Matthews (2000, 1998b), and Scerri (2003).
3. This component is meant to include the associated disciplines and not just the philosophy discipline itself.
4. Whether or not science students themselves should be presented with PS ideas and controversies is still being debated among researchers (Hodson, 2009). One philosopher of education has reversed his earlier standpoint (Davson-Galle, 1994, 2004, 2008a, 2008b).
5. "The contrast between Dummett's and Rorty's views indicates not only the most divergent and antithetical understandings of the accomplishment of

modern and recent analytic philosophy but of the self-understanding of philosophy itself" (Bernstein, 1983, p. 7).

6. Habermas takes Rorty to task for undermining the one role still left to philosophy, being the "guardian of rationality." "If I understand Rorty, he is saying that the new modesty for philosophy involves the abandonment of any claim to reason—the very claim that has marked philosophical thought since its inception" (1987a, pp. 298–299). While Habermas agrees with Rorty, philosophy must give up its assumed Kantian project of ultimate knowledge arbiter (*Platzanweiser*), it must still "retain its claim to reason," taken to mean stand-in (*Platzhalter*) and interpreter.
7. It has also been historically associated with particular schools of thought (e.g., idealism, rationalism, empiricism, existentialism, etc.), hence particular *philosophies*, themselves often associated with individual philosophers (e.g., Plato, Kant, Marx, Nietzsche).
8. This is not meant to discount the next three. Logic has made a renewed appearance in science education under the guise of critical thinking and scientific argumentation; those in ethics intersect with discussions of values and socio-ethical issues (Allchin, 2001; Corrigan, Dillion, & Gunstone, 2007; Witz, 1996; Zeidler & Sadler, 2008); even aesthetics has been considered for the field (Girod, 2007).
9. See the chapters in Bonjour and Sosa (2003) for a concise overview; Section 5.1.1. targets the former.
10. That one must inevitably justify the value of philosophy for teachers and many researchers suggests that a cultural predicament already exists concerning what constitutes "education" in our present age.
11. Which includes essentially their "orientations" toward teaching, identified in science teacher education research as formative dispositions attached to identity (Van Driel & Abell, 2010; Witz & Lee, 2009).
12. Probably the ongoing reality of the academic divide between the "two cultures" maintained as two solitudes in universities to this day (as described by C. P. Snow; Shamos, 1995; Stinner, 1989) contributes to the hostility or indifference since science teachers are not generally required to endure arts faculty courses. All this in combination with the common negative image that academic philosophy is preoccupied with obtuse speculation and unresolved disputes all remote from everyday matters. Certainly quite different, encouraging evaluations can be had (Matthews, 1994a; Nola & Irzik, 2005).
13. They continue: "Too often the selected textbook defines course scope, sequence, and depth implying that a textbook's inclusion of information, in part, legitimizes teaching that content. Textbooks also exert a significant influence on *how* content is taught" (Clough, Berg, & Olson, 2009, p. 833).
14. Many teachers would probably declare "science for all" or "scientific literacy," though seldom with awareness these slogans are replete with ambiguities—the latter goal even suffering inherent incompatibilities due to serious shifts in connotation, and this despite its ultimate prominence in worldwide "standards" documents (Jenkins, 2009; Schulz, 2009a, 2009b; Shamos, 1995). The science for all theme arguably partially appropriate for junior science nonetheless vanishes when specialty upper secondary or tertiary courses are

reached, for here the status quo is maintained as "technical pre-professional training" (Aikenhead, 1997b, 2002b, 2007). In this case an extreme narrowing of the "literacy" notion is found, HPSS aspects are distorted or abused, while the concealment of existent curriculum ideologies remains unrecognized in absence of educational philosophy (e.g., scientism, academic rationalism, "curriculum as technology," or social utility; Eisner, 1992).

15. In his comprehensive review, the categories "Vision I" and "Vision II" were postulated to account for two major competing images of science literacy behind many curricular reforms. The former designates those conceptions that are "internally oriented," that is, towards science as a knowledge- and inquiry-based discipline and including the image of science education as heavily influenced by the identity, demands, and conceptions of the profession. The latter vision, alternatively, is "outward looking," towards the application, limitation, and critical appraisal of science in society. The image is influenced instead by the needs of society and the majority of students not headed for professional science-based careers. Here the question of the "social relevance" of the curriculum is paramount. He claims that while the second vision can encompass the first, the opposite is not true. My contribution, spearheading Egan's ideas, can be considered a third option for literacy: "vision three."
16. For linked views see Anderson (1992); Fensham (2004); and Matthews (2002, 1994).
17. It should not be forgotten that the 17th century scientific revolution introduced "science" as a field of research and study under the academic umbrella of *natural philosophy* to distinguish it from the reigning scholasticism of the universities, hermeticism, and neo-Platonism. Our modern conception of the term and the severance of philosophy from science is of relatively recent origin. The division emerged historically as a development in intellectual thought and specialization, which evolved within European industrial society in the mid-nineteenth century. Whewell, for example, first coined the term "scientist" in 1833.
18. Scientists and philosophers alike have found it necessary to launch important new *sub-disciplines* to address foundational questions and concerns arising from their scientific areas of expertise—notwithstanding those scientists who disparage the study of PS overall (e.g., Weinberg, 1997). Philosophy of physics (Cushing, 1998; Lange, 2002), philosophy of chemistry (McIntyre, 2007; Scerri, 2001) and philosophy of biology (Ayala & Arp, 2009), are becoming established research fields, including philosophy of technology (Scharff, 2002), likewise lauded for teachers today (de Vries, 2005).
19. Unfortunately it appears that science education worldwide and many science teachers themselves have tended not to kept abreast of these advances and what they possibly offer for curriculum design, instruction, and reform efforts. One might hope these sub-disciplines offer, minimally, deeper and improved insights about subject content, but more so, a better understanding of the essence of the discipline, the core of which teachers are required to inspire and impart to their students. Certainly these are less well-known to science teachers, and not canvassed by science education researchers to

the extent of interest shown in the post-structuralist and "science studies" literature. See Allchin (2004); Collins (2007); Hodson (2008); Kelly et al. (1993); Nola and Irzik (2005); Ogborn (1995); Roth and McGinn (1998); and Slezak (1994a, 1994b).

20. With such a faculty teachers could better function in their role as *mediator* between the scientific establishment and their pupils, also between public discourse about science and pupils or adults not conversant with how science evolves or the nature of modern techno-science (see also Hodson, 2009).
21. The term "scientism" can be interpreted in different ways; most construe it negatively (Bauer, 1992; Haack, 2003; Habermas, 1968; Matthews, 1994). Nadeau and Désautels (1984) attribute five components (as discussed in Chapter 4). Irzik and Nola (2009) are careful to distinguish legitimate scientific worldviews from illegitimate *scientistic* ones: "A scientific worldview need not be scientistic. Scientism, as we understand it, is an exclusionary and hegemonic worldview that claims that every worldview question can be best answered exclusively by the methods of science...that claims to be in no need of resources other than science. By contrast, a scientific worldview may appeal to philosophy, art, literature and so on, in addition to science. For example, scientific naturalism can go along with a version of humanism in order to answer worldview questions about the meaning of life" (p. 87).
22. What is being suggested here can be taken to correspond with a key objective of critical pedagogy, popularized by the Marxist teacher educator Paolo Freire (1970), their advance to "critical consciousness."
23. More recent studies have been undertaken to improve teacher education and science classrooms by providing explicit HPS and NoS strategies and resources (Kokkotas, Malamitsa, & Rizaki, 2011). Allchin (2012) provides several examples from the Minnesota SHiPS collection. His (2013) provides even more elaborate examples for classroom use. Höttecke, Henke, and Riess (2012) comment on a large scale funded European HIPST project. Maurines and Beaufils (2013) discuss the situation in France regarding physics courses.
24. In their major review, Abd-El-Khalick and Lederman (2000) mention three major lines of research on NoS in the past 50 years that has proceeded in sequence. The first line investigated students' conceptions of NoS and concluded that their inadequate views were primarily the result of inadequate curricula. The second line, which investigated the design and implementation of curricula (some using HPS) aimed at NoS conceptions, showed significant improvements in post-test scores but tended to ignore the central role of the teacher and their understanding of NoS. As more studies came to indicate the importance of teacher epistemology for influencing student learning, this opened up the third line in the research.
25. They mention (1) the nature of the subject-centered culture of teaching; (2) teachers' lack of skills in teaching HPS, also their contrary attitudes and beliefs; (3) institutional framework of science teaching; and (4) the nature of textbooks and their role as a fundamental didactic support.
26. A glance at science teacher education textbooks does indicate that the community recognizes the value of some PS for teacher preparation. As

examples, both the textbooks by Bybee, Powell, and Trowbridge (1967/2008; U.S.) and Ebenezer and Haggerty (1998; Canada) have one chapter devoted to philosophy of science, mentioning the philosophies of Bacon, Popper, and Kuhn explicitly. Summers (1982) was among those who had earlier addressed the need for explicit PS for science teacher preparation, as had Scheffler (Matthews, 1997; Siegel, 1997).

27. Science education to this day has been unable to resolve the principal dilemma concerning the conflict of the two competing "visions" of its purpose (hence competing conceptions of "scientific literacy"). Roberts (2007) admits the community must "somehow resolve the problems associated with educating two very different student groups (at least two)" (p. 741).
28. Yet despite these disturbing findings, researchers in these newer fields of study (also chemical education research) still struggle uphill for respect and acceptance in their academic departments, where educational studies and research continue to be afforded a low priority (Gilbert, Justi, Van Driel, De Jong, & Treagust, 2004; Hestenes, 1998).
29. The cyclical return of such talk does indicate there is still something deeply amiss with ongoing science education and which both earlier periods of so-called reforms—the scientist/discipline-centered curricular reforms of the 1950s/60s and the STS inspired movement of the 1980s—have failed to adequately address and rightly resolve. For a skeptical look at earlier crisis-talks see Klopfer and Champagne (1990. There are compelling arguments that indicate the widely accepted correlation between improved secondary science education—and hence assumed improved literacy—and the supposed benefits accrued for either national economic performance or increased critical citizenship are largely illusory (Shamos, 1995). An in-depth global study undertaken by Drori (2000) on the widely assumed "policy model of science education for economic development" highlights the many methodological problems associated with the empirical studies and economic models which render any connection between the two "unreliable" (p. 31). The empirical evidence is inconclusive at best, states Drori, regardless of claims to the contrary (and as charged against the competency of teachers and school science throughout the 1980s). And the burden of ensuring a "scientific pipeline" for future supply of specialists falls primarily to post-secondary pedagogy.
30. Such as Aikenhead (2006); Bencze (2001); Donnelly and Jenkins (2001); Fensham (2002, 2004); Roberts (1988); Roberts and Oestman (1998). Laugksch (2000) draws attention to different social group interests in defining "science literacy." Ernest (1991) also identifies several interest groups as determinants of mathematics education.
31. Fensham's 2002 paper "Time to Change Drivers for Scientific Literacy" (movement away from the academic driver to "social" and industry-based drivers) provoked a lively response from researchers about the "educo-politics" of curriculum development, especially about what role academic scientists should play, if any (Aikenhead, 2002b; Gaskell, 2002). Such a suggestion though would re-orientate science education back towards the recurrent (and contentious) "social relevancy" goal and the progressivism of Deweyan-type philosophy (DeBoer, 1991)—whose educational theory is

often concealed. It may even involve a Faustian bargain with industrial- and vocational-driven interests. Gaskell believes the risk is worth it. But given the complexity of techno-science and the great diversity of vocations and business interests today, this leaves one wondering if any sort of meaningful consensus on curriculum is achievable, even locally.

32. Fensham in fact suggests that it is the "dominance of psychological thinking in the area" that attests to why Dewey is *not* cited more frequently among respondents in the U.S. (still the most prominent philosopher of education linked with science education in North America).
33. Important works are Bailey et al. (2010); Blake et al. (2003); Chambliss (1996); and Winch & Gingell (1999).
34. Authors in alphabetical order include: Bailin, Burbules, Davson-Galle, Garrison, Grandy, Hodson, Matthews, McCarthy, Norris, Phillips, Scheffler, Schulz, Siegel, and Zembylas (see respective references).
35. Matthews comments this may be the significant reason why the science education research literature "is dominated by psychological, largely learning theory, concerns" (Mathews, 2009b, p. 23). Others have also cited the domination of psychology and conceptual change research (Gunstone & White, 2000; Lee, Wu, & Tsai, 2009).
36. The typical tendency is to adopt philosophical or ideological views from well-known authors outside the field but often not accompanied by critical appraisal of such authors: "the work of Kuhn, von Glasersfeld, Latour, Bruner, Lave, Harding, Giroux and others is appropriated but the critiques of their work go unread: it is rare that science education researchers keep up with psychological and philosophical literature" (Matthews, 2009b, p. 35).
37. Related to this topic is the question of what worldview(s) science assumes or requires in order to be sustained, hence which one(s) educators need to be supportive or cognizant of (Matthews, 2009a). This further raises the question of the *universalism* of "Western science," whether or not its knowledge and truth claims are necessarily culturally confined, or merely *evolved*. Disputes over the interpretations of "multicultural science" will not be addressed here, but again science educators require philosophical training in order to adequately tackle these controversial topics. Philosophical treatment of this subject can be found in Hodson (2009), Matthews (1994), and Nola and Irzik (2005).
38. The literature on *Bildung* and *Didaktik* is extensive. Some references to its historical development are Barnard (2003); Beiser (1998); Gadamer (1960/1975); and Schiller (1795/1993).
39. Chapter 3, Section 3.1 will be concerned with a more involved discussion of this important subject.
40. Science education and *Bildung* in Germany have been examined by Benner (1990) and Litt (1963).
41. One Canadian study involving science teachers had sought to fuse the *Bildung* ideal with the STS paradigm and cross-curricular thinking (Hansen & Olson, 1996).
42. "In the one, the maturing young person is the purpose of the curriculum. In the other, the teaching of subjects is the purpose. In the one case, disciplines

of knowledge are to be mined to achieve its purpose; in the other, disciplines of knowledge are the purposes" (Fensham, 2004, p. 150).

43. Jenkins (2009) notes the same problem with reform movements and policy documents. This complaint (although dated but still relevant) was earlier attested by DeBoer in his Preface to his insightful *History of Ideas in Science Education* (1991).
44. Roberts (1988) draws attention to where teacher *loyalties* commonly lie: "The influence of the subject community is an especially potent force in science education. In general, the 'hero image'...of the science teacher tends to be the scientist rather than the educator [or philosopher]" (p. 48).
45. Hirst (2008b) has recently complained that in some countries such as England there are now moves afoot to de-list such courses for teacher training altogether. It would not be a stretch to conclude that such a downgrade in the general value of PE cannot fail to negatively impact science teacher professional development.
46. The leading journals of the English-speaking world are: *Studies in Philosophy and Education, Educational Theory, Educational Philosophy and Theory,* and *Journal of the Philosophy of Education.*
47. Some classroom case examples are Hadzigeorgiou et al. (2011), Kalman (2010), and Ruse (1990). Bailin and Battersby (2010), Giere (1991) and Kalman (2002) offer science teacher educators rich material for enhancing science subject-related critical thinking.
48. "Domination, resistance, oppression, liberation, transformation, voice, and empowerment are the conceptual lenses through which critical theorists view schooling and pedagogy" (Atwater, 1996, p. 823).
49. Different kinds of answers are provided by Aikenhead (2006, 2007); Apple (1992); Bencze (2001); Donnelly and Jenkins (2001); Gaskell (2002); Gibbs and Fox (1999); Klopfer and Champagne (1990); Roberts and Oestman (1998); and Zembylas (2006).
50. For examples of teachers caught in the debate see Sullenger et al. (2000) and Witz and Lee (2009). For different perspectives on the debate in the academy see Brown (2001); Giere (1999); Gross et al. (1996); Laudan (1990); Nola (1994); Norris (1997); Siegel (1988, 1987a); and Sokal and Bricmont (1998).
51. Science educators continue to quarrel whether basic NoS statements *can* or *should* be defined, even where a measure of recognized consensus is said to exist—inclusive of those now written into global policy documents. The dispute centers on how to determine "consensus" (among which experts?), or questions regarding disciplinary distinctions, or about NoS cultural dependence on "Western" science and Enlightenment traditions, among others (Hodson, 2008; Irzik & Nola, 2011; Matthews, 1998a; Rudolph, 2000, 2002). Good and Shymansky (2001) make the case NoS statements found in "standards" documents like NSES and *Benchmarks* could be read from opposing positivist- or postmodernist-type perspectives.
52. This viewpoint aligns to an extent with Hodson's view (2009) except for the fact that he ignores relating his desired outcomes to educational philosophy and theory: "In my view, we should select NOS items for the curriculum in

relation to other educational goals ... paying close attention to cognitive goals and emotional demands of specific learning contexts, creating opportunities for students to experience *doing* science for themselves, enabling students to address complex socioscientific issues with critical understanding" (p. 20). On what philo-educational grounds the selection is to be undertaken, we are not told, though he considers students' "needs and interests" (overlap with progressivism?), views of experts ("good" HPSS—the Platonic knowledge aim?) and "wider goals" of "authentic representation" of science and "politicization of students." His lofty ambition for science education (thus his notion of "literacy"), however, includes too many all-encompassing and over-reaching objectives. These must clash and become prioritized (or so it seems) once his three stated criteria for subordinating goals force them under his socio-techno-activist umbrella of politicizing students—the ghost of Dewey beacons.

53. The focus here is on the normative nature of the question (i.e., what do policy documents, researchers, or theorists stipulate?), as opposed to the empirical (i.e., what is going on in classrooms now?).
54. See discussion on the topic of epistemic aims by Adler (2002), Hirst (1974), and Robertson (2009), Phillips (2002).
55. See discussion in Egan (1997) and Pring (2010). Smeyers (1994) discusses the European account.
56. Driver et al. (1996, pp. 16–23) offer five rationales for teaching NoS in classrooms, yet they either assume or overlook their dependence upon different, prior educational theories.
57. See Nola and Irzik (2005), Robertson (2009), and Siegel (2010) for discussion of these subjects.
58. See Brighouse (2009) and Pring (2010) for discussion of these subjects. Donnelly (2006) only scratches the surface of the problem with his defined dual clash between "liberal" and "instrumental" educational aims behind community reforms.
59. This remark also targets research concerning situated cognition models, where it has often been asserted practice was either *prior* to theorizing or *without* theory. See critiques of Roth by Sherman (2004, 2005).
60. He continues: "Too often we try to simply derive pedagogical practices from theoretical positions on learning, or diversity, or language, or the latest research on the functioning of the brain, etc." (Erickson, 2007, p. 33).
61. An example of the confusion that results in science education research when PE is ignored is the paper by Duschl (2008). Here empirical research from the learning sciences and science studies is confused with educational goals, which must be chosen on a normative basis. Such research may tell us *how* students (and scientists) learn but expressly not *why and what* goals they *should* learn. And to argue for a "cultural imperative" is to make a normative claim extrapolated from such research—one is dabbling in PE without its recognition. Moreover, whether the avowed economic, democratic, epistemic, "social-learning" goals, and so on (as they've been historically articulated for the field) can be "balanced" as Duschl simply assumes is by no means obvious—PE debates show quite the opposite (as will be discussed in Chapter 2).

62. So that it may "engage in explorations of what [science] education might be or might become: a task which grows more compelling as the 'politics of the obvious' grow more oppressive. This is the kind of thing that Plato, Rousseau and Dewey are engaged in on a grand scale" (Blake et al., 2003, p. 15).
63. Carr (2010) holds that educational theory might be better suited to ethics (moral reasoning) than with any sort of empirical science, which is not to dismiss the worth of some empirical work: "On closer scrutiny, it seems that many modern social scientific theories of some educational influence are often little more than normative or moral accounts in thin empirical disguise" (pp. 51–52). This deduction leaves unanswered the important question as to what the proper role and value of empirical research for educational theorizing is to be. The topic is controversial and engenders debate in PE. See Egan (2002) and Hyslop-Margison and Naseem (2007) for a negative assessment and Phillips (2005, 2007, 2009) for a positive view.
64. "We have suffered from tenuous inferences drawn from insecure psychological theories for generations now, without obvious benefit" (Egan, 2002, pp. 100–101).
65. To name just some in the Western tradition; Eastern and other traditions have of course their own major philosophers who have concerned themselves with education.
66. A classic source of material for this orientation are the essays in Henry (1955).
67. As those in the *Socio-Scientific Issues* (SSI) reform movement today insist (Zeidler & Sadler, 2008).
68. See DeBoer (2000), Fensham (2002), Gaskell (2002), Jenkins (1994), Roberts and Oestman (1998) for responses to such questions.
69. Recall the ongoing past disputes between "science for scientists" and "science for all" perspectives on curriculum, goals, and policy (Bybee & DeBoer, 1994; DeBoer, 1991). The most recent STEM reform movement in the U.S. can be justifiably accused of redefining science education as "science for engineers."
70. As examples: Roberts (2007), Bybee and DeBoer (1994) regarding item 1; Lederman (2007), McComas et al. (1998), Driver et al., (1996) regarding item 3; Sula (2009), Yager (1996), Layton (1993) regarding item 8; Matthews (2009a), Cobern and Aikenhead (1998) regarding item 9.
71. Gunstone and White (2000) have earlier sought to make the case for science education as a discipline, but they linked it with the criterion that as a research field it was required to show progress. By concentrating strictly on research related to students' content learning and alternative conceptions, they answered in the positive by identifying two key factors that indicated progress: changes in research styles over time and the discovery of several learning principles. They also suggest science education should be aligned with the *social sciences* and as such could not aspire to the rigor of the natural sciences, which can show a development in their investigative processes from general principles to "laws." Science education, they argue, could at best aspire to the discovery of principles.

CHAPTER 2

SCIENCE EDUCATION REFORM, PHILOSOPHY OF SCIENCE EDUCATION, AND EDUCATIONAL THEORY

History teaching still turns a blind eye towards science, the most exciting and noble of human ventures, and science and mathematics teaching is disfigured by the customary authoritarian presentation. Thus presented, knowledge appears in the form of infallible systems hinging on conceptual frameworks not subject to discussion. The problem-situational background is never stated and is sometimes difficult to trace. Scientific education—atomized according to separate techniques—has degenerated into scientific training. No wonder that it dismays critical minds.

—Imre Lakatos, philosopher of science

2.1. SURVEYING THE ISSUES

This chapter is concerned with how science education reform can be made more effective but takes a very different perspective from many present and previous deliberations in the extensive literature on this topic.[1] It seeks to harness insights from research in particular fields (social constructivism, cultural-linguistics, and philosophy of education) to argue that science

Rethinking Science Education: Philosophical Perspectives, pp. 63–107

education both as academic discipline and practice requires foremost an overarching *theory of education* for and about science comprehension.

Although science education research continues to be heavily influenced not only by ideas but especially by *theories* coming from outside the discipline itself (primarily from the fields of psychology, philosophy and sociology of science, socio-linguistics, and cognitive science), little attention has been given to developing in-house specific *educational* theories that could better attend to the needs of science teaching, learning, and curriculum unique to it, and that (as an educational sub-discipline) it more properly shares with others in educational studies. Such insights though as can be gleaned from these other academic fields that can throw light on the complexity of education are naturally to be welcomed, but the point here is that due to the poverty of educational theorizing within much of science education, such imported theories tend to overshadow and mask the need to develop in-house theories that could help resolve conflicts internally generated—those, for example, that revolve around questions of goals, values, content selection, learning, and instruction.

Moreover, not only do these external disciplinary theories usually have various built-in constraints of their own (only occasionally acknowledged), they serve at best to illuminate only partial aspects of educational concerns—for example, offering ideas about development, motivation, memory, learning, and others—and they tend to offer these from perspectives that can have little in common with what either the practitioner or educational theorist is ultimately concerned with, or worse, their application may do outright damage to education (Egan, 2005a; Slezak, 2007; Thomas, 1997).[2] These points cannot be emphasized enough. As will be argued, some of these ideas and solutions embedded as they are within their own theoretical framework (for example, the Piagetian stage theory of development) could very well be in conflict with an alternative, fully developed *educational* theory (Hirst, 1966). In any event, although they may be able to address some aspects and issues, they certainly cannot address the more fundamental ones pertaining to the values, goals, and teaching of science exclusive to it, such as the ongoing disputes over the meaning and assessment of *scientific literacy*, even the recent critical questioning if such a key aim is either reasonable or attainable (DeBoer, 2000; Shamos, 1995). My view concurs with Donnelly (2004) that "science curriculum reform [should] be seen as a self-contained transformation, motivated by an independent vision of the aims and purposes of science education" (p. 778).

Therefore the call is to re-shift the focus of science education, especially to address those within the community (philosophers, curriculum theorists, learning theorists, practitioners, etc.) concerned with the broader questions of science education as an *educational* endeavor including the aims, ideas, and issues respective to it (goals, conceptions of development and learning, curriculum reform). Not generally since Dewey (1916/1944, 1945) has such a particular deemed emphasis been brought to bear in science education—something those within the larger community have basically lost sight of (an exception is Roberts & Russell, 1975)—and as is well known, Dewey's image of science played a key role in his overall educational philosophy, including applying its assumed scientific way of thinking ("method") to enhance the reasoning skills of students and thereby hoping to forward democracy in society. Such principles have now become the hallmark of progressivism, and they continue to influence science educational ideas today (Bybee & DeBoer, 1994; Wong, Pugh, & the Dewey Ideas Group, 2001), especially the STS reform movement and their science literacy conception (Hurd, 2000; Solomon & Aikenhead, 1994; Yager, 1996), although the actual link to an earlier educational philosophy is often overlooked or not clearly delineated (DeBoer, 1991; Rudolph, 2002; Shamos, 1995). The argument put forth here in effect is to turn the focus around: *not* to expand the notion of science education, to be used as a cornerstone for an educational philosophy in general (as Dewey had hoped to do), but rather to explore ideas in education insofar as it is possible to develop an overarching *educational philosophical* framework—or *metatheory*—and to bring this back into the fold of science education to help guide and solve curricular and learning issues relevant to this field.

This is where I believe Kieran Egan's ideas can be of immediate importance (1983, 1986, 1997, 2002, 2005b). He affords us the ability to conceive of education in relatively new, creative ways and with a rather different set of categories and vocabularies ("cognitive tools," "imagination," "narrative understanding") than those heretofore commonly employed in educational discourse ("structure of the discipline," "knowledge forms," "stages of development," "growth," "constructivism," "information processing," "multiple intelligences," etc.). His cultural-linguistic theory of education (concerning *educational* and not *psychological* development) will be introduced and elaborated as one possible over-arching conception of education that can serve to guide curriculum development and learning about science, especially since crucial narrative aspects of his theory already coincide with current work by some HPS reformers (Klassen, 2006; Metz, Klassen, McMillan, Clough, & Olson, 2007; Stinner, 1995a; Stinner et al., 2003). Furthermore, besides the obvious link to the "narrative approach," which is receiving increased attention from several researchers (Norris, Guilbert, Smith, Hakimelahi, & Phillips, 2005; Millar & Osborne, 1998), there are conceptual and theoretical links to the growing body of work drawing on Vygotsky, cultural psychology, and language studies, which focus on discourse and the social-cultural nature of learning and development.[3]

Egan's conception of an *educated mind* certainly aligns itself (in large measure and to various degrees) with those who have always sought to frame science education in terms of broader aims, such as *cultural literacy* (most recently Carson, 2001), or that it needs to be properly located within a *global* (Snively & MacKinnon, 1995), "humanizing," or even "liberal education" framework (Aikenhead, 2007; Donnelly, 2004; Carson, 1998; Matthews, 1994; Shamos, 1995)[4]—and much less so to be confined either to narrow *technical training* (traditional curriculum) or understood primarily as *means* to serve civic and/or socio-political ends (the "socialization" imperative; typically the STS movement but more recently "science education for socio-political action"; Hodson, 2003; Jenkins, 1994, 2000; Roth & Desautels, 2002). Granted, all *three* expansive aims or goals (cultural-personal, technical-disciplinary, socio-utilitarian) have accompanied the history of science education for a century or so, each continues to have force and forceful spokespersons, and each has come to the fore for different reasons in the past (Bybee & DeBoer, 1994). As Mathews (1994) clarifies, these three "curricular orientations" are not necessarily mutually exclusive: "Curricula that stress one, usually include something of the others. What is in contention between the views is the general orientation of the science program, and the goals that it seeks to achieve" (p. 15). Importantly, the connotation of "scientific literacy" has shifted accordingly (DeBoer, 2000).

Indeed, it is time to ask the essential questions once again: "what does it mean to be *educated?*" (Barrow & Woods, 2006), and supplemental to this for our particular profession, "what does it mean to be educated *in science* (and in physics, in geology, etc.)?" This latter crucial question should be kept distinct from—though it often tends to be subsumed by—the question harboring a social *utilitarian* criterion: "what do we educate people in science *for*?" Depending upon the stakeholders involved with science education (students, teachers, parents, business community, academic scientists, science education researchers, etc.) one expects very different kinds of answers to these questions, while science education history bears special witness to the array of answers provided to the dominating *utility* question in particular with respect to the science literacy debate (DeBoer, 1991, 2000).

What is normally done is to start with the utility question first, to look "outside" to the demands of either society or the professional science community and then proceed backwards to define goals and structure curricula according to the needs and diverse demands defined there. Egan offers us the choice of starting differently, from a potential autonomous and "inside" educationalist perspective (and subsuming the utility question). Although Egan himself is not considered in some circles to be a professional philosopher of education, he nonetheless addresses issues pertinent to educational philosophy and seeks to help bridge the disciplinary gap between philosophy of education and genuine educational (curriculum) theory. Moreover, as an *educational theorist* he has long argued for the establishment of education as an autonomous discipline, for it to be better demarcated from other fields of study (especially psychology; Egan, 1983, 2005a), and his reasons here serve equally well to further the self-conception of science education as an independent sub-discipline (a noteworthy topic that can only be touched upon here). The question to be asked, and the solution sought, is whether his *comprehensive* educational theory ("metatheory" or "grand" theory) can contribute in significant ways not only to enhancing science learning (the narrower use), but also if it can help resolve at long last in some sort of definitive way (assuming many in the community would accept it) the enduring impasse in the self-conception of school science education due to its central goals being at cross-purposes (the wider use). These are questions addressing different audiences, the former focusing on teacher pedagogical knowledge and learners, the latter focusing on the wider research community with its evolving history and identity.

With these solutions in mind, this chapter has been organized into five further sections and will be concerned with the important theme of a self-reliant *philosophy of science education* (PSE) for teachers and the research community, whereas the following chapter will address Egan's educational metatheory directly. In the next (or second) section, a succinct overview of the quandary of science educational reforms is presented to clarify the deadlock in the major goals, including reference to the lack of consensus concerning scientific literacy conceptions. These problems will lead us to ask, in the *third* section, whether or not a turn towards more philosophical-based reflection and educational theory in the discipline is required, with the proposal that, among others, a metatheory may offer an answer to the dilemmas we face. The *fourth* section will then address the question: "what is a metatheory?" The *fifth* and final section will ask: "why does science education need one?," notably addressing the options resulting from the "scientific literacy" debate and dilemma.

2.2. OVERVIEW OF SCIENCE EDUCATION REFORMS: A RECYCLING OF COMPETING IDEAS?

2.2.1. Three Problems of Science Education

Around the world, science education is in a period of intense self-reflection and self-criticism as it moves into the 21st century. For a while now several authors have even described the situation as one of "crisis" (Duschl, 1985; Gibbs & Fox, 1999; Matthews, 1989; McDermott, 1991; Reiss, 2000)—a word, however, that appears to be more *slogan* than description, having been used off-and-on for decades. Nonetheless, educators both at schools and universities are troubled about declining enrollments (O'Meara, 2006),[5] the closure of physics and chemistry departments in countries such as Canada and the United Kingdom (McKee, 2005), poor performance on international tests (OECD, 2004: *PISA, 2003*; IEA: *TIMSS, 2003*), and the general unpopularity of lecture-based courses. The abandonment of science by numerous college students after taking introductory courses in physics and chemistry has been a cause of concern for many years (McDonnell, 2005; Rigden & Tobias, 1991). Among students themselves there is widespread perception that the content, especially as presented in physics and chemistry classes, is largely formalistic, dogmatic, unexciting, and irrelevant, as evidenced in several countries (Canada: Brouwer, Austen, & Martin, 1999; U.S.: McDonnell, 2005; Germany: Reiss, 2000; Norway: Angell, Guttersrud, Hendriksen, & Isnes, 2004; Denmark: Nielsen & Thomsen, 1990).

This identifies the first of three problems central to science education that foreground this chapter. I shall call it the *problem of meaning as related to relevance and non-transferability*: the common inability of students (and also citizens) to transfer their classroom or previously attained scientific knowledge to issues and problems related to their everyday lives. This is the view from the *inside*, from those within educational institutions: the teachers and students affected by curricular demands and designs, teaching and learning styles, as well as an educational system long dominated by traditional instruction based on "structure of the discipline," "textbook science," and standardized testing (although it comes as no surprise to science educational researchers who have instead emphasized the need for *science-technology-society* (STS) context and curricular themes for over 30 years now; Roberts, 1982; Solomon & Aikenhead, 1994; Yager, 1996).

This predicament is not unique to science education, of course, and has been identified as a general dilemma with public schooling (Egan, 1997; Gardner, 1991; Ungerleider, 2003). In brief, Egan (1997) understands the issue to be a combination of at least three conflicting goals of education in schools and society (with their underlying educational theories), along with a failure to consider appropriate age-developmental (emotive-cognitive) understandings in students. Gardner (1991) also sees the problem associated with a combination of factors: the failure to recognize students' preconceptions that they bring to classrooms, narrow focus of schooling on abstract *knowledge* at the expense of true *understanding*, and elevating the cognitive domain at the expense of developing a variety of other intelligences, especially the emotional domain. Ungerleider (2003) zeros in on the *curriculum* as the primary culprit:

> The problem is the curriculum. The curriculum of the public school has become bloated, fragmented, mired in trivia, and short on ideas. It does not

> demand that students connect what they learn with anything else. It does not challenge them to reach their limits. The curriculum stifles curiosity. Although it demands effort, it does not reward deep thought. (p. 105)

Identified here, albeit in simple terms, are important issues relating to meaning and use, knowledge versus understanding, and the proper status and location of disciplinary knowledge (or a teacher's CK) within science teaching and curriculum that could possibly enable such understanding to take place, all of which will need to be addressed further. Let us move to the subject of understanding for the moment and examine this briefly as it comprises another problem of science education (while recognizing its ties to the previous issue of meaning and use). It is true—and accepted as self-evident among teachers and the wider public—that science education of the specialized disciplines at the upper secondary and undergraduate levels does indeed impart degrees of *technical content proficiency*. But what is not so well-known among teachers and the public—although it has been disseminated for decades in science educational research journals—is that such acquired *formal knowledge* remains largely theoretical, without context and "inert" in students' minds when they leave their science classrooms.

The view from *within the science education research community* has established without a doubt "the [predominant] failure of science teaching to disturb ingrained beliefs" among learners. Students, in fact, "often live in two worlds, one for the science exam, another for everyday life. This is alarmingly illustrated in recent American surveys showing that belief in astrology is very little effected by completion of an American science degree" (Matthews, 1989, p. 5). This "two minds outcome," as it is sometimes referred to—and one that would surely disturb teachers and cause much unease among the public if it was more broadly recognized—is the inability of students trained within the same discipline to solve basic problems in different contexts and attain deeper conceptual understanding of the topics commonly taught not only in classroom science,[6] but in other subjects as well (Gardner, 1991).

Secondary- and college-level physics and chemistry students in numerous studies have exhibited this phenomenon in the past 30 years, and some examples will illustrate this by now widely accepted problem of science learning (Duit, 2006; Duit, Goldberg, & Niedderer, 1992; Garrision & Bentley, 1990; McDermott, 1984; McDermott & Redish, 1999; Nakhleh, 1992). Physics students are able to successfully solve mechanics problems using Newton's second law (F = ma), and even obtain high scores as measured by standardized tests, but most tend to revert back to "common sense" reasoning and Aristotelian-type conceptions of forces and motion once outside of the "text-test" framework (DiSessa, 1982; Hestenes, Wells, & Swackhammer, 1992; McCloskey, 1983; Palmer, 1997). In chemistry, researchers have found that profound misunderstandings persist throughout schooling, from elementary through to undergraduate, and occasionally even into graduate school, about atomic and kinetic molecular theory. A significant portion of students continue to hold to ideas that appear to be based on their everyday experiences, such as insisting that the atoms of metals themselves expand when a metal rod is heated, or that the bubbles rising in a pot of boiling water are comprised of air or hydrogen (Nakhleh, 1992; Osborne & Freyberg, 1985). Some of these conceptions bear remarkable resemblance to the Aristotelean-type ideas of early scientists in the history of chemistry, for instance that gases are weightless substances (Mas, Perez, & Harris, 1987). This situation can be traced back to the apparent pedagogical failings of educational institutions and academic science departments to properly prepare teachers. According to an early study by McDermott (1984), the research clearly indicates

> that many students emerge from their study of physics or physical science without a functional understanding of some elementary but fundamental concepts. Similar problems exist at all levels of education, with little difference between high-school and college students. Of particular concern is the apparent failure of universities to help precollege teachers develop a sound conceptual understanding of the material they are expected to teach.[7] (p. 31)

Thereto, one could argue that post-secondary instructors have equally not been adequately prepared for the obstacles raised by the pedagogy of teaching and learning science content as structured by science academic disciplines in introductory undergraduate courses (Hake, 1998; Hestenes, 1998). These serious impediments to true understanding are due in large degree to; the limitations of teaching de-contextualized, abstract knowledge along with an algorithmic (solely numerical) problem-solving approach; ignoring both everyday or techno-social problems and the *nature of science* emphasis, at the same time assuming a trouble-free transmission or "Morse code" *notion of language*, so typical of textbook-dominated science learning at both the high school and undergraduate stages in physics and chemistry teaching (Mestre, 1991; Nakhleh & Mitchell, 1993; Niaz, 2005; Stinner, 1995b). Furthermore, as may have become evident, this problem is directly connected to—and also supported by the research literature amassed over the years on—students' "misconceptions" and "alternative frameworks" that they invariably bring with them to classrooms (and rarely discard) and that are too often ignored by teachers during traditional instruction.[8] I shall call this *the problem of context and conceptual understanding* in the physical sciences.

A third problem associated with science education has come to be identified by those viewing education primarily, but not exclusively, from the *outside*, those concerned with the *public understanding of science* (which now has its own academic journal by that name): scientific institutions, think tanks, government agencies, business and corporate interests, and concerned citizens in general. This is the *problem of science literacy*—although science educators as well have for decades been preoccupied with this problem for their own reasons and concerns.[9] A reasonably high level of scientific literacy among citizens has come to be seen by many within modern societies—both Western and non-Western, whether developed or developing—as of vital importance for maintaining both economic and medical health but also democratic policy-making and environmental sustainability, because of the intimate ways these are wedded to, and dependent upon, techno-scientific innovations (McEneaney, 2003). It has, in fact, been understood as a principal objective of science education since the 1950s—by some arguably as early as 1900 (DeBoer, 2000)—and is often touted in the media as the key component in today's "knowledge economy." Unfortunately the very low levels of scientific literacy as evidenced by the public—around eight to 10% (although highly dependent upon how it is *defined and measured*)—in numerous surveys taken in several Western countries has remained remarkably and disappointingly unchanged over the past several decades (Miller, 1996, 1998).

So one can safely conclude that regardless from which of three current important perspectives science education is viewed (whether from inside institutions themselves, or from within the research community itself, or from outside interests), there is a common perception of an unsatisfactory state of affairs and hence renewed calls for significant reform(s). Unfortunately, any view from the *present vantage* tends to the myopic; it is rarely adequate to recognize either the roots or the wider dimensions of fundamental problems. What is required is broader *historico-philosophical analysis*.

2.2.2. Crisis-Talk and Historic-Philosophical Analysis

Why research within science education generally tends to have so little impact on classrooms is due to various factors and continues to occupy the community as a serious problem (Boersma et al., 2005; Jenkins, 2001; Millar et al., 2000). Another important reason, cautiously suggested here, is the omission of developing any kind of metatheory for the discipline. Such an *educational failure* could help account for why curricular reforms are particularly vulnerable to the *political* whims (or "ideologies") of various stakeholder groups, an enduring situation several researchers have taken notice of (Aikenhead, 2007; Donnelly & Jenkins, 2001; Fensham, 2002; Gaskell, 2002).[10]

What *has* transpired, and been reluctantly admitted, is that several conflicting redefinitions of the *meaning* of "scientific literacy"—long seen as *the* fundamental goal of science education (both within and without the community)—have occurred over this extended time period, though unfortunately with little consensus on the part of the community itself, "with one exception: everyone agrees that students can't be scientifically literate if they don't know any science subject matter" (Roberts, 2007, p. 735). This has been *the* major contributing factor to the ongoing disputes surrounding curriculum programs, content selection, and the choice of suitable targets and their evaluation when constructing appropriate measurement instruments, whether for standardized tests or for public understanding surveys (Bauer, 1992; DeBoer, 2000; Laugksch, 2000; Shamos, 1995). This lamentable state of affairs, although acknowledged, cannot be ignored indefinitely. Jenkins (1992) has made the observation:

> In essence, therefore, scrutiny of the notion of scientific literacy raises in an acute form the question of what counts as science education. It is perhaps worth recalling, therefore, that, in a democracy, the answer to this question is socially, rather than academically or theoretically determined. (p. 237)

One can go along with the first part of this view while taking serious issue with the second. Though one can readily acknowledge that since school science exists both as a public institution and as an educational undertaking it must be responsive to its socio-cultural environment and the interests of stakeholders, it does not therefore follow that science education or literacy must be primarily *socially* determined—on the contrary, this is to undervalue, if not to undercut, its educational merit. A major argument of this book is to undermine such a notion (because such a determination usually results in an impasse) and insist on the converse, that the answer to the question must *first* be addressed *educationally*, that is to say, to be educational-philosophically determined, and afterwards socially negotiated.

Internally the research community has had considerably more success (in large measure because it has limited its focus to localized empirical studies) in addressing the "two-minds" outcome of domain-specific subject topics in contemporary instruction relating to canonical knowledge (e.g., kinematics, chemical bonding, or diffusion)—where many students enter and leave classrooms with their alternative conceptions typically intact[11]—through various reform tactics: "interactive classrooms," new "inquiry" techniques, conceptual change strategies, micro-computer based simulations (MBL), and others. And although the research on conceptual change is now extensive and formidable, much still remains to be answered about how to improve science learning (Duit, 2006; Erickson, 2000; Scott, Asoko, & Leach, 2007; Wandersee, Mintzes, & Novak, 1994).[12] Indeed, the majority of science education research work seems focused on such content-based "technical" and learning studies (Gunstone & White, 2000;

Hodson, 2003; Lee et al., 2009[13]). Indeed, Gunstone and White have made the case that science education has "progressed" as a *discipline* because of several *principles* discovered through research work related to alternative conceptions and student learning. As necessary and important as these are individually, there still exists considerable criticism concerning the *image of science* that underlies such studies with their focus on formal, decontextualized scientific knowledge—the mainstay of most school and college learning—and they serve at best as *intermediate* and not final aims, of ordinary chemistry and physics education, in particular, and of science education in general. Nor can they assume to have skirted the wider literacy question; quite the contrary, such studies implicitly commit themselves to a very narrow conception ("technical training"), for they are satisfied with a restricted (and *ahistorical*) understanding of formal content knowledge within isolated and specialized disciplines—a conception that has come under noteworthy attack for decades now.[14]

Related to this, one can identify two more criticisms that immediately come to the fore: the limitation with respect to *context* in which the subject content knowledge is presented and assumptions about the *nature of content knowledge—the former a pedagogical issue and the latter an epistemological one*. The issue of "context" has only recently come to concern physics and chemistry researchers (Finkelstein, 2005; Gilbert, 2006; Redish, 1999). Klassen (2006) has provided *five contexts* that are important for learning (theoretical, practical, social, historical, and affective) and has developed a story-driven contextual model. As discussed later, the issue of *context as narrative*, especially as explicated by an educational metatheory, will be seen as paramount for successful learning. Furthermore, such metatheories will purposely *problematize* the disciplinary-based content knowledge for educational purposes. (The epistemological problem will be dealt with separately in Chapter 4.)

The interplay of internal- versus external-motivated reforms in science education is long and fascinating, involving a complex and controversial relationship among diverse social groups (claiming disparate interests, goals, and definitions of literacy), the character of which has overlapped with the conflicting and unresolved wider issues of the *value and aims of public education*, the complexity of *techno-scientific impact* on society, as well as the enduring academic quarrels over the *nature of science* (NoS). These conflicts are probably inevitable, and pertinent questions remain to be answered as to whether or not they are in fact intractable given the intersection of several other equally substantive dimensions, especially the following:

- the fact that science is historically and culturally dynamic and exhibits itself to us *three-fold*, as an accumulated body of knowledge, but especially as a knowledge generating and techno-social enterprise—the impact on societies and cultures of which has both positive and negative consequences (different reform "waves" have laid stress on one or the other of these aspects to be taken as the key factor in defining "literacy");

- the implicit *educational project* (inherited from progressivism and the liberal education tradition, which remains active to some degree), geared at enabling learners to develop their potential as autonomous individuals, including cultivating creativity and critical thinking skills (stressed as the "humanistic aspect" at different times in reforms);
- the ever-present desire by diverse social interest groups to orientate education towards either socio-political or socio-economic useful *ends* (utility arguments)—what can be called the *socialization imperative*.

Looking within the science education community (and ignoring for now those arguments from social groups *without*; Laugksch, 2000), this last imperative is usually understood in two predominant ways: either as preparation for professional and technical careers (the "pipeline" argument, behind the conventional curriculum) or, more commonly, for "civic responsibility" and "science for all," with their assorted *interpretations* (those involved with HPS, STS, SSI reforms, "science for social action"; also the stated intentions of current "standards" documents).[15] This imperative has dogged science education since its inception (which Bybee and DeBoer, 1994, p. 358, label as the "social efficiency and effectiveness" and "national security" justification)—a required demand of science teaching more so than any other school subject—and is one that several notable critics have countered is misguided and strains its capabilities as an educational endeavor (Bauer, 1992; Jenkins, 1997; Shamos, 1995). Yet it is one many educators are not willing to let go.

In any event, what seems to have been overlooked in its entirety is that *three* longstanding yet venerable and operative *ideas* in education—themselves inexorably embedded within science education (as sketched above)—could be undermining each other. As Egan (1997) has persuasively argued, schools in the West as *educational projects* are ineffectual primarily because they are caught among three chief objectives (or rationales) that effectively serve to check or undercut each others' *intended aims*: whether to teach science for (1) intellectual development (knowledge), or (2) individual fulfillment (character), or (3) socio-economic benefit. (The first can be associated with the original knowledge-based educational project of Plato, the second with Rousseau, and the last is a cross-cultural and timeless expectation of most societies.)[16] In effect, three educational philosophies are competing with each other for a defining role in schools. When educational goals are examined historically, these three are ubiquitous; they persistently present themselves, albeit in different guises,[17] and they certainly can be identified throughout science educational reform history (Bybee & DeBoer, 1994).[18]

Now no one normally holds exclusively to one or the other, although usually one or the other is emphasized over the other two at a given time (depending upon the defined "crisis" at hand and under influence of respective social group interests), and the modern school and indeed many curricular so-called "standards" documents aim at a sort of *balance* among them.[19] Egan maintains, however—and I think this is the correct assessment—that the attempts to achieve "balance" are illusory, because one cannot "balance" incommensurate educationa philophies (see footnote and previous footnote 16).[20] If this is indeed the case, then the impasse must encompass science education as well, whether one wishes to concede this or not.[21] What is required as a way out of this stalemate, argues Egan, is for education to mature and develop its own discipline-specific theory. This would, in turn, go a long way (perhaps not all the way), in resolving the internal-external tensions among diverse social group ideologies and related curricular paradigms that have shaped, and still shape, the contours of the science educational endeavor.

The *history and philosophy of science* (HPS) reform movement has a history over a century old too and has taken part to differing degrees in the debates defining the aims of science education, including *scientific literacy*, and hence revising curriculum and instruction accordingly (reviews are given in Duschl, 1994; Matthews, 1994). It has had only partial success, however, since its agenda must still compete with alternative and well-entrenched curricular "projects"—though "paradigms" is a better word—within the actual *schooled* community (the main ones being traditionalist, progressivist, STS, standards movement).[22] Although this term as popularized by Kuhn (1970) has some drawbacks, it serves the purpose for which it is used here (Shulman, 1986).[23] What is missing in all science *education* paradigms though, are *defining high-level educational theories that could inform and drive them*.[24] More recently a host of other reformist research approaches (e.g., to integrate social activism, or incorporate gender, multicultural, and global issues, among others) have been demanding increased attention while pressing for their own goals and targeting their preferred issues. And although the latest school standards documents (insofar as they are *implemented* policy documents) in the English-speaking nations show considerable overlap on several topics, and in particular have emphasized the significance of incorporating the NoS for a broader and more accurate understanding of science (to the relief of many HPS reformers; McComas & Olson, 1998), this NoS goal nonetheless often runs at cross-purposes to others—including competing "visions" for science education (Roberts, 2007). Some have even questioned the value of NoS instruction, including the epistemic basis of so-called "Western science" (Rudolph, 2000). As a consequence, both a conceptual and value-based impasse among the alternative paradigms has come about.[25] This tends to deadlock attempts at practical implementation, affords traditionalism the default position of the status quo, and at minimum it undermines the potential of what HPS reformers seek to accomplish.

This becomes noticeable if we briefly compare HPS to its two major curricular "competitors": STS and the conventional specialist paradigm. While there has been some fruitful intercourse in the past between STS and HPS reformers (especially their mutual humanistic emphasis), those within the two paradigms now seem to be drifting apart given the divergence of views on NoS, the goals of science education, and the notion of literacy (Aikenhead, 1997b; Turner & Sullenger, 1999)—especially with the relocation of some former researchers within STSE shifting towards a kind of "political activism." Some are even flirting with relativist, postmodernist, and "postcolonial" ideas (Aikenhead, 1997a, 2002a; Loving, 1997; Roth & Désautels, 2002; Schultz, 2007; Weaver, Morris, & Appelbaum, 2001; Zembylas, 2006).[26] It would seem that a critical mass of incommensurability has been reached. There are researchers who now argue, in fact, that since STS has not lived up to its promise of properly engaging students with science and society concerns, it is time to move "beyond STS" (Zeidler et al., 2005).[27] In essence, STS "may be an underdeveloped idea in search of a theory" (Zeidler et al., 2005, p. 358).[28] Shamos (1995) had already made this charge a decade earlier.

Comparing HPS with the traditional academic (or specialist) paradigm, it is quite apparent that at the *upper levels* this paradigm (social group, curricula, textbooks, literacy conception, and practices) remains deeply entrenched, where an outlook prevails that continues to undervalue the epistemology and history of science (also the social practices of contemporary science), while overestimating the efficacy of lab-based "cookbook" inquiry[29] and content acquisition in decontextualized settings (see Lederman, 1998). Although the new worldwide HPS movement is now over twenty years old and some inroads are occurring at the secondary level (as presentations at International History and Philosophy of Science Teaching Group conferences attest to), at the tertiary level little appears to have changed (Galili & Hazan, 2001; Mason & Gilbert, 2004). Yet here too an educational theory is missing, indeed an educational *philosophy*, to justify the practices and curricula in these classrooms—although everyone seems agreed that its main purpose is induction into academic science (Fensham, 1997; Jenkins, 2000; Shamos, 1995).[30]

What is characteristic of both traditionalism and STS, speaking generally, is that each paradigm embraces social groups holding competing conceptions of science literacy that, taken alone, are exclusive and form oppositions—which Roberts (2007) has perceptively described as "Vision I" and "Vision II," respectively. Put in practical terms, this stark contrast has several negative consequences for science teachers, for they are forced to choose between options they would prefer to avoid: the choice of mainly teaching science either as "knowledge of" (Platonic project) or as "knowledge for" (socialization); preparing their students primarily for science-oriented vocations or for "active citizenship." Moreover, it forces a choice in their *allegiance* to two vital social groups, either to the academic science community or to the science education community (that is, to one particular sub-group *within* that community). Unfortunately by identity, by training, and by calling (by appeal of their *specialty* and by request of their profession as *educators*) this divorce is not possible, nor should educators be placed in such a quandary.

Again, a "balance" does not appear an option with exclusivist-type paradigms.[31] Yet this kind of dilemma seems destined to occur when educational theories are lacking, when teachers are not philosophically equipped to grapple with competing curricular visions, and when external social groups on top of well-intentioned science education researchers (all beholden by conviction to their own educational ideologies) unwittingly initiate with science teachers (Pedretti et al., 2008).[32]

2.3. PHILOSOPHY OF SCIENCE EDUCATION AND EDUCATIONAL THEORY

At issue in my view, and as has become apparent so far, is there exists a much greater and fundamental problem plaguing school science education as a discipline. It goes deeper than "crisis-talk," disagreements over conceptual change strategies and learning theories, debates over literacy, goals, and how to balance competing group interests, including the disparate perspectives on the worth of "Western science" in terms of questioning its epistemology, objectivity, and universality. At bottom, what should have become apparent but remains hardly noticed is that the HPS movement suffers from not having developed an overarching *conception of the educational endeavor* (or "philosophy" if you prefer) behind its intended reforms—something it shares with all other reformist projects and their various and changing curricular emphases (Roberts, 1982, 1998). Hence, while it partakes in the debates (intrinsic versus instrumental goals; the meaning of science literacy or constructivism, curriculum content choice, etc.), it remains hampered and cannot rise above them. My accent on "educational theory" therefore equally entails the conviction that science education requires a conceptual reorientation—it needs to re-shift its theoretical discourse towards the *philosophy of education* (PE) in fundamental ways, in particular through a sub-discipline *philosophy of science education* (PSE).[33] This new field of inquiry begins in a consideration of what Ernest (1991) had already outlined previously in his work for a "philosophy of mathematics education" (although some have argued that his project is more about sociology than philosophy).[34] It can be considered a truism that modern science education, especially science teacher education, has tended overall to ignore philosophy for psychology, especially its theories of learning and development that still dominate education research (DeBoer, 1991; Duit & Treagust, 1998; Fensham, 2004).[35] It is high time for science education to get its philosophical house in order, and hence, to return to thinking about its foundational problems and goals.

When examining the two latest *International Handbooks of Research* (e.g., Abell & Lederman, 2007; Fraser et al., 2012) as well as the two earlier defining and comprehensive handbooks on science education (Fraser & Tobin, 1998; Gabel, 1994), which together can be taken to represent the "best thinking" in the discipline, one is amazed to discover a complete lack of reference to philosophy of education, or in fact any discussion related to educational philosophical *theory* (as mentioned before). Fensham's important book *Defining an Identity* (2004), which surveys researchers and the scope of their work over four decades (also discussed previously), illustrates this glaring omission best. Likewise, even a cursory perusal of individual reform-minded literature shows a near complete lack of acknowledgement of terms like "theory of education" or "educational philosophy."[36] This is equally true of all reformist-type "standards" documents in both the United States (e.g., American Association for the Advancement of Science, 1993) and Canada (Council of Ministers of Education, 1997). (Clearly, statements of standards, a list of goals and principles and mandated policy documents, while *useful,* cannot in the end substitute for an educational philosophy. Such items though usually contain implicit philosophical messages and telegraph explicit educational values.) This also holds for those interested in HPS reforms,[37] although there have been exceptions: Matthews (1994) notes that the influential U.S. National Science Foundation had already commented back in 1980 when criticizing school reforms that there existed a lack of "a sense of direction and a theory and philosophy which should provide guidance to curriculum development and instruction" (p. 31, quoting their report). Nevertheless, what surprises is that one might expect a movement that has rightly focused on the need to integrate philosophy *of science* issues into science education would have been cognizant to a much greater extent of the need to inquire into a philosophy *of science education* (PSE) as a framework within which to do the necessary education of the subject at hand, considering their *educational* focus. In almost all of those cases mentioned, one proceeds as if philosophy of education (PE) and its related topics and issues are essentially irrelevant.

In having stated the premise that the discipline needs to partially shift its research focus towards more philosophical-based reflection and analysis, this was not meant to imply that philosophy as a subject has been completely neglected. Such a view is obviously mistaken. Science education is known to have borrowed ideas from pedagogues and philosophers in the past (as examples, from Rousseau, Pestalozzi, Herbart, and Dewey; DeBoer, 1991), yet most science teachers and too many researchers seem little aware of, or even concerned to know about, this rich and extensive historical background. It has been said that the most significant philosophical impact, at least in North America, has come from the pragmatism of John Dewey, especially in the pre-World War II "progressivist" era, and some authors continue to champion him today (Kruckeberg, 2006; Wong et al., 2001). Dewey's ideas have also been presented at previous IHPST conferences (Briscoe, 1990; McCarty, 1990).[38] There have certainly been other intermittent forays into philosophical terrain, some *indirectly*—for example, Stenhouse (1986) and Lemke (1990) who have looked to Wittgenstein and language studies for insight and reform ideas—and some more *directly*, such as Eger (1992), who has insightfully championed the worth of Gadamer's (1960/1975) *philosophical hermeneutics* for science educational reform (discussed further in Chapter 5). No doubt the most extensive and ongoing discussion that *does* include philosophy is the valuable but nonetheless limited focus on relevant themes in the PS for science education.[39]

Roberts and Russell (1975) were the first, however, to explicitly articulate the need for a *shift* from psychology towards philosophical analysis (drawing on the earlier work of Peters, Oakeshott, and Scheffler in the PE, and Nagel and Toulmin in the PS). They argued for the necessity of science classroom teachers to select and develop philosophical-based "theoretical perspectives" when scrutinizing aspects of their science teaching—like the nature of authority, the kinds of arguments used in classrooms and the image of the nature of science—as well as the epistemology and metaphysics ("worldviews") behind curricula. Their arguments were not limited to simply stating rationales, for they referred to several research cases at the time that exemplified such perspectives, and one can easily recognize how these different research emphases have grown and are being actively pursued today (especially by HPS reformers), although without awareness of their previous claims and examples. Since then Anderson (1992) has repeated the worth of including "philosophical perspectives" as one vital dimension (in a multiple perspective analysis) when discussing how to make curricular reforms more successful. Certainly earlier there has also been Scheffler (1970/1992), who had stressed the importance of the direct inclusion of philosophy into aspects of science teaching and teacher training (as mentioned previously). Finally, Siegel (1989, 1992) and others have long argued for the value of enhancing critical thinking and "educating reason"—another research direction that today has grown and gained importance, emphasizing "practical reasoning" as well as *scientific argumentation* (Bailin, 2002; Brickhouse, Stanley, & Whitson, 1993; Kuhn, 1993; Matthews, 1994; Osborne, Erduran, & Simon, 2004).

In sum, part of the present shift in focus is to remind those in science education: (a) that since it belongs to the sub-field of *education* it avoids educational philosophy at its own peril, and (b) to insist on a greater awareness and appreciation of its own neglected historical philosophical roots. Lastly, (c) to insist on the value and need to *explicitly* develop a philosophy "of" and "for" the discipline itself. As presently envisioned, such an internal reflection, as part of its self-conception and "research program" (Lakatos, 1970), would seek to formulate appropriate educational theories, possibly a "grand" or "metatheory."

2.4. WHAT IS A METATHEORY?

We have previously examined the historical development of *educational theory* itself in Section 1.4.5. The original emphasis on the explicit requirement for a metatheory in education had been discussed by Aldridge et al. (1992) following the proposal first put forth by Egan in the early 1980s encompassing his critique of "scientific psychology" and the demand that educational studies stake out independent territory (Egan, 1983). Simply put, a metatheory is essentially a worldview or paradigm that seeks to formulate a coherent account of explanations *of* and prescriptions *for* a given range of phenomena within its specified conceptual framework; it has pre-established criteria for empirical interpretation and judgments, and it directs research efforts along given lines within scientific or scholarly communities. In other words, it serves several roles: as the *lens* through which a community observes and reports data; as a *semantic-net* that provides meaning and conceptual order; and finally (characteristic for some social and applied sciences rather than for the natural sciences), as providing *prescriptions* for improvements and action. In the social sciences they can be classified into broad categories according to their makeup: as metaphysical, as political and economic, or as socio-linguistic. For our purposes, a metatheory in the disciplines of psychology and education "gives the big picture or may be described as the umbrella under which several theories of development or learning are classified together based on their commonalities regarding human nature" (Aldridge et al., 1992, p. 683).

These authors suggest that *four* different psychological metatheories continue to influence educational practice in various ways depending upon how the individual-environment nexus of development and learning are described and evaluated: *organismic* (or biological; examples are Piaget, Werner, neo-Piagetians, psycho-analytic theories); *mechanistic* (Skinner and the behaviorist school); *dialectical* (influenced by Marx, Hegel, and Vygotsky); and *contextualism* (based on the pragmatism of James, Dewey, and others).[40] None of these though, are true educational metatheories, the authors insist. Nor is, it seems to me, *cultural psychology*, a relatively recent school of thought (Cole, 1996), and currently of lesser influence (although it goes back to the "second psychology" ideas of Wilhelm Wundt in the 1920s). It can be taken as a *fifth* metatheory, or alternately, may represent the convergence of two metatheories, those of socio-linguistics and dialectics.[41]

What is important to realize is that any *educational* (meta)theory must needs be a normative one, for it seeks to *prescribe* an educational process to ultimately yield a certain outcome or *aim* (Hirst, 1966). This is usually a kind of person or the ideal of what an educated individual should aspire to become given the values and dispositions to be cultivated and methods employed in the specified program (Frankena, 1965). Further, it is in the worth of that final aim that the pedagogical methods of the educational project are justified, which traditionally have themselves been framed within the values and aspirations a society has deemed of ultimate importance: "The *value* of this end-product *justifies* the stages that lead toward its realization. Becoming a Spartan warrior justifies training in physical hardship. Becoming a Christian gentleman justifies exercise in patience and humility" (Egan, 1983, p. 9, original italics).

In Western civilization a succession of diverse *aims* or ideals have historically followed since the time of Ancient Greece, and some of the greatest Western minds have been preoccupied with formulating various philosophies of education to define their respective ideal and suggest ways to realize it (Lucas, 1972): Plato, the (philosopher-king) man of knowledge; Aristotle, the "good" or "happy" active citizen; Augustine and Aquinas, the Christian saint; Locke, the successful Christian mercantile gentleman; Rousseau and romanticism, the natural development of self-actualization; Kant, the autonomous individual, self-ruled by moral "good will"; and Dewey, personal and social "growth" through ever-changing experience, as the basis for developing critical democratic citizenship.[42]

Frankena (1965) insists that any PE must ask itself three basic questions: *what* dispositions (or "excellences") should be cultivated, *how* to cultivate them, and *why*?[43] When examining the position of the educational theorist Kieran Egan (1983), he does appear to have these same generally in mind, though as I take it, he reformulates and generalizes them with a slight shift in accent. Instead of using terms like "dispositions to be cultivated" and "ideal," he talks in terms of "end-product" and "aims" while explicitly raising the important fourth component of *development*: it is of the essence of an educational theory, he writes, that it answer four key questions: *what* to teach (curriculum), *how* to teach (instruction), *when* to teach it (stages of learner development), and, most importantly, *why* to teach it (specification of the end-product, aim, or ideal). That said, the similarity in questions and intent is obvious.

Certainly these four questions appear on the surface to be matter-of-fact curricular questions and are asked by any competent teacher at the *classroom level* of practice; there, however, the "what" is predetermined by the state- or department-sanctioned syllabus, the "how" is usually learned *in situ* with various instructional strategies, and the "when" is often shaped by precepts taken from developmental psychology and, increasingly today, cognitive science. The "why" is taken to be, at best, a restricted focus on building a "solid foundation" and presenting correct knowledge of science products and processes (Roberts, 1982), or at worst, simply as the hoped-for achievement of "knowledge outcomes" based on the standardized test at the end of the course. (Although I believe most educational practitioners are more thoughtful about the objectives of their courses.) At the *intermediate level*, these questions are also occasionally asked during professional curriculum planning within science departments and state ministries, where STS- and HPS-inspired reformers have also sought to spearhead policy change. Too often, though, policy is driven by professional or economic interests (Donnelly & Jenkins, 2001).

But to ask these questions at the higher *theoretical (and foundational) level*—the level on which any philosophy of education will direct us to concentrate, and which I suggest must be addressed by anyone wishing to critically reflect on curriculum—is to seek answers of a qualitatively different order. It is to ask basic questions of the quality and legitimacy of the education program itself—including the courses embedded within—the justification of aims and content, the kind of educated mind or individual formed by such courses, who constructs and benefits from the project so defined, and so on. It is at this level that I maintain science education has been found wanting in not framing aspects of such a discussion in terms of the development of metatheory, instead of, say, "knowledge outcomes," "multicultural science," "science literacy," or "power-ideology" analysis (*critical pedagogy*; e.g., Apple, 1992; Barton & Yang, 2000; Friere, 1970). Moreover, Egan asserts that because of the nature of the questions addressed at the higher level, psychology has little, if anything, to contribute to education, and the sooner educational practitioners become aware of the autonomy and uniqueness of their discipline, the better off and more successful their enterprise will be.[44]

The term "theory" itself is not unproblematic of course, for it has a range of meanings when comparing its usage in the natural and social sciences—which is not surprising and not necessarily an issue given the purpose and domains of the different disciplines—but in education it is found to be especially ill-defined and often contradictory, if not outright misused (see critique by Thomas, 1997). Likewise in science education the conception and usage of "theory" is quite varied, and it is better to speak of theory in two senses, as "theory lenses" or "theoretical frameworks" as normally applied to conventional research (Abd-El-Khalick & Akerson, 2007). (But, as mentioned, there is little to no direct reference here to *educational* theory). Even this given conception, though, is of a different *kind* and much more narrow or "lower level" from the sense I have in mind—that is, its conception as a "grand" theory *of education* and its link both to the PE and curriculum studies. A typical (albeit older) example is Shymansky and Kyle (1992, pp. 760–761). In discussing how to re-think curriculum research and practice, they can only identify "two theoretical frameworks" that "appeal as being suitable for research on science curriculum" (p. 761), being those of "radical constructivism" and "knowledge constitutive interests" (the latter referring to Habermasian ideas of knowledge and interests). Yet neither of these is in any way adequate as a metatheory of education for the discipline, since the former is confined to questioning the epistemologies of science, the teacher, and learner,[45] while the latter is confined to examining the nature of three different types of knowledge (technical, practical, emancipatory) in relation to helping ascertain the appropriate social project of schools—yet it should seem clear what is first required is to specify one's educational theory as a prerequisite to address and answer these kinds of issues and problems.

Interestingly, Fensham's book (2004) *Defining an Identity*—as discussed, where he canvasses the views of prominent science education researchers worldwide—once again offers an important insight about the science education community, this time concerning the role of theory (his Chapter 7). He admits that the development of theory is a significant indicator of a discipline's *advance as a research field*:

> If the existence of theory and its development is a hallmark of a mature research field there is some evidence that the research in which the respondents have been engaged in science education has reached this point. On the other hand, the role that theory plays in the respondent's remarks was so variable that it is not possible to attach this hallmark in a simple way to much of their research. (Fensham, 2004, p. 101)

With that admission he acknowledges that the use of theory is constricted and there was little interest on the part of researchers to develop their theory of choice further. What is significant though, is the range of *borrowed* theories that the researchers have relied heavily upon.[46] The spectrum stretches from social anthropology, ethnology, and cultural theory to psychology, cognitive science (e.g. information processing; schema restructuring), and PS (e.g., conceptual change theory).[47] He notes that those researchers employing a "political framework" to curriculum, or concepts of power and ideology, shift the common focus of science education onto entirely different factors that influence science teaching and learning. Essential PE-type questions like "what counts as science education?" or "How are ideological meanings reproduced in science education?" are raised, but surprisingly not addressed with that perspective or discipline in mind. One observes, rather, that in all cases educational theory and PE nowhere make an appearance.

Fensham also mentions the topic of "grand theory" (p. 107), one of the very few who do so. He writes that only one respondent had admitted to theorizing on this scale, namely the educator Joseph Novak, who had earlier published *A Theory of Education* (1977). This book, however (as is familiar today), is based explicitly on the psychologist Ausubel's quasi-neural theory of meaningful learning in combination with Toulmin's philosophy of science, and principally restricted to learning theory. Novak (1977) has today continued to hold to the value of this theory and the belief that "theories in science education would be developed that have predictive and explanatory power" (p. 106), a belief that seems to closely align educational theory with theories in the natural sciences, an arrangement both Hirst (1966) and Egan (1983, 2002) explicitly reject.[48]

We see that another important distinction that must be made clear is how the term "theory" is used with respect to other disciplines. It is in the nature of an *educational* metatheory foremost understood as *theory* that the notion of "theory" as used in this discipline (to prescribe a kind of *educated person* or *mind*) should be clearly distinguished and not limited to a tighter, prior formulation of the term as commonly applied in other fields—the natural or social sciences, philosophy of science, or even in those psychological schools who wish to define themselves exclusively as "scientific" or empirical disciplines (e.g., behaviorism, Piagetian theory, neuro-psychology).[49] It is to the credit of the ongoing development of science education as an independent discipline that some researchers have recognized this distinction, at least with respect to theories in science: "A failure to consider this fundamental difference between the nature of scientific theories and the theories of science education is damaging to science education" (Gunstone & White, 2000, p. 294).[50] The distinction, however, seems more difficult to make out when researchers draw upon theories from other disciplines. An important example is the paper by Norris and Kvernbekk (1997), and one of the few to address the vital subject of educational theories and their application in science education, although confined to *constructivism* (the typical subject when this topic is broached in the literature). They use the semantic notion of theory as discussed in epistemology and philosophy of science as a lens to analyze and critique the usefulness of the constructivist theory of science *learning* as developed by the late Rosalind Driver and her associates. This remarkable and original paper, while offering insight into how goal-directed theories function and the problematic relation of any theory in the social sciences to practice (that the connection always remains an *indirect* one, through mediation of auxiliary hypotheses and by interpretations and judgments of practitioners), nonetheless bears inherent limitations because of their use of such a semantic conceptual lens, as Driver (1997) in response had pointed out. I would argue, along with Driver, that "normative goal-directed theories" as *defined by* the semantic notion of theories is not possible at the level of generalization of a metatheory, although it may be possible at a lower level, and that the "analytical techniques used by [the authors] to analyze the theory actually distort the core features of the theory" (p. 1007). Further, I would add that one must be very attentive to the context when using such terms as "normative goal-directed theory."

As discussed, any educational metatheory must be both normative and goal-directed but not in the combined sense of the prior use of these terms in the semantic-epistemic conception, especially if it detracts from the explanatory power of the theory. Moreover, as I understand it even Driver's useful constructivist account would itself serve at best as a sub-theory to a broader educational metatheory, for it functions exclusively as a theory of learning and instruction and expressly not as a theory of either *cognitive development* (which at least the stage theory of Piaget and other neo-Piagetian theories have articulated; Niaz, 1993) or *content selection.*[51] And as Osborne (1996) has correctly pointed out, the omission of these two latter important aspects ("when" and "what" to teach) must rule out any form of constructivism as a viable candidate for metatheory status, irrespective of the grand claims of some enthusiastic endorsers (Matthews, 2000).[52]

2.5. WHY DOES SCIENCE EDUCATION REQUIRE A METATHEORY?

In having followed the discussion one could hope that the merit of metatheory has become palpable: *science education does not have one*. The inconsistencies and incompatibilities in its major educational goals bear this out (as argued in Section 1.4; also Bybee & DeBoer, 1994), the confusion in the conceptions of science literacy (based on alternative "visions") attests to it (Roberts, 2007), and the disputes (sometimes strident) over constructivism give witness to it (Matthews, 2000). It requires a metatheory for reasons comparable to what Aldridge et al. (1992, p. 684) claim is necessary for education: (1) that "the purposes of a psychological theory are different from an educational theory," thereto, (2) the implementation of such theories is usually problematic, and (3) the "relationship of development and learning [of science] is unique in educational settings." These three reasons (which are inter-related) aim to detach education from psychology, and so may be considered as *separation* arguments. They go on to list another two, but these are based more so on the actual inadequacy of the discipline itself—a *curriculum*-based argument and a *grounding* argument: (4) "education does not have a unifying metatheory which incorporates both curriculum development and assessment procedures," and (5) "education does not have a metatheoretical basis" (Aldridge et al., 1992, p. 684). We will see later that Egan considers the first three arguments as particularly forceful, especially since he holds that applications of disciplinary theories outside of education far too often do serious damage to it (1983, 2005a). Moreover, the fourth curriculum-based argument does not go far enough in its conception of curriculum; any useful curriculum must also seek to incorporate content and learning in correlation with the maturation of the learner at age-appropriate stages. Such a learner-based *developmental* theory that correlates maturity with curriculum, learning, and assessment would truly qualify as a "unifying metatheory" and be exceptional. The question whether Egan's metatheory qualifies will be addressed in the next section. If so, it has serious ramifications for how people can come to be "educated in science," as to what that can *mean*, and thereto, possibly help ground science education as a discipline.

> Psychology as a discipline is new, hardly more than 110 years old. Psychologists have generated, developed, or expanded at least four world views and multiple subtheories. Education has been around for centuries. We have no metatheory. We have no educational theories at all. (Aldridge et al., 1992, p. 687)

In writing that "we have no educational theories" I cannot imagine the authors should be taken to imply that educational theories have never existed or been proposed in the past. As mentioned, such theories go back at least as far as Plato (*Laws*, 1970; *Republic*, 1974, *Meno*, 1975) and Aristotle (*Politics*, 1962/1981), but these were primarily philosophers, as were Locke (1693/1964) and Rousseau (*Emile*, 1762/1979). While the latter two were occasionally employed as private tutors (as was Aristotle to Alexander the Great), their foray into education was limited, and they remained mainly outsiders, although Rousseau's ideas or versions thereof (in progressivism) are spread widely, in various guises (Darling & Nordenbo, 2003). Plato and Aristotle can be considered both educational theorists and practitioners, given their establishment of the two famous ancient institutions of higher learning (the Academy and Lyceum, respectively). How far they attempted to actually incorporate their theoretical ideas into practice in those places is unknown, yet it does appear plausible that Plato considered his Academy as contributing to the development of a state-guardian class in Greece at the time. Be that as it may, the point here is not to judge the educational credentials of some well-known thinkers of the past, rather, to pronounce along with the cited authors that for quite some time education has suffered from the characteristic lack of such theories formulated *by* and *for* those directly involved in educational studies.[53]

Considering the current controversies about prioritizing goals in science education (Bybee & DeBoer, 1994), one may be surprised to learn that even educational debates have a long history. Once again, PE can offer insight into long-standing science educational dilemmas. Aristotle records the following about education in his day:

> But we must not forget the question of what that education is to be, and how one ought to be educated. For in modern times there are opposing views about the task to be set, for there are no generally accepted assumptions about what the young should learn, either for virtue or for the best life; nor yet is it clear whether their education ought to be conducted with more concern for the intellect than for the character of the soul. (1962/1981, p. 453)

It is remarkable to contemplate how his discussion mirrors the debate of values and aims that has steered science education since its inception in the 19th century. Consider if you will the conflicting meanings (post-World War 2) of "science literacy,"[54] still identified as the overall objective of science education as discipline and practice: whether it is to be primarily understood as personal self-fulfillment (i.e., "virtue" as its own intrinsic worth), or for "critical citizenship" in a democracy (i.e., as instrumental worth, "the best life," STS), or rather solely for development of "mind" *per se*, as mastery of subject-based formal knowledge and as a tool for developing inductive (later redefined as "critical") reasoning (i.e., "intellect" development, science "processes," Traditionalism, "scientific argumentation"). Lastly, whether it should encompass foremost moral development when arguing "socio-scientific issues" (SSI) or "science education as/for socio-political activism" (i.e., "character of the soul"—always seen by Aristotle in terms of socio-political *activity*).

Note as well that the *three fundamental goals* underlying education (as elaborated above) can be identified here and roughly mapped onto the corresponding conceptions of literacy and onto existing school science curriculum paradigms. Some critical observers had thus come to the conclusion that already by the late 1980s the usefulness of the literacy concept had exhausted itself:

> It was confusion over the differing connotations of scientific literacy, the attempt to *subsume all under a common heading*, from mastery of basic knowledge to national technological superiority, from science viewed as a cultural imperative to that of social responsibility, from science content to science attitude, that finally led to serious questioning of its real purpose. The goals were simply too fragmented and not accepted with universal enthusiasm by the science education community. (Shamos, 1995, p. 85, italics added)

Regardless of how one wishes to characterize it, I believe one can without much controversy state that "science literacy" nonetheless still remains (for good or ill) the prime goal or aim of science education,[55] certainly in the English-speaking world, but increasingly in other parts as well (Jenkins, 1997; Roberts, 2007)[56]—although Shamos has insightfully argued that its common conception tied to citizenship is fundamentally flawed, that the community is chasing a utopia, that it continues to refuse to accept the grounds why it has failed in achieving it, and finally that many rationales typically put forth to justify it are a *myth*.[57]

In this sense the discipline *has* decided, no matter how contested, on its "why," and to differing degrees also on the "what" and "when" (depending on the prior construal)—however, these are merely mechanical because the "why" has nowhere been embedded within an educational metatheory (as argued previously). Further, it seems clear that some advocates construe the literacy aim to be a *kind of person* (the wider notion: the images of either the professional scientist or the active citizen), whereas others construe it to be primarily a *kind of mind* (the narrower notion). Hence, the desired or constructed curriculum is dependent in large measure upon one's prior paradigmatic and incommensurate commitments (traditional, multi-culturalist, HPS, STS, SSI, "socio-political action," etc.) *based on the background goal*. It thus serves more as *slogan* than as definition. I submit that as long as this state of affairs persists the controversy surrounding scientific literacy is not resolvable—nothing less than a scandal. *We have a situation here where a discipline cannot agree on the most fundamental purpose and goal of its educational endeavor* (DeBoer, 2000; Roberts, 2007; Shamos, 1995)[58]

One can therefore conclude, given this consistent mode of discourse about "science literacy," that the community is placed before one of *three* choices:

- *Exclusivist* option: one chooses either an already given or hoped for curricular paradigm; this could be the knowledge-based, specialist "Vision I" literacy conception (the given: traditionalism) or, at the other end of the spectrum, opting for an "extreme" form of "Vision II" by redefining literacy as "collective praxis," such as the (hoped for) image held by Roth and Barton (2004), according to Roberts (2007, p. 769).

- *Inclusivist* option: one agrees instead to hold fast to as many conflicting meanings as possible (e.g., Hodson, 2008). Along with DeBoer (2000) one simply accepts that the term stands for "a broad and functional understanding of science for general education purposes" (p. 594) and "because its parameters are so broad, there is no way to say when it has been achieved. There can be no test of scientific literacy because there is no body of knowledge that can legitimately define it. To create one is to create an illusion" (p. 597). Rather, only specific goals can be achieved in a piecemeal fashion, where his historically identified nine different conceptions are chosen as in a smorgasbord, attentive to the context of school culture and society wishes, and where "schools and teachers need to set their priorities" (DeBoer, 2000, p. 597). With this option, divergence is chosen. It is then assumed that "consensus about one definition throughout the worldwide science education community is a goal not worth chasing" (Roberts, 2007, p. 736).
- *Abandonment* option: one chooses to reject the term as both useless and meaningless for educational purposes, along with Solomon (1999) and Shamos (1995).[59]

Option two, although seemingly attractive on the surface, does not seem to me to be viable, and one can imagine numerous problems associated with it. Just mentioning one, it assumes a degree of autonomy for schools and teachers that they generally lack, and that in the climate of "accountability" and standardized testing and under the influence of powerful outside social groups would seem to check their ability to make the kind of choices DeBoer would like.[60] A reversion to option one would in all likelihood result, namely, the default traditionalist position.[61]

In any case, if an educational metatheory is to be of service to science education it must also acknowledge and address these options in the deadlock. It may also put into question the assumptions and scope of the discussion and even the entire character of the discourse that has heretofore been conducted (Schulz, 2011; Witz, 2000). As mentioned previously, *the essential merit of metatheory lies in creating curricular coherence, properly transposing subject content knowledge for the learner, and defining and steering educational aims.*

Now some may wish to argue that such a plurality of meanings is not necessarily a bad thing and, on the contrary, indicates the health and "maturity" of the discipline, where typically a variety of viewpoints are allowed to be expressed, criticized, and defended, and where a diversity of research programs are in play. "It might be argued that science educators collectively do not see curricular reform as a unified 'project' at any level. Nor should they, some might say: there is merit in diversity" (Donnelly,

2004, p. 764). It was in large part in reaction against the perceived trend of an all-embracing "one-theory-view" of constructivism within the science education research community that several authors responded that "consensus" can be an unattractive and debilitating quality. Because "'consensus' is at most a transitory feature of scientific progress both in science and education" (Niaz et al., 2003, p. 790), the authors reasoned instead that it was the critical appraisal of competing research programs that had facilitated what can be called "progressive transitions" in the discipline. Some may therefore protest that my suggestion for the need of consensus around "metatheory" may similarly represent a regressive rather than a progressive move for the community.

While in general agreement with the view expressed by the above authors regarding constructivism, I would still like to point out a few things: first, that most kinds of research programs are not usually informed by any sort of high-level *educational theory*; secondly, that debates concerning constructivism occur at a lower level with regard to *learning* theories (but where the values and principles of an educational metatheory as can impact the discussion is missing), and thirdly, that debates concerning issues of goals and science literacy must be transported to and answered on a metatheoretical level. Surely such a discourse is long overdue (Anderson, 1992). Here is precisely where issues in PE come to the fore.

There is no doubt that diversity is to be welcomed and "while accepting the partial legitimacy of this view," I argue, along with Donnelly (2004), "[we] yet maintain that curricular change in science is, to a degree which is counterproductive, an inchoate search for rationales, justifications, and specific innovations in the face of perceived deficiencies" (p. 764). Certainly one must always be vigilant against any kind of "groupthink"—as was perhaps threatened by constructivism and about which the authors correctly raised concerns—that talk of metatheory usually brings with it. But science education has shown itself time and again to be vulnerable to an extraordinary degree with its reliance on extra-disciplinary metatheories (psychological and economic), on the one hand, and has not managed to develop any explicit educational metatheories of its own, on the other. Moreover, one should point out that especially with respect to the natural sciences (a mirror I likewise hold up, for it is favored by too many science educators when formulating arguments and making comparisons for their discipline), even the physicist Lee Smolin (2006), (in his trenchant critique of the conceptual and sociological problems plaguing contemporary theoretical physics) still insists that in order for a community to thrive it requires not only a diversity of research programs and contrary ideas, but equally at times "we can say that science progresses when scientists reach consensus on a question" (p. 295). The obvious inference being, based on an argument of parallelism, that this should hold as well for our educational community.

At this stage in the discussion, and not by accident, a different sort of objection concerning the assumed need of "metatheory"—and indeed "theory" in general—could also be raised, but on different grounds, being the *question of the relationship between theory and practice*. What is of interest for education, so goes the oft-repeated stance, is not a matter for "theory" (at best it is only narrowly concerned with theory) because the discipline is predominately concerned with *practice*—this, as it turns out, is a basic critical rejoinder that has earlier come to the fore in associated literature. Things have certainly not been quiet regarding the relation of theory to practice in the social sciences, in curriculum studies, or the philosophy of education; still, here is not the place to indulge an elaborate discussion of the issues involved (Blake et al., 2003; Hiley et al., 1991; Thomas, 1997). Nevertheless, a few comments seem unavoidable. With the attack and undermining of foundationalism (see Section 5.1.1), and hence the rejection of the claim of the universal validity of theory, there are those who have sought at one extreme to solve the theory-practice problem by hoping to considerably diminish the role if not dispose of theory altogether—at whatever level. They have been encouraged in this regard from several quarters, both critical and practical: a return to Aristotle and his ideas of habituation and *phronesis*, the hermeneutic tradition, Feyerabend's "anarchistic" criticisms of the philosophy of science, Lyotard's attack on "totalizing metanarrative," as well as a revival of interest in Deweyan pragmatism. There has come the general realization of the considerable difficulties entailed when applying *any* theory in the social sciences to practice in useful ways.[62] This is the most radical and strongest position and, if successfully challenged, carries away the weaker ones in tow.

To meet this challenge, *five* counterpoints will be brought to bear: The first point to be made here is that "theory" can and has survived the collapse of foundationalism, and such a wholesale and radical rejection is therefore not warranted. For already with Hegel "theory" was taken as systematic without being construed as foundationalist (Rockmore, 2004).

Secondly, the need for metatheory must be maintained as a critical bulwark against several alarming tendencies, of which I cite *two*. The first tendency is an *academic* one, which tends to cognitive and moral relativism, only partly as a result of the "linguistic turn" (Rorty) after the collapse of foundationalism, but more commonly under the strong influence of a misguided anti-science attack (Gross & Levitt, 1994; Gross, Levitt, & Lewis, 1996). For science education to ignore educational philosophy and theory is to leave it directly exposed and vulnerable to those who would seek to reshape the discipline by importing theories based on radical social constructivist views and epistemological-relativist perspectives (Foucault, 1980,

1989; Loving, 1997; Weaver et al., 2001). An example here is the attempt to use ideas from Lyotard to transform the discipline (Schulz, 2007; Zembylas, 2000). The other tendency is *socio-economically* driven, and described as the new "managerialism" with its emphasis on "accountability," testing, and proper socialization, required, so it is argued, in order for students to navigate the emerging global techno-economy. This kind of socialization argument, though, is only a more recent manifestation of one that has pursued science education for decades (as discussed earlier), and by taking the form of an extreme reduction to educational instrumentalism (*economic utility*; Apple, 1992; Jenkins, 1997) can only be effectively challenged on the grounds of an alternative vision offered by a metatheory.

A *third* argument is the fact that all perceptions and practice are unavoidably "theory-laden," a phrase brought to popular parlance because of the developments in the philosophy of science, a fact well-known to those in the HPS movement. The position that science education requires a high-level theory to inform its practice and clarify its aims is basically to acknowledge that "theory" should "come to the surface" and be explicitly recognized. Otherwise one inevitably runs the risk of using some theory of which one is not cognizant, but which nonetheless guides practice and covertly channels the discipline—which seems to be the ongoing case with the field and its buckling under to the socialization imperative, itself often shaped by either economic or social-utilitarian philosophical metatheories (e.g., Dewey). This equally and markedly holds true at the local classroom level. Bruner (2006b, p. 161) has stressed that there already exists a folk psychology and folk pedagogy among both teachers and learners, and that it is their "folk theories" that engage the practice of teaching and learning, and not necessarily in the best interest of the learner, that the pedagogical theorist must in some way also take into consideration.[63] In short, to assume "theory" can be avoided is deceptive.[64]

This fact is linked to the *fourth*: science education in particular and education in general (let me repeat) tend to default to external disciplinary theories according to the newly discovered maxim that "educational development abhors a theoretical vacuum." It is commonplace to defer to metatheories in psychology, but even here educators as a rule often disagree which sub-theories of development, learning, or motivation to apply (Bruning, Schraw, & Ronning, 1995). Egan has shown by a series of cogent and careful arguments that when education defers to metatheories in psychology, whether (as examples), directly via behaviorism or Piagetian theory (1983, 2005a), or indirectly via progressivism (2002), the debasement of practice and a wide-ranging impoverishment of the field must necessarily result.[65] One suspects that science educators' mix-and-match approach with regard to, say, psychology, "science studies," or, lately, language theory, is predominately dictated by *conditions* of immediate practice (including classroom culture) and not informed by any sort of educational theory or a more encompassing educational *philosophy* behind their instruction—the latter necessity, as stated before, Scheffler (1970/1992) had stressed some time ago.[66] The popularity and conflicting strands of constructivism itself bears witness to this (Geelan, 1997). Such a complex, incoherent, and chaotic state of affairs I cannot imagine will contribute moving the field towards a more academically grounded or "mature" discipline (Kyle et al., 1992).

Hence, *lastly*, a metatheory (as previously mentioned) is required to advance the field into a suitable discipline standing on its own merits (as Piaget had opined could be termed a "science of education," yet Egan would fervently eschew such a notion, as no doubt Dewey would have done). It is needed to "develop and coordinate curricula congruent with assessment procedures" and helps decide "what to teach, when to teach, and how to teach." Barring this, "we will continually take the bandwagon or pendulum-swinging paradigm as our process and adopt the next psychological theory that comes along" (Aldridge et al., 1992, p. 686). It would also bring science education back into the fold of how "theory" has come to be positively reappraised again in the philosophy of education, so that it may truly "engage in explorations of what [science] education might be or might become: a task which grows more compelling as the 'politics of the obvious' grow more oppressive. This is the kind of thing that Plato, Rousseau and Dewey are engaged in on a grand scale" (Blake et al., 2003, p. 15).

Educational Philosophy and Science Education as "Socio-Political Activism"

One pertinent example of the third argument stated above is the contemporary reform movement (spearheaded by some international researchers and popular with some policy advocates), namely "science education as/for sociopolitical action." It has been articulated with implicit philosophical aspirations. It could reasonably be interpreted as a rudimentary sort of "*philosophy of*" science education (PSE) as here elucidated (granted, not formulated in this fashion), though not openly and explicitly. The position that science education *should* be oriented (if not exclusively so) to perform socio-political action is a normative claim argued on philosophical grounds, justified because of the apparent promise/claim of enhancing critical-minded citizenship and forwarding democracy. It patently stipulates categorical answers to the key questions "what counts as scientific literacy?" "what counts as science education?" and "what is it for?" Whether or not such a muscular and singularly focused PSE can do justice to the other historically identified aims associated as central to science education (including the *aesthetic* component of science; DeBoer, 2000; Girod, 2007), and therefore is the best option for policy deliberations and reform, is a matter for some dispute—although a considered debate, especially one involving PE, is surprisingly lacking to date.[67]

That this sort of politicized PSE represents a "radical program" to challenge common school science education is understood (Jenkins, 2009; Levinson, 2010). Here our focus is to ask: is such a "program" an adequate PSE?[68] Science education, for example, could plausibly "do" socio-political action at times, while rejecting "as" and "for." In any event, does politico-social activism as put forth substitute ideology for philosophy?[69] Does it presuppose educational metatheory? The present author would argue it must (although this feature is seldom articulated; i.e., social reconstruction). Stepping back, must *any* methodical PSE presuppose metatheory (of some kind)—or can it be gone around for, say, a list of rationales, principles, and exhortations? That debate has not yet begun but would be welcomed.[70]

One of the fundamental responsibilities of a PSE at the research level would be to expose educational theories (especially metatheories), as well as better clarify the relationships between such theories in PE and theories in other (empirical) disciplines (as to their nature, value, and limits), whether one of independence or inter-dependence.[71] In other words, a philosophical appraisal of several domains, such as: conceptual clarification and the validity of borrowed ideas; scrutiny of epistemic and/or moral and political aims—their character and prioritization; analysis of the theory-practice dilemma; also the character, quality, and significance of kinds of assessments or tests employed (range of usefulness), and so on, and hence the question of boundaries, applicability, and relevance.

2.6. SUMMARY

This chapter has so far attempted to ascertain why science educational reform has been so intractable a problem for so long in its historical development. The "conventional wisdom" still seems to hold that, ideally, a proper scientific mind should be a probable, if not an inevitable, result of our contemporary educational system, preoccupied as it is with the mastery of textbook-based formal knowledge (content and processes), stylized laboratory work, and with instruction and knowledge organized according to the disciplinary structure of the separate sciences. Yet in the past several decades numerous studies in science education research in general, and physics education and chemical education research in particular, have established that all three of these components are fundamentally inadequate to meet this outcome. That is, the curriculum remains too narrowly specialized, school laboratory work distorts the image of actual scientific inquiry, and instruction has chiefly overlooked the psychology and contexts of learning, including accommodating students' prior conceptions (or misconceptions), which create considerable barriers to comprehending complex and abstract scientific schemes.

Although the conventional wisdom is identified as an immediate obstacle to reforms, as an established school educational paradigm it was linked to a broader impediment located at the nexus of a confluence of several factors. The first factor is competing school-based "paradigms" with their own largely exclusivist conceptions of scientific literacy. These are themselves strongly influenced by social groups and their educational "ideologies" (both external and internal to the community), including several science education-based research programs. Secondly, there is the ongoing neglect of generating discipline-specific educational theories (not restricted to learning theories, and which could help clarify curriculum and goals), in conjunction with, thirdly, the prevalent disregard of philosophy of education. Crisis-talk has commonly tended to mask the competing interests of diverse social groups, including the lack of consensus over "science literacy" and division within science education research groups. It has been inferred that the fundamental reason why there continues to be inherent difficulties for prioritizing the main goals in science education is reflective of the reality that three essential aims exist that underlie general education, which themselves undercut each other's potential and which science educational reform history bears witness to but which it has not openly acknowledged. Attempts at "balancing goals" were judged ineffective if not illusory largely because the aims are articulated differently by contrasting educational philosophies, while they themselves often remain concealed.

To help resolve the reform and science literacy quandary it has been suggested that the science education research field requires a partial conceptual and theoretical re-orientation towards the neglected field of PE, and thereto, that a new theoretical-based research field should be broached entitled "philosophy of science education" (partially in comparison with Paul Ernest's work outlined for a "philosophy of mathematics education"). To that end it has been further advised that an educational metatheory should be developed that could help ground science education as an independent discipline, help distance it from various metatheories in psychology, and finally help it better navigate the demands of group interests with respect to instrumentalizing the educational endeavor as a desired socio-utilitarian project. Viewpoints that argued that entertaining a metatheory for the discipline could be counter-productive, or holding the position because education was primarily concerned with practice, "theory" was either superfluous or debilitating, were rejected based on five counter-arguments. In addition, it was argued that a metatheory could offer a possible solution to the "three-options" deadlock science education is currently embroiled within regarding the conceptions of *scientific literacy* by offering an educational discourse coming from an unusual but progressive perspective.

NOTES

1. See Jenkins (1992, 2000, 2003, 2007, 2009); Duschl (1990, 2008); Sunal, Wright, and Day (2004); Hodson (2003, 2008, 2009); Fensham (2002); Gaskell (2002); Aikenhead (2002a, 2002b, 2007); Roth and Désautels (2002); Donnelly (2004, 2006); Donnelly and Jenkins (2001); Hurd (1994, 2000); DeBoer (2000); Solomon (1999); Millar and Osborne (1998); Lederman (1998); Bybee and Ben-Zvi (1998); Claxton (1997); Millar (1997); National Research Council (1996); Yager (1996); Shamos (1995); Matthews (1994); Bybee and DeBoer (1994); American Association for the Advancement of Science (1993); and Shymansky and Kyle (1992).
2. For example, the concept of *learning* as defined by computer-based artificial intelligence models in cognitive psychology could be necessarily at odds with what educationalists mean by "learning." Some still recall the deleterious effects on education resulting from importing ideas of behaviorism (Matthews, 1994). Slezak (2007) has also come to question the relevance of cognitive science claims for science education.
3. See especially Carlsen (2007); Duit and Treagust (1998); Sutton (1998); Kozulin (1998); Moll (1990); and Lemke (1990). (Chapter 5 will examine this research tradition further).
4. This would include the *Bildung* tradition, which has influenced educational ideas and curriculum in Germany, Central Europe, and Scandinavia (Sjøberg, 2003). Duit, Niedderer, and Schecker (2007), writing in a recent review of physics education, have commented that the common notion of science literacy in the English-speaking nations is burdened with an overly instrumental conception and generally neglects to emphasize the importance of developing the individual personality, which is vital to the *Bildung* conception.
5. Figures comparing the percent changes in full-time undergraduate enrollment by field of study in Canada between 1991/1992 and 1998/1999 showed a 21% decline in physics, a 5% decline in chemistry, and a 32% decline in mathematics enrollment. Yet during this same period there was an overall 3% increase in total full-time enrollment. (Trends in the United States showed similar declines). In contrast, other departments for the same period showed remarkable enrollment increases—computer science: 94%; forestry: 73%; geology: 56%; biology: 33%; nursing: 20%; engineering: 9% (O'Meara, 2006). A study of BC grade 12 provincial exam participation rates over a 7-year period (1997–2004) showed a consistent low rate of 14% (physics), 23% (chemistry), and 29% (biology) of the total student body (Nashon & Nielsen, 2007). (An overview of physics education in Canada in general is described by McFarland & Hirsch, 1992).

6. "Examples include: F = MA (for use in schools) versus 'motion implies force' (for use in the real world); natural selection versus creationism; and the kinetic-molecular theory versus the caloric theory of heat" (Wandersee et al., 1994, p. 190). Not all learning will necessarily result in this kind of unintended "two mindset" outcome, though it is pervasive. A "mixed-outcome" where the student reconciles the two views in a hybrid is also possible.
7. Note that although this insight is several decades old, there has been little improvement in the situation (McDermott & Reddish, 1999; Mestre, 1991), although Lillian McDermott, her Physics Education Research group (PER), and others (Hake, 1998) have made significant research contributions towards alleviating this ongoing problem (McDermott, Heron, Shaffer, & MacKenzie, 2006).
8. See Wandersee et al. (1994); Duit, 2006; Duit, Goldberg, and Niedderer (1992); Hestenes et al. (1992); Mestre (1991); Zoller (1990); Osborne and Freyberg (1985); and Driver and Erickson (1983).
9. Laugksch (2000) has identified four major interest groups concerned with this very important and controversial topic: science educators, public opinion researchers, sociologists of science, and those involved with informal and nonformal science education (science museums and exhibitions, zoos, science journalists, writers and film makers, etc.). The agents I have listed should therefore be identified as a *fifth* group (this diverse, powerful and influential group could possibly be labeled "socio-economic"), although there are overlapping interests with groups two, three, and four. My preoccupation is with groups one and two.
10. "We must not forget that curriculum decisions are first and foremost political decisions. Research can *inform* curriculum decision-making, but the rational, evidence-based findings of research tend to wilt in the presence of ideologies, as curriculum choices are made within specific school jurisdictions, most often favoring the status quo" (Aikenhead, 2007, p. 880, original italics).
11. Indeed some studies suggest that students' conceptions of motion, heat, light, electricity, genes, atoms, and many others are only rarely adjusted or replaced by canonical ones after traditional instruction—although it can be admitted that the degree of the tenaciousness varies considerably according to level, subject, and the alternative kinds of instructional strategies employed (Wandersee et al., 1994).
12. As an example of the scope of what has gone on in physics education research (PER) alone over a 20-year period—and moreover to spotlight as evidence for the critique that follows—see McDermott and Reddish (1999). An exception is Kalman (2008). See also the earlier critique by Cushing (1989) of tertiary physics education targeting its ongoing neglect of incorporating important aspects of NoS into curricula.
13. According to their comprehensive survey of research topics 2003–2007 for three common journals (IJSE, SE, JRST), the two categories "learning-conception" and "learning-contexts" made up the top two at 38.8%. If one includes the category "teaching" (13.9%) these top three give 52.7%. (Although the authors note a decline in conceptual change publications compared with

their previous search 1998–2002.) The "philosophy, history, nature of science" research category showed a paltry 8.2% of publications for these journals (a noteworthy decline from the previous 4-year period of 16%).

14. Because the ongoing perception of science education at the upper levels worldwide is almost exclusively seen as *technical pre-professional training* (TPT) rather than a broader conception of education either *about science* or about science *and society*, science teacher training, along with classrooms at secondary and tertiary levels, continue to ignore the epistemology, social practices, and histories of science. They are almost universally failing, as others have pointed out, to bring to fruition in students' minds a more *authentic* view of science: the development of how scientific ideas change; the nature of scientific reasoning and inquiry (also *modeling*); the methodologies of science and its epistemology—also termed *nature of science* (NoS) discourse (as discussed later in Chapter 4).
15. Today, with the push from parents and the business community stressing the need for creating a techno-scientific trained workforce for the competitive global "knowledge economy," the external pressures are again impinging upon science education and could possibly drive it towards a new "career focused track" in ways most in the community (whether in traditional, STSE, or *Bildung* circles) may find highly objectionable. Here one is wise to heed Apple's (1992) concerns about the reduction of educational aims to *economic utility*. This is an example where internal/external interests could clash. Alternatively, the current concerns about global warming and environmental awareness among the public and business and the desire to transform curricular objectives accordingly, may represent a case where internal/external interests could overlap—as those within our community arguing for literacy as "socio-political action" have stressed (Hodson, 2003; Jenkins, 2000; Roth & Barton, 2004). The most typical case of congruence still continues to be the one between the interests of secondary science teachers and those of the external academic science community, by simple virtue of academic teacher training. The close "ideological" relationship between these two social groups is occasionally bemoaned by some researchers, for example, by Fensham (2002).
16. In brief: socialization conflicts with the "Platonic" (knowledge-focused) project because the former seeks the conformity to values and beliefs of society while the latter encourages the questioning of these; Socialization also conflicts with the "Rousseauian project" since the latter argues that personal growth must conflict with social norms and needs. It sees growth and hence education in *intrinsic* terms instead of as utility for other socially defined ends. (Here exists the principal tension between the *Bildung* tradition and the dominating utility view of education and science literacy of the English-speaking world.) The Platonic and Rousseauian projects conflict because the former assumes an epistemological model of learning and development and the latter a psychological one. In the former "mind" is *created* and the aim is *knowledge*, in the latter it develops *naturally*, requiring only proper guidance, and the aim is *self-actualization*.

17. A history of biology education clearly illustrates this: "A major theme throughout ... has been the continuing debates about its primary goal: whether it should be a science of life and emphasize knowledge or whether it should be a science of living and emphasize the personal needs of students and the social needs of society" (as cited in Roberts & Oestman, 1998, p. 5). Roberts though, does not recognize that there are three underlying educational conceptions at odds here; he defines instead the concept of "companion meanings" associated with curricula, and further, sees here only an example of two at odds.
18. As examples, the authors identify that "throughout the 19th century the goal of personal intellectual development [Rousseau] competed with the goal of learning science facts and information [Plato]" and that this competition "is evident in two curricular models that became popular" (Bybee & DeBoer, 1994, p. 365). Later, during the "progressive era," "there was considerable lack of agreement on the goals," whether "the knowledge goals" [Plato] or "application of subject matter to the lives of students" [social utility] (p. 369). Here the community became "polarized" between extremes and how to organize the curriculum. Needless to say, a surprisingly similar debate flared up again during the 1980s "curricular wars" and remains with us today.
19. Walker and Soltis (1986/1997) reach a similar conclusion when assessing these three chief conflicting goals of education that drive curriculum reform efforts. Roberts (1988), too, holds that "balance" is both desirable and achievable during public policy curriculum deliberations.
20. It is to assume one can apply them as principles while ignoring their underlying theoretical frameworks. "The trouble is, each of them has significant problems separately, and together they do not blend into a coherent curriculum. We are so used to mangled curricula, however, that their fundamental incoherencies are accepted as necessary 'tensions' produced by the competition of 'stakeholders' " (Egan, 1997, p. 206).
21. The example from Bybee and Ben-Zvi (1998) is typical: "Using the term 'scientific literacy' implies a general education approach for the science curriculum. General education suggests that part of a student's education that emphasizes an orientation towards personal development and citizenship ... [it] suggests that one should begin the design of a program by asking what it is that a student ought to know, value, and do as a citizen" (p. 488). They seem unaware that "personal development" and "citizenship" can seriously conflict. They further appear to assume that instrumentalized education for "citizenship" as both definition and goal does not entail considerably difficulties in its own right. (Here one only need peruse the philosophy of education literature! For example, see Mitchell, 2001). They continue: "We contrast this approach to designing a science curriculum with the initial effort in which individuals ask what it is about physics, chemistry, biology ... that students should learn" (p. 488). Here they explicitly devalue the Platonic project. As seen later, "literacy" is here defined in accordance with conceptions beholden to the STS paradigm.
22. One must tread carefully here. It is helpful to distinguish between "research programs" (Lakatos, 1970) within the science education *research* community

and "paradigms"—often implicit and hidden—that encompass the beliefs, instruction, culture, and curricula of practitioners in classrooms. The latter can be influenced by the former community in different ways. Another distinction should include *social groups* and their "ideologies," which inhabit either university-driven research programs or classroom communities, although an alliance between the two can certainly exist (e.g., STS researchers and STS teachers). For a useful discussion of social groups and their interests, see Ernest (1991); for ideology analysis see Säther (2003). For a discussion of the dominating community research programs see Anderson (2007), Tsaparlis (2001), and Erickson (2000).

23. Namely, as a *tradition* influenced by key theories, texts, and authorities that guide how groups of scholars, scientists, or practitioners see evidence, solve problems, perform research, and communicate with each other over a time period. His revised term for it as a "disciplinary matrix" serves equally well. But I do not follow him in his strict elaboration of the term defined as a single dominating one for the sciences. Shulman has argued that due to the richness and complexity of fields like education and the social sciences their "mature stage" of development might instead be characterized by several leading and co-competing paradigms.
24. This omission does put into serious question my use of the term. No matter in what field where the term is commonly applied it is generally understood that *at least* one high-level theory characterizes the paradigm of the discipline, although a few may be in competition (e.g., Newton, Einstein, Heisenberg in physics; Darwin in evolutionary biology; Skinner, Piaget, and Ausubel in psychology; Marx, Weber in sociology, etc.).
25. See Good and Shymansky (2001) for how statements concerning NoS in both the major U.S. "standards" documents *Benchmarks* and *NSES* can be read from opposing modernist or postmodernist perspectives.
26. It is one thing to say a broader and more authentic science education should include the occasional critical discussion involving the oft neglected nature of "frontier science" and related socio-scientific issues, and that the literacy definition could possibly comprise some aspect of this; it is quite another to *demand* that the discipline and literacy should be *defined by and structured around* frontier science, socio-technological issues, or even socio-political action exclusively. Such a radical approach, as Shamos has insightfully remarked, usually skirts the core question of how the *science* itself should be taught and learned. And there is certainly no need here to repeat the earlier and still unresolved debate of the 1980s "curricular wars," the conflict between two major curriculum models: between those who sought (and still seek) to organize curriculum around STS-type themes and those who resist any such context not heavily tied to the subject disciplines and knowledge structures (see Bybee & DeBoer, 1994, pp. 378–380). Besides the fact that this debate looks suspiciously like the older version of the Platonic academic program (e.g., Hirst, 1974) versus socialization, it could certainly have used some philosophical-historical insight.
27. Those within the recent *science-societal issues* (SSI) approach argue that STS has failed to properly account for an explicit NoS discussion, for scientific

argumentation and for ethical and cultural values connected with techno-scientific impact issues, and, more importantly, it lacks a theoretical framework.

28. Turner and Sullenger (1999, p. 15) also note that the STS camp has suffered from internal tensions because of the inability to articulate what its proper goal should be, torn among four different approaches of cultural, practical action, technocratic, or interdisciplinary issues-based emphases or projects. One notices again the consequences of neglecting to first specify one's *educational theory*.
29. Often seen in either *naïve* inductivist or falsificationist terms (Bauer, 1992). See also Claxton (1997).
30. This induction does include three sub-purposes as "curriculum emphases," which Roberts (1982) had identified as the requirement to: (1) build a "solid foundation" (logic of subject topics in succeeding years), (2) give "correct explanations" ("products"), and (3) ensure "scientific skill development" ("processes"). These "emphases," furthermore, can be directly linked to Eisner's (1985) "curricular orientations," being influenced by "academic rationalism," "curriculum as technology," and "development of cognitive processes" as three of his five predominant orientations that have heavily influenced science curriculum decision-making since inception of the 1960s traditionalist, academic paradigm (see Duschl, 1990, p. 67).
31. It might be objected that teachers normally attempt to address valid aspects of both paradigms and allow for individual differences and needs. While this may be true in general, it fails to acknowledge that curriculum is prescribed and that traditionalism is by far the status quo—hence there is often little maneuverability between desired and dictated curriculum. Exactly for these reasons Jenkins (2000) has called for a paradigm shift in science education. However, as I see it, even if this was successful, the traditionalist paradigm would nonetheless remain as an effective competitor. ("Paradigm" is here understood not in the strict Kuhnian sense but, again, according to Shulman, 1986).
32. The paper by Pedretti et al. (2008) is revealing on several levels, not least of all what is exposed here about their own social group "ideology," predisposed to the validity of the STSE paradigm (especially prevalent among researchers in Canada), as the only legitimate educational game in town.
33. Roberts and Russell (1975) had much earlier argued for the need of philosophical *informal analysis* for examining common concepts (like "teaching" and "authority").
34. He foregrounds a useful analysis of educational "ideologies," including their connection to five kinds of social groups. (Note *Table 6.3*, Ernest, 1991, p. 138). The "philosophy" he offers is primarily social-constructivism.
35. This is not meant to imply that there has not been some important research work that has sought to clarify the differences between the two for science educational purposes, although such literature tends to favor conceptual change research, where the "philosophy" in question usually refers to philosophy of science (Duschl & Hamilton, 1992; Duschl et al., 1990).
36. Indeed, there exists little reference to "philosophy of education," "educational theory," or just plain "philosophy" (Hurd, 1994, 2000; Jenkins, 1992,

2000; Millar et al., 2000; Shymansky & Kyle, 1992). When scanning for the reference of such terms in the community's established research journals (e.g., *Science Education, Journal of Research in Science Teaching*) it quickly becomes apparent that when, for example, "philosophy of education" is cited the sense is rarely linked to that discipline and its concerns.

37. One need simply scan the subject indexes of some important publications: Bevilacqua, Giannetto, and Matthews (2001); Matthews (1994); Duschl (1994).
38. A comparison of Egan and Dewey represents a fascinating clash of two contrasting conceptions or philosophies of education, one I cannot address here (see instead Polito, 2005; Egan, 2002). In passing, one can disclose they are as incommensurate as are the fundamental differences that Bruner (2006a) has identified between Piaget and Vygotsky in the field of psychology, both of whose developmental ideas have been raided by science educators to augment their *learning* theories (Duit & Treagust, 1998).
39. Earlier arguments are Duschl and Hamilton (1992, 1998); Matthews (1994); Burbules and Linn (1991).
40. In Bruning et al. (1995, p. 218; a standard textbook in teacher preparation courses for the use of cognitive psychology in instruction), the authors explicitly admit that three different kinds of *constructivism* ("exogenous," "endogenous," and "dialectical") are beholden to three different metatheories (mechanical, organismic, and contextualist, respectively).
41. Many of these are characterized by descriptive and prescriptive elements (either implicit or explicit).
42. It should be noted that Dewey's aim is among the least predetermined of the others, although it could reasonably be argued that Kant's ideal is also dynamic insofar as he allows for education's dual aim, the "perfecting" of man *qua* man plus the improvement of society and "the human race." In addition, Frankena (1965, p. 156) also notes that such a dual aim in Dewey could considerably conflict—that the expected growth of the individual and society may clash—in anticipation of Egan's critique, which claims the clash is inevitable insofar as modern schooling is molded according to progressivist precepts. Alternatively, for Dewey, but also for Aristotle and Kant, such a possible conflict was thought to be reconcilable in principle.
43. Such questions are actually the purview of what is demanded of an educational *theory*. Philosophy-of-education (PE) properly understood is a much broader field of inquiry that encompasses an analysis of such theories and questions (Peters, 1966), as discussed in the last chapter, which today usually overlaps with curriculum studies. Frankena seems to have been working with a constricted conception at the level of theory.
44. Egan (1983) asserts, in fact, that the weight of theory goes all the way down, because "every consideration relating to education—whether the organization of furniture in the classroom or matters of local policy-making, so far as these are educational rather than socializing matters—must be derived from an educational theory. That is, there can be no such thing in education as distinct lower-level theories—whether of classroom design or instruction or motivation or whatever—but only general, comprehensive educational

theories with either implications for [such] things or direct claims about such things" (p. 123). Alternatively, Hirst (1966) had argued against the view of education as an autonomous discipline.

45. This is to critique the theory on educational grounds. It has of course been vigorously critiqued on philosophical and epistemic grounds (Kelly, 1997; Phillips, 2000). Abd-El-Khalick and Akerson (2007) further mention its critique on empirical grounds.
46. "This borrowing can have the healthy effect of bringing new insights to bear on the problems of science education, but it can also lead to superficial descriptions that do not seem to be pushing for deeper understanding" (Fensham, 2004, p. 101). He fails to mention a *third* possibility, that outside theories can do outright damage to education, as Egan (1983, 2002) argues for the cases of behaviorism, Piaget, and progressivism. The presumed relevance of cognitive science has lately come into question as well (Slezak, 2007).
47. Reliance upon psychology is clearly predominant, primarily Bruner, Gagne, and Piaget in the 1960s and 1970s and the significant role they played marking the revolt against behaviorism.
48. It is admitted that Novak's book and subsequent paper (1978) did help offer an important counter-theory in support of the growing dissatisfaction with the dominance of Piagetian theory arising in the late 1970s (although some science educators continue to hold neo-Piagetian views). With the growth of conceptual change and constructivist research in the 1980s and the influence of philosophy of science ideas, this dominance was gradually displaced in the science community—but on the other hand, Erickson (2000) cautions there is much common ground between Piaget and the newer constructivist theories. Egan's cultural-linguistic metatheory (1997) is inclusive of learning theory but goes beyond it, and outright rejects Piaget.
49. A fundamental difference is that such a theory is not empirically testable in practice (at least according to some presently applied research conceptions and methods), which is not to imply it will have no empirical consequences. See here Hirst (1966) and Driver (1997). One may speak loosely, for example, about the "falsifiability" of such a theory if its methods do not contribute in large measure to the aim/ideal sought; however, because of the complexity of educational phenomena even here caveats abound. Again the problem is that as an independent discipline education must spell out a useful notion characteristic of the discipline and not beholden to previously framed conceptions specific and useful in other fields, which are hardly transferable. On the surface the magnitude of this difficulty for education should not be expected to be of the same order, nor create the same dilemmas as it has for "scientific psychology," which explicitly seeks to emulate the natural sciences. Egan (1983) writes: "The study of education is [to be] engaged in so that we may construct better educational programs, and prescribing how to construct such programs is the function of educational theories. Such theories are clearly unlike theories in physics or psychology, and we might well debate whether the differences are so great that the term 'theory' should not be used to refer to them" (p. 119). He goes on to argue that the term serves

an important purpose and, in any event, the question should be decided on pragmatic grounds, and hence should be chosen.

50. Hirst (1966) had made this distinction decades earlier, but it bears repeating again: "Educational theory is in the first place to be understood as the essential background to rational educational practice, not as a limited would-be scientific pursuit" (p. 40). He goes on to criticize those who would "fall back" on their "scientific paradigm maintaining that the theory must be simply a collection of pieces of psychology." Both Hirst and Egan stress the normative character of such a theory. From an Aristotelean point of view (and I would like to *interpret* Hirst here accordingly), the fundamental difference lies in the nature of the reasoning involved—*phronesis* ("practical discourse") instead of *episteme* ("knowledge that is organized for the pursuit of knowledge and the understanding of our experience"). Originally I had thought to have discovered an implied recognition of this distinction in his paper, but he has admitted of late to having "failed" to acknowledge this in his earlier works (Hirst, 2008a, pp. 119–120).
51. Norris and Kvernbekk (1997) do not seem aware of this, nor is there any reference to philosophy of education.
52. This would also hold true for *conceptual change theory* (CCT), which Abd-El-Khalick and Akerson (2007, p. 190) have pointed out is probably the only original "educational theory" (used in the restricted sense) so far explicitly developed by science educators.
53. This fact has not gone unnoticed, notably by Piaget in 1977 (quoted in Aldridge et al., 1992): "The general problem is to understand why the vast array of educators now laboring throughout the entire world … does not engender an elite of researchers capable of making pedagogy into a discipline, at once scientific and alive, that could take its rightful place among all those applied disciplines that draw upon both art and science" (p. 685).
54. The term itself first came into use in the late 1950s. Initially broadly framed in terms of science, culture, and society relationships, it soon came, however, to mean learning technical, subject-specific knowledge: "This emphasis on disciplinary knowledge, separated from its everyday applications and intended to meet a perceived national need, marked a significant shift in science education in the post-war years. The broad study of science as a cultural force in preparation for informed and intelligent participation in a democratic society lost ground in the 1950s and 1960s to more sharply stated and more immediate practical aims" (DeBoer, 2000, p. 588). By the 1980s the phrase had become commonplace: "Yet despite the problems of definition, by the 1980s scientific literacy had become the catchword of the science education community and the centerpiece of virtually all commission reports deploring the supposed sad state of science education" (Shamos, 1995, p. 85).
55. The "standards" documents in both America (American Association for the Advancement of Science, 1993) and Canada (Council of Ministers of Education, 1997) make this explicit. Unlike in the U.S., the Canadian document has legislative force. Bybee and DeBoer (1994) have concluded: "Scientific and technological literacy is the major purpose of K–12 science education. This purpose is for *all* students, not just for those destined for careers in

science and engineering" (p. 384). That science education should also be saddled with the added burden of trying to specify, teach and *aim* at "technological" literacy seems rather odd considering the debate that still revolves around "scientific" literacy. Unfortunately both terms are too often conflated whereas in some countries they are tied to separate courses (Layton, 1993). And certainly while all parties may agree that science literacy is for *all* students, that slogan explains little, neither does it consider the fact that many students have no interest in science, nor does it contribute to resolving the quandary concerning criteria for *content selection* (Fensham, 2000).

56. Roberts' (2007) comprehensive review of the term (also noting the occasional conflation of scientific and technological literacy) illustrates its increasing reference in handbooks and at science education research conferences worldwide.
57. Most importantly, his demolition of the two standard rationalizations of the common literacy conception—which Fensham (2002) acknowledges and terms the *pragmatic* and *democratic* arguments—effectively serves to undermine what has become the grounding assumptions of the STS literacy rationales.
58. The various "standards" documents have sought to achieve a kind of "balance" among these goals all-the-while feigning a consensus that is non-existent. And as stated earlier, they run at cross-purposes according to Egan's critique (see DeBoer, 2000 for a list of *nine* competing goals). I would add, to square intrinsic aims (science for personal development, for aesthetic appreciation and cultural literacy of the value of science) with extrinsic utilitarian aims (socialization) is most difficult and probably unworkable since it is to square a *deontologic* with a *teleologic*, to seek something for its own good versus to seek it as utility for another end. We may seek to do this with aspects of our lives, but to attempt to establish curriculum and goals on these conflicting principles, especially in the confines of a real school schedule and practice, seems destined to fail. Hence, the "standards" documents taken alone seem to be engineering their probable curricular gridlock as a working project and attempted *practical* resolution in classrooms of the science literacy debacle (Fensham, 2000; Roberts, 2007; Rudolph, 2000; Shamos, 1995).
59. Long a popular advocate of the STS conception of literacy, Joan Solomon would now drop this word for what she calls "scientific culture." Shamos alternatively would substitute science "literacy" with "awareness." He does describe three kinds of science literacy, however, those of "cultural," "functional," and "true," all of which Roberts nonetheless locates in the "Vision I" category.
60. The example of what has happened with the *NSES* document during implementation efforts in California supports this assertion (Bianchini & Kelly, 2003).
61. Option two would also seem to allow for a *two-tiered* kind of science education system (along with its respective science literacy conception): allowing a school to choose a type of "Vision I" for the science-bound student (traditionalism) and another stream for the general non-science student with its "Vision II" focus on either cultural or technological literacy or social action—chosen

as desired. Shamos has suggested something along these lines as a way to resolve the impasse. Roberts (2007) admits that although "this approach is at odds with the majority of the science education community," it is "a stark and forthright acknowledgement" the community must "somehow resolve the problems associated with educating two very different student groups (at least two)" (p. 741).

62. "Increasingly it has been claimed that education is itself practice with its own internal rationality, mediated by tradition, which does not need to be informed by external theory from the 'disciplines of education,' including philosophical value-theory, and that practical action should not be conceived on a technicist model of the application of high-level generalizations to particular cases" (Blake et al., 2003, p. 7).
63. "Stated boldly, the emerging thesis is that educational practices in classrooms are premised on a set of folk beliefs about learners' minds, some of which may have worked advertently toward or inadvertently against the child's own welfare" (Bruner, 2006b, p. 163).
64. This censure is meant to include authors such as Thomas (1997), who when criticizing "grand theory" (he mentions Piaget, Chomsky, and Habermas) as a "totalizing discourse" constraining both creativity and thought would opt instead for "ad hocery," for practitioners' reliance on "reflective thinking" and "craft knowledge" learned in situ. Although these are undoubtedly necessary, they are hardly sufficient.
65. He argues: "psychology has generated much knowledge that is properly of interest to education. Knowledge by itself is mute, however; it is made articulate by being organized into a theory.... Knowledge about the *psyche*, about learning, development, motivation, or whatever, becomes psychologically articulate when organized by a psychological theory. The same knowledge may become educationally articulate only by being organized within an educational theory. This is not, I think, a trivial point. It means that apart from an educational theory no knowledge and no theory have educational implications. Even knowledge about constraints of our nature becomes educationally useful only when it has become incorporated within an educational theory" (Egan, 1983, pp. 122–123). Hence, the controversial conclusion: "psychological theories at present have no implications for education" (p. 125).
66. Some may object here and argue that science education (and education in general for that matter), can be considered a mere "field of interest" upon which other disciplines have rightful bearing and thus should *not* be considered a discipline in its own right. Hence, the argument has been ignored why education, when being conceptualized primarily as a field of *practice* that it, at best, could rightfully accept an amalgam of theories from other disciplines, or use different theories for different purposes, such as ethical theories to define the "why," and social and psychological sciences for the "what," "when," and so on. This position ignores both the logical and historical reasons why there are separate disciplines. Disciplinary theories frame the kinds of questions and answers relevant to them: using a psychological theory to guide your questions will get you psychological answers; if you use a theory from sociology, you get sociological answers, and so on. Disciplines

(surely) are basically imprecise methodologies for answering certain kinds of questions. Even *if* education is only a "field of interest," you need questions, methods, and answers appropriate to that field of interest. The record of ignoring the distinctiveness of education in research does not give grounds for confidence that this is serving us well. Calling education a "field of interest" does not prevent one forming a "high level theory" about it—as long as that theory is appropriate to the nature of the phenomenon and can guide one into asking appropriate educational questions, and suggesting possible answers, about the phenomena.

67. Leaving aside questions if its individual educational claims are either warranted or empirically validated. Strong advocates for this kind of politico-social activist PSE (just naming some researchers) are Hodson (2009) and Roth and Désautels (2002). Criticisms leveled against it are provided by Hadzigeorgiou (2008) and Levinson (2010).
68. Does it fully take into consideration the three dimensions of the synoptic framework shown in Figure 1.1?
69. Roberts (1988) had earlier cautioned the research community about the "*individual ideological preference* of professors of science education" that can "indoctrinate science teachers into believing that what counts as science education is the ideology of a single curriculum emphasis (or perhaps a few emphases)" (p. 50, original italics).
70. It seeks as well to address the common blurring of lines between "descriptive" and "normative" research work, the expectation that classroom research *should* change classroom teaching and learning, as Sherman (2005) points out, but strictly in accordance with a specified (ideological) program. This academic conflation may indeed be due to our culturally inherited situation, in other words, "if we can't be objective, we'll be openly ideological" (p. 205), but regrettably real "openness" is rare. The argument here in a nutshell is that science education avoid (c)overt ideology for candid philosophy.
71. Such a conversation can be considered an extension of one already discussing the difference between epistemology and psychology (Duschl et al., 1990; Matthews, 2000; Southerland et al., 2001), or critiquing the assumed validity of cognitive science theories for science education (Slezak, 2007).

CHAPTER 3

PHILOSOPHY OF SCIENCE EDUCATION AND KIERAN EGAN'S EDUCATIONAL METATHEORY

Engaging the imagination is not a sugar-coated adjunct to learning; it is the very heart of learning.

—Kieran Egan

3.1. LOCATING *PHILOSOPHY OF SCIENCE EDUCATION: BILDUNG*, EDUCATIONAL METATHEORY, AND PEDAGOGICAL CONTENT KNOWLEDGE (PCK)

The central importance of metatheory discussed so far raises the related question as to whether or not educational systems in other countries have also considered the need of such theories for their use as a coherent approach to the problems associated with curriculum aims and implementation. This has indeed been the case with the *Bildung*-centered *Didaktik* tradition of central and northern Europe. The *Bildung* conception and its tradition functions as a metatheory of education in ways very similar

Rethinking Science Education: Philosophical Perspectives, pp. 109–138

to Egan's ideas and intention as to what an "educational theory" should comprise and achieve. Hence it can serve as a useful parallel theory for comparison. And not only that, it equally serves to draw a distinction to the common Anglo-American "curriculum" tradition prevalent in the English-speaking nations. What is of even more immediate interest to our concerns will involve not just sketching out the contrasts so described but more so in locating the position and value of a *philosophy of science education* (PSE) among these educational theories and traditions, especially in light of some recent awareness of these differences in science education research.

In the 1990s groups of scholars on both sides of the Atlantic had already begun a cooperative project to better understand similarities and differences between the two major educational traditions.[1] Fensham (2004) states that the catalyst for this dialogue can be traced back to Lee Shulman's (1987) original discussion of *pedagogical content knowledge* (PCK). It was the recognition that the problematizing of subject content for its appropriate integration in *pedagogy* "attracted the interest of those in the *Didaktik* tradition for whom subject content had always been a central issue and integral approach to research" (Fensham, 2004, p. 146).

What also became apparent during the dialogue were the obstacles encountered with foreign concepts like "Didaktik" and "Bildung." The closest term to the former is "didactics" in the English-speaking world, but this has a separate and primarily a pejorative connotation,[2] whereas the latter term has no equivalent at all and bears difficulties in translation. *Bildung*, a concept that has developed out of the early 19th century German romantic movement, encompasses an array of ideas often employed separately in the English-speaking world when one focuses on key educational aims: the personal development of the individual and coming-of-age; the cultivation of intellectual and moral powers through subject matter to advance critical thinking and enhance the state of freedom for individuals and society; or viewing the general role of education to foster "an ideal of self-determination, the formation of character, or exercise of autonomy, reason and independence" (Vásquez-Levy, 2002, p. 118). All these labels and qualities are intrinsic to the concept:

> *Bildung* embodies a double process of inner-developing and outer enveloping, what Germans call *Allgemeinbildung* and *Ausbildung*. On the one hand, the concept *Bildung* describes how the strengths and talents of the person emerge, a development of the individual; on the other, *Bildung* also characterizes how the individual's society uses his or her manifest strengths and talents, a "social" enveloping of the individual. (Vásquez-Levy, 2002, p. 118)

On this formulation, one quickly recognizes the link with the "humanistic tradition" of liberal education mentioned earlier as one of the three vital goals behind the purposes of science education. Given this interpretation one could in fairness associate the values of the *Bildung* tradition with two prevalent "curriculum ideologies" in American education as identified by Eisner (1992), being "rational humanism" and the "personal" stream within progressivism.[3] It should be additionally mentioned here that Gadamer (1960/1975) takes *Bildung* to be one of four fundamental guiding concepts (next to *Sensus Communis*, judgment, and taste) that characterize the significance of the humanist tradition for the humanities (*Geisteswissenshaften*) and which therefore clearly distinguishes them from the so-called "method" and disciplines of the natural sciences. Gadamer presents an elaborate discussion of the concept (including the roots of the idea and its conversation in Herder, Humboldt, and Hegel), which is steeped in Rousseauian notions of *self-formation*, but which also later came to emphasize the vital role played by culture in this process, in particular Herder's linguistic-cultural re-interpretation as "rising up of humanity through culture" (quoted in Gadamer, 1960/1975, p. 9). Rousseau's influence weighed heavily on Goethe's and Schiller's views on education and the Romantics in general, and the further expansion of the self-formation idea by Herder, according to the Canadian philosopher Charles Taylor (1991), provided necessary contributions to the modern notion of the self as unique and authentic.[4]

With the idea of *Bildung* just described, one may even so come straight away to identify some serious difficulties with the conception. The first point of issue is that *Bildung* appears as to be caught in a paradox, being defined as both *means* and *ends*, as the process of learning and personal growth as well as the final aim at which such growth or education itself terminates; between becoming educated and being educated (*Gebildet*). It shares this paradox with the ancient Greek conception of education as *paideia* as it developed in late Hellenism.[5] This should not be entirely surprising since German classicism relied heavily on the ideals of the Greeks and saw in *paideia*[6] a useful analog. (Other parallels can be drawn today: Eisner identifies the "Paideia Project" of Adler, 1983, as another example of curricular "rational humanism.") Secondly, *Bildung* was initially taken as inclusive of education *for* the state, but not as mere means to that end; rather self-realization was taken to be an end unto itself, and where the state could serve as the means (Beiser, 1998). Yet an inherent tension arises here between self-realization (Rousseau) and the socializing demands of citizenship (including perhaps the wishes or commands of the state), as was sketched in Chapter 2 with respect to Egan's critique of these two essentially incompatible educational ideas. Vásquez-Levy in her *Essay* (2002) does not recognize the tension(s) when she writes:

> This means that, through education, human beings are gradually opened to and connected with the world both as a historical process and as a natural

> and social environment. Education prepares us to participate as responsible citizens in the *polis*, the political and social order, so that we have some control over our own destiny with others. This approach is certainly what Socrates had in mind: the telos of education is free citizenship.... Thus, *Bildung* is the process of developing a critical consciousness and of character-formation, self-discovery, ... an engagement with questions of truth, value and meaning. The education of individuals is, therefore, a recapitulation of the cultural development of the world and the practice of freedom and work towards higher liberation (pp. 118–119).[7]

This encompassing description of *Bildung* as given above, moreover, goes even further, and appears to ascribe to this idea an amalgam of mismatched educational ideas, counting Rousseau, the Platonic knowledge-based project, as well as (using Eisner's 1985 terms) both social adaptation and social reconstruction—the last two without a doubt mutually contradictory. It seems the conception (taken in its widest sense) is burdened by too many demands, even encumbered by incompatible ideas. But in actual practice choices must be made when decisions about curriculum orientations are considered; it is then no coincidence that when criticisms of the *Bildung* paradigm arose from those within critical theory (of the Frankfurt school) starting in the 1950s/1960s, we find them targeting the differing aspects of its interpretation as manifested in schools: The philosopher Adorno at first arguing that the notion had been "stripped of its normative content, its relation to a good and just life" with its critical, emancipatory edge questioning societal rules and expectations, for a shallow adaptation of capitalist norms and competence to function in the social order (Blake & Masschelein, 2003, p. 40); afterward, other adherents of the School closer to our time had even come to impugn "the modernist ideal of authenticity and questioned the limits of autonomy"[8] (p. 45). It hardly surprises today that this paradigm has come to suffer from attacks similar to ones that postmodernists have leveled at the Enlightenment tradition, those that have come to question the entire notion of the induction of children into a cultural tradition, one "which conceived this enculturated maturation as a form of emancipation" (Blake et al., 2003, p. 9), and one that placed a heavy stress on the role of reason and rationality in the process. What is of issue here, however, is that Adorno's complaint can be taken as representative of the tensions between the "social adaptation" conception and a knowledge-focused Platonic-type tradition vying for dominance in the implementation of the *Bildung* interpretation, while the later critiques have put in doubt the essence of Rousseauian child-centered self-determination at the very core of the paradigm (to be replaced by a theoretical focus on education for "social reconstruction" purposes).[9]

My digression on analyzing the conception of, and problems attached to, the *Bildung* paradigm serves a *three-fold* purpose. Besides the usefulness in clarifying the idea itself for an audience unfamiliar with the concept was the chief objective to illustrate that *Bildung* as an overarching conception of the educational endeavor has by no means escaped the tensions and incompatibilities associated with the *three fundamental ideas* underlying general education as discussed earlier; on the contrary, these inconsistencies appear to be inherent to its inception and interpretation, of what it *means* and *how* it is to be implemented, and this is no doubt mostly due to its Rousseauian pedigree. This brings us to the second purpose, the comparison of this tradition with the better-known Anglo-American "curriculum" tradition. On many levels, the two traditions form a stark contrast, but as Reid (1998) has commented, the dialogue between the two camps had been furthered by a mutual recognition of the apparent strengths of the other based on the perceived lack or weakness in their own. Yet this dilemma too seems to me to surface once again the incompleteness of the two traditions with respect to the *three* underlying educational ideas identified before, because the advocates of each respective tradition have come to cherish that one fundamental *idea* seen as missing in their own educational project but alive and appealing in the other.

For example, in northern Europe, especially Scandinavia, there has been a revival of interest in Dewey's ideas of schooling as a form of socialization into "differentiated democracy" and hence, coming to grips with the paradoxes of the aims of education that American education has already wrestled with for a century: "how can notions of utility be accommodated to the traditions of liberal education? How can authority in curriculum matters co-exist with a diverse population, not all of whom would see themselves as sharing a common cultural heritage? How can a common curriculum be made accessible to all students?" (Reid, 1998, p. 22). On the U.S. side (especially during and after the Reagan years of the 1980s and their concern for establishing national "standards" of academic excellence due to anxieties about global competitiveness) has come an admiration for the *Bildung/Didaktik* ability "at maintaining coherence, at fostering achievement in the disciplines, and at retaining public confidence in standards of performance" (Reid, 1998, p. 23). In other words, what the U.S. admired about the *Didaktik* tradition was its accomplishment at "academic rationalism" (Eisner)—or the Platonic-knowledge based "academic" program (Egan)—while the Europeans admired the American pragmatic approach using schools for fostering socialization of its student population in a multicultural context. Once more one notices the recycling of competing ideas not only as a recurring problem for science education but spanning two very different educational traditions and cultural contexts. While there certainly is a willingness to learn and adjust one's tradition in light of the other, there seems to be (as of yet) little appreciation of the impasse provoked by the educational projects of both.

We now come to the third purpose of our contrast, which is to demonstrate how the advantages of the *Didaktik* tradition over the "curriculum" tradition, also discussed in some latest science education research circles (mostly in Fensham's 2004 *Identity* book), offers insight into positioning a *philosophy of science education* (PSE) and Egan's metatheory. Both Vásquez-Levy (2002), speaking generally about education, and Fensham (2004, Ch. 10), specifically addressing science education reform, admit the advantages of the *Bildung* paradigm over the standard Anglo-American "curriculum" tradition in allowing for: (1) a coherent *theoretical* approach to curriculum design and implementation insofar as it serves as an *educational criterion* to assess subject content; thereto, and related to it, (2) problematizing content for the learner. It places more emphasis on teacher autonomy and pedagogical knowledge to assess content for educational aims and purposes, and hence, forces them to wrestle with the nature and significance of the subject/discipline as consequent to improving their pedagogy and how pupils can/should learn the subject content in question.

When comparing the new 1994 system-wide "standards" documents for science curriculum reform in Norway and Australia,[10] Fensham (2004) comments on the "stark and striking difference" between the two, and how the *Bildung* idea (as a metatheory) in Norway affects the rationale of education and steers the curriculum (his Table 10.1, p. 151):

> In the one, the maturing young person is the purpose of the curriculum. In the other, the teaching of subjects is the purpose. In the one case, disciplines of knowledge are to be mined to achieve its purpose; in the other, disciplines of knowledge are the purposes. (Fensham, 2004, p. 150)

One notes here how one educational idea (the Rousseauian project) dominates and is prioritized over another (Platonic knowledge-based project). What offers the advantage though, is that teachers come to be more concerned with pedagogy and the learner, and not, as is often the case (at secondary and tertiary levels), see themselves chiefly as subject specialists taking a mere *instrumentalist stance* on curriculum, to expand knowledge in students' minds. (We will see in Chapter 5 this presupposes a rather trouble-free "transmission model" of language and knowledge acquisition.) Thereto, Vásquez-Levy (2002) emphasizes the importance of *Didaktik* analysis as linked to this project:

> *Bildung*-centred Didaktik keeps analysis of teaching events/classroom work within the realm of pedagogy and, simultaneously, can engage teachers in deep levels of reflection concerning the what, how, and why of teaching and learning. In this way, *Bildung*-centred Didaktik prevents the teacher from being consumed by purely institutional concerns, which may be antagonistic to students' *Bildung*. (p. 120)

This in fact can help contribute to avoiding the situation where the teacher's role itself can be reduced to, or predominately taken to be, an *instrumental* one; that is, the teacher as mere "instructor" commanded to implement prescribed curriculum—as is often the case in the curriculum view[11] and school science education generally. The Didaktik tradition can suffer this charge, too, if it is reduced to mere instructional methodology. Englund (1998) described such a narrowing to have occurred in Sweden, where later a "broader didactic" based in curriculum theory and philosophical inquiry brought about a renewed sense of didactic analysis. Curriculum theorists then widened their horizon and asked questions about "factors determining *educational* content [of school science subjects] and ... *why* certain content was chosen" (p. 15, italics in original). *These kinds of concerns and the type of philosophical inquiry involved are exactly the purview of PSE.*

Egan's metatheory serves the exact same purpose as (1) above (but is not subject to the tensions of internal educational inconsistencies inherent to *Bildung*, as previously discussed). In addition, part of the critical aspect of a PSE for teachers equally serves the purpose of (2) above, acting as a type of *Didaktik analysis*—the involved process by which teachers critically examine how the content can meet the concerns of learners and learning progression when fulfilling the requirements of the educational metatheory (whether *Bildung* or Egan).

Fensham (2004) comments that both established traditions (curriculum and *Bildung*) have now come to understand this procedure of concurrent content and curriculum scrutiny as encompassing what Shulman had earlier come to portray as a teachers' pedagogical content knowledge (PCK).[12] Unfortunately it tends to be mainly undervalued in the typical science education "curriculum" tradition, especially science teacher education programs:

> This interaction between content and educational process—a central aspect of PCK—is certainly an issue in *Didaktik*, but the first stage of the *Didaktik* analysis, transposing the disciplinary science knowledge itself to the purposes of school science, is often not recognized by those in the Curriculum tradition. An alternative meaning for PCK is to regard it as the way a teacher's knowledge of science content is modified by the experience of teaching it. (Fensham, 2004, pp. 155–156, italics in original)[13]

What primarily distinguishes the two traditions then is the way they conceive of, and organize, school science education:

> In the *Curriculum* tradition the teaching/learning stage is most obviously problematic, whereas in the *Didaktik* tradition, the *transposition* stage and the teaching and learning stage are both problematic. In the former tradition, the science content itself is essentially given. In the latter tradition the

> science content is initially the site of the problem, because of the decisions teachers should make about it, and then this content is intimately involved in the problems of its teaching and learning. (Fensham, 2004, p. 152)

For comparison, again, Egan's cultural-linguistic based metatheory equally problematizes both stages, and he explicitly mentions that the teacher needs to place the subject content for "transposition" in question as well, as a means to satisfy the prior specified metatheoretical educational purposes, including the need to determine the *emotional potential* of the content for the learner when lesson planning (Egan, 2005b).

Fensham, moreover, goes even further when discussing the *nature* of science content knowledge (CK) as found in textbooks and mandated curricula, and his insight here is noteworthy. When he places both school traditions side by side he identifies a common shared *failing*: both traditions regrettably still assume that the subject content (discipline-based knowledge) is itself unproblematic. (The *Bildung* tradition only problematizes the content *form* or *context* for pedagogical *transposition* and its usefulness in relation to the essential educational aims of the metatheory.) Certainly *Didaktik,* it can be admitted, does readily acknowledge (in a way that "curriculum" does not, although Dewey and Schwab stressed the difference) that the two contexts of *knowledge structures* are quite different: when CK is organized *either* in subject specialization (i.e., in disciplines like physics, chemistry, or biology), *or* organized for learning for *pedagogical* and schooling purposes as required by the demands of an educational metatheory. "In other words, the knowledge in these sciences is not automatically in a form that makes it meaningful or worthy of a place in schooling committed to education as *Bildung*" (Fensham, 2004, p. 147). This fact distinguishes the fundamental difference between the two, to be sure. (Hence, the crucial role of PSE, or, on the other hand, *Didaktik* analysis, as was stated.) However, what is still completely missing in both traditions, argues Fensham, is a deeper questioning of the nature of the formal content itself, as typically structured, sequenced, and offered by the science disciplines or specialties themselves. This implies that another vital role of PSE is to help teachers scrutinize their own, and indeed their science discipline's CK base, and thus to take it for granted no longer:

> Despite recognition in the *Didaktik* tradition of the need for *didactical analysis* or *transposition*, both traditions have, I believe, hitherto, held strongly to the idea that the content for school science subjects should be determined by what is accepted as lying within the content of the corresponding disciplinary science. (Fensham, 2004, p. 158)

At this juncture I would stress that this is exactly where the merit and requirement of history and philosophy of science (HPS) awareness plays a fundamental role. It assists those specialist teachers to move beyond the strict confines of the disciplinary knowledge structures they have been so accustomed to—and been enculturated into—in their academic training, by providing them with wider, deeper and in many cases, more truthful perspectives on the sciences (informing their *image* of science and *epistemology*). Moreover, this offers them another criterion to assess the adequacy of their CK for pedagogical purposes, that is to say, in structuring their PCK for better science subject comprehension (see the earlier comment by Matthews, 1994, p. 204). This requires science teachers in particular to look beyond the organized confines of their *textbook*-based knowledge and to secure "the 'deepest objective substance' of the subject matter and know its educative aspects, and understand students' internal and external capacities." Only then is "he or she free to reflect on approaches to teaching" (Vásquez-Levy, 2002, p. 124).[14] Egan (2005b) expresses a similar expectation for teachers to dig deeper into their subject disciplines (CK) to uncover (or recover) the *imaginative* potential of the subject knowledge for learners, which can be seen as another legitimate interpretation of PCK.

One therefore concludes that the content must be seen as problematic for *two* reasons:

1. it needs to be *transposed* into an age and culltural appropriate form accessible to the learner, and
2. it needs to be *broadened* as a knowledge base, to include NoS studies.

With respect to the first, a significant part of the *transposition*[15] will include shaping the context by using different, appropriate approaches (like storytelling, narratives, drama, border-crossing, argumentation discourse, etc.) and assisted by PE awareness, including metatheory (for example, Egan's metatheoretical ideas and frameworks, 2005b). With respect to the second, the next chapter will address the issue of teacher's CK and NoS studies.

When one examines Figure 3.1 (which contrasts the two traditions), the position and purpose of a PSE for a teacher's PCK in the schema becomes apparent.[16] Also illustrated is where the *nature of science* factor ("science studies/HPS") must be included to allow for a re-evaluation of curricular and disciplinary-based CK. These aspects are shown in square brackets. The *Didaktik* tradition expects of the specialist teacher to weigh the educational merits of disciplinary CK with the desired aims of how *Bildung* constitutes the formation of learners as educated human beings and citizens. Moreover, whereas the *Didaktik* tradition draws upon and relies on the disciplinary sciences only as a knowledge source for one component of a teacher's PCK, the curriculum tradition of science education by contrast is fundamentally rooted in the disciplines themselves since they normally structure both school curriculum and teachers' CK (especially at the upper levels).[17] In addition, one notes that educational metatheory (hence any sort of thinking regarding educational philosophy) is here markedly absent. Notice how Egan's metatheory can be situated more easily within the preconceptions of the educational system of the Central-Northern European tradition, while posing as a rival to the *Bildung* paradigm.

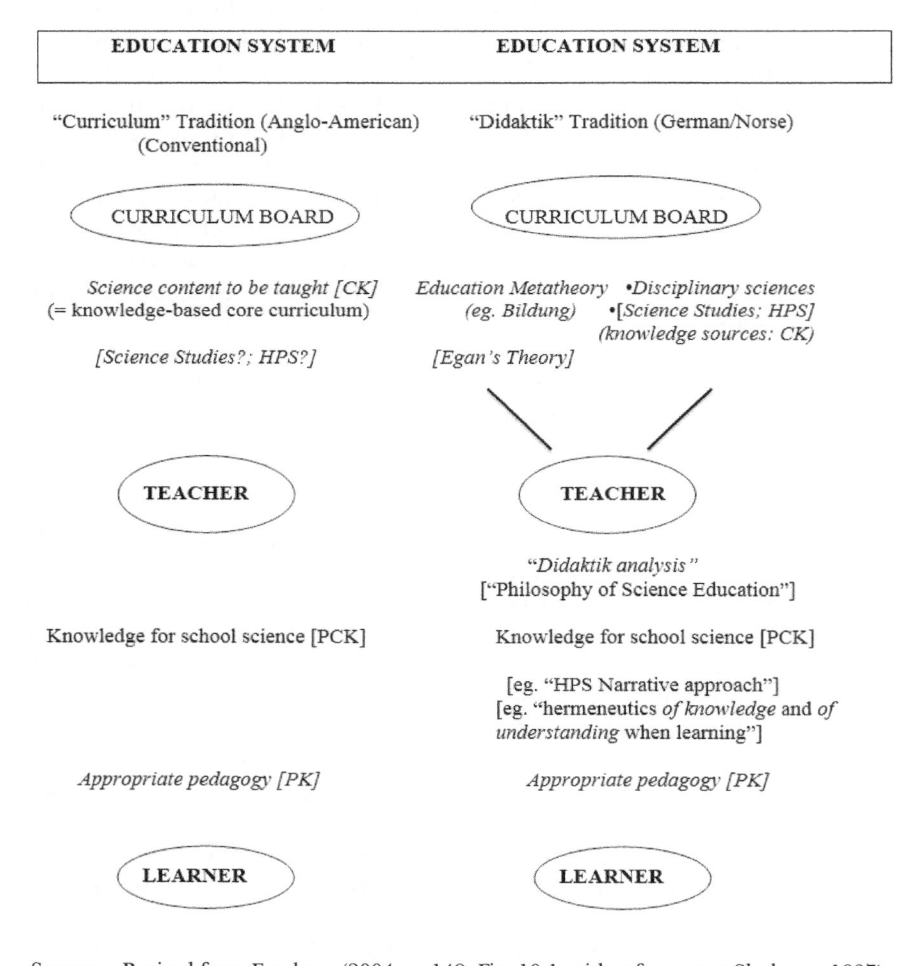

Source: Revised from Fensham (2004, p. 149, Fig. 10.1, with reference to Shulman, 1987).

Figure 3.1. Two curriculum traditions.

3.2. KIERAN EGAN'S EDUCATIONAL METATHEORY

Egan's ideas on imagination, learning, development, and curriculum that underpin his educational theory have been developing for several decades and been described in previous books. The most complete articulation of the metatheory—occasionally referred to as "imaginative education"—is found in his (1997) *The Educated Mind: How Cognitive Tools Shape Our Understanding*. Polito (2005) has recently (and correctly) discussed how his metatheory is linked with the cultural-linguistic development of humanity, especially by portraying its features in sharp contrast to Dewey's ideas, and Hadzigeorgiou (2008), and Hadzigeorgiou, Klassen, and Froese-Klassen 2011) have emphasized its value for humanizing physics education. Thus, while Polito helps paint the broad picture, Hadzigeorgiou and colleagues exemplify with the concrete, and I have chosen along with the former to contribute with a further sketch of the wider canvas.[18] Many researchers in and outside of the HPS movement are already familiar with his earlier ideas stressing story-telling and narrative as a vehicle for providing curricular context to enhance instruction and learning (Egan, 1986; Matthews, 1994; Stinner, 1995a; Stinner, Mcmillan, Metz, Jilek, & Klassen, 2003).

Egan's construction of the metatheory has grown from an awareness of the perceived carnage brought about by the misapplication of psychological theories in education (1983, 2002, 2005a)—including a large measure of skepticism towards the ability of research findings to inform educational theorizing[19]—bringing forth various (aforementioned) *separation* arguments. Since these arguments equally serve to underpin and shape his alternative metatheory, let me therefore first focus on these before examining Egan directly. He holds that psychological and educational metatheories must pass one another by because of the inherent differences with respect to what each seeks to describe, explain, and apply. His insight about not grounding an educational metatheory on psychological theories is based on two big arguments.

The *first* big argument (also cited by Aldridge et al., 1992) insists that for education, and unlike for psychology, there exists no natural educational process to describe or explain, rather "an educational process exists only as we bring it into existence" (Egan, 1983, p. 3). It is the business of educational theories to be prescriptive, while for psychological theories it is to be descriptive of an assumed "natural development" of human beings at various stages, with the aspiration of becoming explanatory and normative. At most then, psychology's descriptions of cognitive development, say, can serve as *constraints* to the educational project of what is deemed of value for developmental stages and goals. While this may seem sensible, and while this assumption is indeed taken for granted in most educational practice (see Gredler, 1992 as typical), Egan himself is skeptical and cautions against "psychological description constraining educational prescription."

He gives examples from Piagetian-influenced researchers who have shown that young children lack those concepts necessary to make meaningful sense of abstract ideas like "religion" or "history" (and I would add "science"). While valuable information, it unfortunately provides the educator with no basis to decide "what," "when," or "how" to teach ideas that will eventually contribute to the development of such important concepts. *How such knowledge is to be evaluated and used is precisely the task of an educational theory.* Sadly, too often the deduction has been made that such subjects are themselves irrelevant until the appropriate "stage" has been arrived at by the learner and hence should not be taught—a conclusion Egan (1993) refers to as the "psychological fallacy." Its pervasive influence "is a tribute to the power of psychology over education, and the theoretical poverty of education" (p. 12). It follows that common terms of use such as "learning," "motivation," and "development" could very well have different connotations in the two disciplines.[20] Furthermore, not only are psychological theories contested—including the isolated, artificial, and controlled conditions in which many empirical results have been undertaken—one cannot even be sure what exact constraints of our nature are supposed to be described by them, given that when considering human behavior it is difficult if not impossible to separate nature from culture (Cole, 1996).

This brings us to his *second* big argument. There need be no reason why cognitive "development" as elaborated by cognitive psychology should resemble cognitive development as described and prescribed by an educational high-level theory. It belongs to the aspiration of a scientific-oriented psychology modeling itself on the natural sciences that among the criteria for its scientific credentials belongs the assumption of an underlying "nature" common to all human beings that can be unearthed and abstracted beneath the domineering layers of a socializing culture. Indeed, Piaget's stage theory of cognitive development is premised upon exactly such a postulate and "is derived from his analysis of the biological development of certain organisms" (Gredler, 1992, p. 222), what I had previously referred to as the organismic metatheory in psychology (Kitchener, 1992).[21] This viewpoint is actually an old one and goes back to Aristotle, and it runs through progressivism (Egan, 2002). One could call it the age-old quest for the "biologized mind." Egan, following Vygotsky's socio-cultural outlook, is diametrically opposed to such a *conception* of mind—which he considers a fiction—and rejects as a false belief the corollary that children's minds have a preferred natural kind of learning that can be isolated and explained by psychological research. Rather, it may be precisely the *culture forming conditions of mind* that an educational theory may wish to foreground and as such may prove invaluable for the educative process, the very thing a scientific-oriented psychology hopes to minimize if not in fact to eradicate from its research program.

> Much of what is most distinctly human in learning and development has been suppressed by the search for the biologized nature of the mind. That search has avoided the cultural stuff that seems to constitute mind and is not particularly amenable to study by research methods devised to deal with the natural world ... there is no naturally preferred form of human intellectual maturity. We are not designed, for example, to move in the direction of "formal operations" or abstract thinking or whatever. These forms of intellectual life are products of our learning, "inmindating," particular cultural tools invented in our cultural history. (Egan, 2002, pp. 113–114)

It goes without saying that should the central assumption of the existence of an underlying nature common to human cognitive development turn out to be false, it would have serious ramifications for the self-conception of a scientific psychology. This is not our concern here, though, nor does the merit of Egan's argument require its overthrow. It just requires that such a nature remain if not irretrievably buried beneath, then at least irrevocably shaped by, human cultural history (as Bruner, Cole, and others working within the socio-cultural school of psychology now argue is indeed the case; Cole, 1996; Kozulin, 1998).[22] This brings up the significant question to what extent *mind* can be considered a legitimate scientific object of study. There is currently much debate on this subject, with "mind" being located somewhere between the cerebral cortex and culture (Leahey, 2005). Egan's sympathies clearly lie more with those who hold what has been described as the "strong culturalism" view (Bakhurst, 2005).[23]

With these arguments in view, and supposing such a take on "mind," Egan has constructed his metatheory using insights gained from many diverse fields, drawing upon anthropology, cognitive science, philosophy, history, linguistics, and classical studies, and following in the tradition of the Russian socio-historical approach to psychology and education (Cole, 1996; Kozulin, 1998; Moll, 1990; Vygotsky, 1978; Wertsch, 1985). In sum, its hallmark features the significance of the three central ideas of *imagination*, the *mediation of socio-cultural cognitive tools*, and *recapitulation*. Each one of these three complex and historically-loaded ideas is carefully explicated in his works, and for brevity I can only sketch them here. Yet for many educators, and especially for science educators, such ideas may sound strange and radically new, accustomed as they are to the more conventional discourse, one dominated by language fashioned by knowledge structures ("subject disciplines," "curriculum content") and psychology ("cognition," "learning as information processing," "stages of development," etc.). And yet his three main ideas are themselves part of a rich historical and educational background, which for the case of imagination alone is considerable, running through Western thought as it has and stretching back to its early differing appraisals both in Ancient Greece and the Judeo-Christian religious heritage.

Egan's metatheory is grounded on the fact of the *historicity* of language in human anthropology and cultural development and how this has managed to shape—albeit in ways not yet entirely understood—*both* the brain and the mind. "Without the historicity of language, human nature and the human mind remain essentially unchanged in history" (Polito, 2005, p. 486). This position stands in sharp contrast to Dewey and his philosophical anthropology, which undergird his philosophy of education:

> There is no historicity of language in Dewey; therefore there is no historicity of consciousness. For Dewey, man's consciousness has remained fundamentally the same since the birth of its existence to the present. The form that consciousness takes is to inquire about the outer world in relation to the needs of self-preservation ... it is this technological advance which drives language development.... Like Vico, Egan believes that we could not have acquired the capacity to think in consistent and controlled manners proper to rational adults except by first being able to think imaginatively and associatively.... For Egan, language and reality develop in constant interaction with each other. Language responds to the world's changes and, in turn, changes the world. (Polito, 2005, pp. 485–486)

One can assert without much controversy that there has been a general cultural progression of the human race from plain mimicry and artifact construction (common to our primal *homo sapiens* ancestors), to oral language use, literacy, and finally to more complex forms of language symbolism and use, including theoretical thinking, as noted by many—though, granted, such a development has been largely uneven and not universally realized by all cultures. Even today, although a few remaining indigenous oral cultures have managed to withstand the test of time, they too, will inevitably enter the cognitive and physical space of literacy, which will, in turn, significantly shape their culture and mind for good or ill, as it has for all other human societies—including as examples, the early Maya, the Arabian tribes (which gave rise to Islamic civilization in the seventh century), and Germanic and Slavic oral cultures, which, when schooled in literacy by the Greeks, Romans and Judeo-Christians, would transform (and themselves be transformed by) the Ancient world and *create* early medieval Europe. The exceedingly long historico-cultural development since our early hominid prehistory, which appears to be neither inevitable nor "progressive" (in the older 19th century evolutionary sense), has nonetheless brought with it the discovery and invention of both *physical* and especially *cognitive* tools, which according to their own sequence and time have wrought technological advance as well as expanded the human capacity to reason.

What is of importance here is that with each major *cognitive transition* (body-centered mimetic to oral language to literacy and numeracy) have come an array of other specific cognitive "languaged" tools that have correspondingly formed the mind and broadened human understanding to grasp and make sense of the world, and that Egan further claims to have identified and catalogued. Where the knowledge of this cultural-linguistic developmental framework becomes useful for *education* is as a *recapitulation premise*: that children while growing up are themselves remarkably going through a similar bodily and linguistically-based cognitive development in their personal growth, and Egan argues it is the task of an educational theory to help learners recapitulate—that is, become aware of and maximize—the use of the mental tools at the appropriate stage that comes along with such a development. And as with human cultural history, so with students, the use of the imagination is crucial in bringing about the success of such a *cognitive recapitulation;* in effect it drives the process.[24] Egan's metatheory in short is an extended argument *for making possible the recapitulation of the co-evolution of human cognition and culture as an educational undertaking*. This has never been articulated as an educational project before, certainly not in any PE of which I am presently aware.

Egan has proposed a detailed five-stage progression that represents five different kinds of human "understanding": *somatic* (or body-mimetic), *mythic*, *romantic*, *philosophic* (or theoretic), and *ironic*. (These bear resemblance to Merlin Donald's study of human cultural history in his *Origins of the Modern Mind*, 1991, which suggests a three-stage cultural succession of *mimetic*, *mythic*, and *theoretic*—brought about by the tools of the body's mimetic skills, oral language, and external symbols, respectively).[25] Notice the emphasis here, in contrast to traditionalism, is not focused on knowledge structures *per se* (although these are of value) as an enabling of a fuller comprehension of subject matter due to the mental make-up or *assembly of mind* already at hand (and in process of moving on to the next assembly). He has further distinguished the five by presenting and analyzing an inventory of "sub-tools" characteristic of each (a few shown below). In addition he has suggested *approximate* ages for when the learner can be expected to enter the particular stage, when the cognitive tools are first stimulated:

- *somatic* (sub-tools: bodily senses, rhythm and musicality, gesture, reference); pre-linguistic and extra-linguistic thinking; body use to represent thought, action, and communication; appears before humans invent oral language; (children ages 0 to 2).

- *mythic* (story, metaphor, binary opposites, humor, images, sense of mystery); oral language stage; associated with oral societies; (children from 2 to 7 years).
- *romantic* (extremes of reality, heroes, wonder, revolt and idealism, hobbies); literacy stage generates a "new consciousness"; (children from 7 to 14/15 years).
- *philosophic* (drive for generality, general schemes and anomalies, search for authority and truth); "theoretic thinking"; associated with academies (adolescence).
- *ironic* (limits of theory, reflexivity and identity, radical epistemic doubt); reflexive use of language; poetic and meta-cognitive thinking; (early adulthood).

These five kinds of *understanding* (or "super tools") to all intents and purposes need to be *recapitulated* in the classroom, according to Egan, in order for education to work best.[26] The understandings are not to be viewed as hierarchical, rather as intellectual-concentric, with one coalescing into the other over time—though there are some losses along with the accumulated gains (*memory*, for example, is much more powerful in oral language cultures). The last four above all can be taken as our "languaged engagements" with the world and represent increasing "degrees of culturally accumulated complexity in language" (Egan, 1997, p. 30). The connection between "cultural development in the past and educational development in the present" (p. 27) are the elements and stages of the culturally mediated cognitive tools common to both. What had begun long ago but now surrounds the growing child, the socio-cultural entrenched cognitive tools become gradually internalized in successive stages.[27] So Egan makes broad use of Vygotsky's original notion of cognitive tools but exploits and extends them in novel ways unique to the educational context.

To a large extent, Egan admits, the degree to which the "understandings" are maximized and which sub-tools are stressed over others is itself culture dependent. Moreover, they are only imperfectly developed in the normal course of human maturity, partly because the "understandings" and their sub-tools have not been made *explicit* and largely because educational systems have either undervalued or ignored them. Some come more easily with maturity and socio-cultural embeddedness than others, although they are poorly harnessed for effective instruction and learning in typical school settings.[28] Because "mythic" and "romantic" thinking are intimately linked to our emotional selves, they are ubiquitous in the everyday world: in the media, the entertainment industry, pop culture, and political rhetoric. "Philosophic" understanding (or perhaps better put as "theoretic thinking"), on the other hand, is acutely fragile; it represents a fairly late flowering of human history and civilization and requires institutions (schools, universities, research bodies, art and music colleges) for its achievement and advancement.

The *aim of education* is thus redefined in this conceptual framework to produce a kind of "five-fold mind" (although this label can be misleading), one that is conscious about and can make *use* of the five cognitive-cultural "toolkits" (along with their various listed sub-tools) when exploring and thinking about the world, be it created by nature or constructed by man. The "how" and "when" of teaching any curriculum is answered by *organizing* subject topics around the cognitive tools and sub-tools available at the corresponding stage of development of the learner—which would engage their imagination, emotions, and interests in a natural way, because these are the dominating "thinking" tools available to them at that stage. Another way to phrase it, the student's "mind" is to become ever more sophisticated by developing five cognitive "layers of understanding" when dealing with the world—this goes greatly beyond the perennial concern in science education to develop better problem-solving or "critical thinking" skills of mind (Bailin, 2002; DeBoer, 1991), although such skills or habits are undoubtedly involved.

3.3. EGAN'S METATHEORY AND HPS SCIENCE EDUCATION

The relevance of his educational philosophy to science education can now be answered and sketched, pertinent to *two levels*, the general and the specific curricular level.

Let us focus on the *specific* or curricular classroom level first. Because it serves primarily as a theory of (educational) *development*, it bears directly upon how learners can be expected to come to *understand* at appropriate age-grade levels and liberates science education from the domination of ideas in developmental psychology (Egan, 2005a; Duit & Treagust, 1998). For example, all in the community are agreed (regardless of paradigmatic commitment) that it makes no sense to present science to learners in primary school from the "logic of the discipline" perspective as is done in senior secondary. Along with other developmental theorists, Egan also seeks to explain *why* this is the case. In marked contrast to how some developmental theories in psychology *attempt to explain* the quality of children's reasoning ability (with their narrower focus on personal cognition and the logical-mathematical category), the explanatory power of his metatheory arises from the changed nature of the discourse, with the shift of focus towards social constructivism and children's cultural-linguistic mental assembly—it is because their developing minds are inherently shaped at this younger stage by "mythic" cognitive tools, and *not* because they are "pre-operational" (or whatever). And at middle school, where the "logic" approach begins to encroach increasingly, Egan explains why this approach must also miss the mark in capturing students' interest and study since their "romantic" make-up of mind inherently searches for meaning and knowledge within that frame. Any amount of motivational coaxing or suggestions from psychological learning theories can have but little effect.[29] Even at the upper levels Egan suggests that the logic of the subject matter could be better understood and appreciated if molded according to their developing "philosophic" frame of mind and its sub-tools. That the logic of subject-discipline organization is a very poor place to *begin* to learn a science subject (as so much PER and CER work continuously confirms), because it represents the *end-stage* of knowledge generation and storage, had already been cited by Dewey (1916/1944) almost a century ago—yet, sadly, such an approach remains ubiquitous.[30] Why not work with what cognitive-emotive tools the learner already has, though in latent form?

Now I would like to list a few brief examples of how Egan's metatheory could impact the science classroom. He has illustrated how the life-cycle of the butterfly could be taught in an imaginative way within the *mythic* framework to foster and develop those tools for making this topic better understood and more engaging for learners in primary classrooms (Egan, 2005b). He has also provided instructional frameworks to develop "philosophic" *understanding* at the upper levels for learning calculus, Newton's laws, and simple harmonic motion, so there is no need to repeat these here. What I would like to underscore is how he has interwoven an emphasis of *narrative* as a thread throughout the tool-using and learning stages. At the "mythic" stage learners are engaged emotionally with story-telling, as is well-known, but he goes on to show how this becomes shifted to a narrative format in "romantic" understanding and on to metanarrative at the "philosophic" stage. One can, of course, still tell engaging stories at subsequent stages and grades, but Egan shows how to structure the subject topics around these important tools in effective ways for instruction.[31] What he is doing is drawing out the richness and meaning-making capacities of story as the mind develops and begins to seek out more complex and varied forms of understanding through narrative (Lyle, 2000). This is precisely where the proper use of the history of science can enter, enrich the narrative format, and deepen comprehension of the curricular subjects. Egan can in fact provide a practical and pertinent metatheoretical *justification* for the contextual-historical *case study approach* as presented by the Stinner research group (Stinner et al., 2003) and as applied by them to age-developmental curricular topics for the early, middle, senior, and college years.

As Klassen (2002) has clarified, the use of the history of science in the curriculum is commonly rationalized as the dual purpose of providing *context* and of enhancing cultural literacy. Especially at the upper levels, students usually do not come to "feel the excitement" of original scientific discovery (e.g., of the electron, DNA, or the Copernican revolution). Rarely are they presented with the "big picture" of the theory that commands a discipline in terms of how it came to be originally formulated (what problems or other theory it confronted), was debated, perhaps was suppressed, even challenged and occasionally overthrown because of recalcitrant anomalies and argumentative discourse (Allchin, 2013; Niaz, 2009, 2010). Instead, students in senior and college courses are drowning in specialized content—in a sea of facts, descriptions, laws, and equations. Egan insists that some sort of *metanarrative* that provides an overall explanatory frame of scientific knowledge, change, and advance should be presented to students at some point in their science education at the upper level. Should there not be any time allotted in the curriculum for those cases where rival high-level theories have clashed in respective sciences (see here Duschl, 1990; Kalman, 2002, 2010)?[32]

Egan argues that due to the learners' *sub-tools* (listed earlier) and their emotive link, the "general schemes" should be up front at the start. The power of theories for explaining and generalizing over many phenomena, for instance, coincides with the cognitive sub-tool of the "drive for generality"; this feeds into their sense of meaning connected to the "search for truth" and the "lure of certainty." And the histories of physics, chemistry, biology, and geology are all well positioned to present to them how "general schemes" (theories) remain incomplete and the important role of debate and anomalies in the scientific community when theories change. This hooks directly to student fascination with the discipline and its evolution; it humanizes the content and opens up questions about the *nature of science* (including the epistemic status and role of models)—all of which allow for better content acquisition. It would certainly help get behind the dry, dogmatic, and static "textbook science" and begin to expose the wonder and dynamism of the historical and cultural development of the scientific enterprise—all components HPS reformers have argued for decades. Certainly some socio-scientific issues (SSI) that are consequent of "frontier science" could also be of relevance here for enhancing the "tool-kit" for *philosophic* (theoretic) understanding (Bailin & Battersby, 2010; Zeidler & Sadler, 2008).

Finally, at the *general level,* the impasse in the *three major goals* of science education (as argued in Chapter 2)—and which lies beneath the contested science literacy conceptions (as stated at the start of this chapter)—is mostly resolved, or so is claimed: insofar as the discipline would *no longer be preoccupied* with debating whether and how specialized content knowledge should be stressed at the cost of teaching and learning about wider sociotechnological issues as preparation for active "critical citizenship," or how both these aims should accommodate (in some way) individual learners, to develop their autonomy and creativity, free from societal constraints, values, and demands. Instead, the *overarching aim of science education* would be (as simple as it sounds, but as difficult as it may be to implement) to assist the learner to master the socio-culturally determined "thinking and emotive tools" available at the appropriate stage of development[33] for the curricular topics at hand. This offers a *fourth*, and completely different alternative as to "why, what, and when" one should educate. This does not so much unravel the long-standing deadlock between the socialization or knowledge-based aims, as removes them both from the *direct* task of educating, on the one hand, and shifts the perennial "social relevancy" requirement (DeBoer, 1991) to a different plane, on the other. The claim is made that when teaching the science content with the tools included at the start in the curricular and instructional planning, students' imagination will become engaged and connect immediately and more intimately with the knowledge and themes presented to them. Knowledge is still very important in this metatheory, but *imaginative education* helps unlock the long discovered and encoded forms and fills them with new meaning. Imagination is not to be taken as the enemy of reason, as the older romantic school and popular opinion would have it, but preferably "Imagination is Reason in her most exalted mood," as Wordsworth (*The Prelude*, XIV, line 192) so sublimely put it, some time ago.

> We can easily forget that learning symbols in which knowledge is encoded is no guarantee at all of knowing. All knowledge is human knowledge; it is a product of human hopes and fears and passions. The primary trick in bringing knowledge to life from the codes in which we store it is through the emotions that gave it life in the first place in some other mind. Knowledge, again, is part of living human tissue; books and libraries contain only desiccated codes. *The business of education is enabling new minds to bring old knowledge to new life and meaning.* (Egan, 2005b, pp. 96–97, italics added)

This knowledge, now filled with fresh significance because it is placed within a *narrative context*, will be better remembered and truly understood, and, so goes the argument, can be better assessed and used as the individual finds necessary—whether for personal-aesthetic intrinsic aims or social-oriented utilitarian ends. In particular, and to the point of socialization, a true *five-mindedness* which *has* mastered the cognitive-emotive tools by the end of schooling should be able to accomplish what the *external* social forces so often demand of science education with their vocational push for "productive citizenship" and the many *internal* (science education) reformers demand for "critical citizenship"—when "socio-scientific issues" are raised, whether at points linked to science curricular themes, or after school and later in life (and as is widely accepted the traditional academic-oriented curriculum with its "Vision I" literacy view constantly fails to accomplish). And the demand to allow for personal "growth" and individual creativity would also be met by this metatheory, precisely because it *foregrounds* student creativity and imagination in learning in various ways, at different age levels. By having the school educational enterprise now focus on developing these "mind-frame" tools *of understanding*, a cognitive and affective dynamic is established that, in principle, would allow these other decidedly important aims to effectively "get off the ground" and be met, since they would come along as by-products during the fulfillment of the educational project. Tools are tools for a reason; they can be put to work in various ways and for different ends.

In short, science education would now no longer be solely preoccupied with formal knowledge mastery *or* socialization *or* personal growth *per se* (as ultimate aims), as it would be focused on tool development and mastery *by which* these others could then be more effectively and eventually accomplished. And science education could not hope do all this on its own; that would be expecting too much of it, but only within the larger scope of the educational endeavor of the school and college curriculum—one that itself is hopefully immersed in *imaginative* education to some degree. In any event, such a metatheory *directly ties educational means to educational ends rather than, as is usually done, educational means to socialized or psychological or some other ends*.

3.4. SUMMARY

Those who wish to structure science education to aim for the exclusive development of a *kind of person* I argue must overshoot the boundaries of what it is feasible to do and aim for in the discipline. And it is to forget that we do not labor alone, as if isolated within educational institutions, in our disciplines in schools and colleges (although many of us operate as if we do, and when scanning the science education literature one easily receives this impression). It is in the purview and value of an educational philosophy for its characteristic metatheory to mark out the ideal, the kind of person, it is true, but we will be more successful at developing and marking out a kind of "scientific" *mind-set* in cooperation with others when working towards that wider humanistic and global goal within the defined educational aims of a metatheory—when at last such an all-embracing framework is drawn up and eventually accepted by the community. (Or possibly one or two defining metatheories). Egan's metatheory has been suggested here, developed exclusively *by* and *for* educators (as Piaget himself had suggested should be done some time ago), as one possible and substantial contribution in that direction. This assumes that such a high-level theory is not only essential and beneficial, but equally a prerequisite to establish science education upon a firmer foundation as an independent academic discipline (but within educational studies)—notwithstanding critical voices who disagree with that position and reject its very premise.

In the interest of avoiding misunderstanding, the argument that science education requires a metatheory is not made on the grounds that *simply because* education—in the Anglo-American "curriculum" tradition—does not have one, *therefore* Egan can be suggested as a theorist to help fill the gap; rather on the grounds that because of some serious problems with science education, a gap was noticed, one that could (in fact should) be filled for educative reasons that can address these problems and better serve science education—and *that* is something worthwhile pursuing. It is not about filling gaps for gaps sake. Nor is the argument saying that *any* metatheory will do, so long as the gap is filled; in point of fact, when comparing the German-Norse *Didaktik* tradition and educational system—one that actually *does employ* a metatheory—it was argued that *Bildung* as a viable alternative is insufficient for several reasons. In short, what is being said is: "this particular metatheory will help (i.e. Egan), and it can stand on it own merits" (whether or not it happens to cohere in the mind of the critic).

The problem regarding the deadlock of the *three* venerable older educational projects with their three key goals time and again vying for priority is not so much "solved" (one cannot "square the circle") but instead put aside for a fresh alternative, which conceives of education in completely new terms of developing the language-based age-appropriate cognitive-emotive tools. We can contribute significantly to helping students at various stages of their development in learning about and using, to their best abilities, the socio-cultural tools through the medium of science, its history, epistemology, and socio-technological impact. Science, in turn, will be better understood and appreciated when approached through the socio-cognitive tools, imagination, and narrative framework, or so it is argued. (The validity and viability of the metatheory, it is agreed, must be appropriately substantiated in classrooms; the recent study performed by Hadzigeorgiou, Klassen, & Froese-Klassen, 2011 has provided an important empirical research instance of support.) Indeed, that is in the purview of *our* knowledge and training, and fully within our area of expertise, and would distinguish our unique contribution. Such a restricted aim for our discipline is quite achievable and we would not be saddled with unreasonable expectations and burdens we could not hope to meet, especially with regards to socialization. Surely, it is the business of society to socialize, but the proper job of education is to *educate*. Regrettably these two are often conflated (Nyberg & Egan, 1981).

Science education (and its newer reform approaches) has been largely (but thankfully not completely) remiss on this issue and has for too often and for too long failed to adequately consider this distinction and flesh out its relevance for the discipline. Instead it has become bogged down in focused disputes concerning constructivism and epistemology or many debates regarding goals and defining science literacy, especially with the well-meaning but flawed primary (if not sole) aspiration to make science students better *citizens*. It is, further, precisely on this point that the hoped for ambition of achieving widespread public literacy of science (above 20% say)—and hence the wider and by now popular meaning of "science literacy" with regards to socialization (along STSE lines or primarily "socio-active citizenship," as versions of "Vision II")—is unachievable and unnecessary, as Shamos (1995) has forcefully argued. This is chiefly, but not solely, because such a goal is inexorably meshed with other socio-cultural forces, interests, and values (themselves time and culturally constrained) *external* to school science and the research community over which they have but little influence and even less control. What is more, this goal is intimately tied to aspects of a hidden philosophy of education (based on Deweyan and by now suspect progressivist principles) that continues to mark out the principal role of schools and the nature of the science curriculum mainly on socio-utilitarian grounds, an underlying philosophy often more implicit than explicit. The hazards are well-known but easily overlooked. These grounds at best tend to downplay the personal development of the individual (e.g., liberal education and *Bildung* traditions) as they strain the social allegiance of science teachers to academic science, and at worst, tend to diminish the value of knowledge and the aesthetic, creative side of science (Girod, 2007). They can just as easily fall victim to economic utility rationales in the face of powerful political forces that too often seek to bend policy in their own interests.[34]

With regard to Egan's philosophy of education and its metatheory, on the other hand, an opportunity is at hand to have school science education stand its own ground and, first of all and finally, argue the validity and purpose of the discipline on merits derived solely from an "inside" educationalist perspective. With this metatheory taken as an *over-arching aim* of science education, a kind of "five-fold tool-equipped scientific mind" (for want of a better description) could equally serve as an *intermediate aim* in the wider educational project suggested. *Scientific literacy* could then be, I suggest, plausibly and practically reinterpreted to mean the creation of a dynamic science mindset imbued with its five kinds of *understanding*, all developed sequentially. This is to choose none of the three earlier mentioned options that currently confront and confound the community, but rather to put the stalemate and the entire discourse aside for a new interpretation and a substitute philosophical educational discourse—conceivably a "Vision III," appropriating Roberts' (2007) terminology. Science education would then be (in part) reoriented to an internal discussion of what should constitute a proper "scientific mind" along these lines—

among those of us concerned for developing a PSE (including debating the very conception)—no doubt a cross-cultural and critical one. (It would certainly be fair to consider my proposal to be linked to the traditional, extended conversation on "mind" within the discipline as uncovered by DeBoer, 1991, but with a different accent and orientation, and a greatly expanded conception based in part on Vygotskian socio-cultural ideas.) It would, at least as presently construed, involve significant study into the epistemology, social practices, and history of science (intellectual and social), along with socio-technological and environmental issues related to scientific discovery, advances, and applications. Here is exactly where HPS reformers can contribute vital and exceptional insights based on a "philosophy of science education" as a specialized sub-field of inquiry. It is to boldly state that the field of science education research should be significantly broadened, as Jenkins (2001) has suggested, and become more philosophically attuned, as Roberts and Russell (1975) and more recently Anderson (1992), Matthews (1994), and Nola and Irzik (2005) have argued. It too, could be partially charted—with critical qualifications—along lines Ernest (1991) had first pioneered for his own field, being "philosophy of mathematics education."

NOTES

1. This dialogue manifested itself in a special edition of the *Journal of Curriculum Studies* in 1995 (no.1, vol. 27) and in a series of books, especially *Didaktik and/or Curriculum. An international dialogue* (1998), edited by Bjørg Gundem and Stephan Hopmann, and the essay review of books by Vásquez-Levy (2002).
2. "The word *Didaktik* is the semiotic indictor of this discontinuity of conversation and, accordingly, of unshared appreciation of each group's work. What is clearly valued as a noun by one group has a derogatory meaning for the other as the adjective, *didactic*, with its association in English with transmissive, instructional teaching" (Fensham, 2004, p. 145; see Lijnse, 2000).
3. Eisner (following Cremin) identifies two related but separate streams within progressivism, one "rooted in the nature of human experience and the development of intelligence ['growth'], the other in social reform" (1992, p. 311); the first which emphasized the personal, the other the political, although Eisner admits that Dewey himself would never have allowed for such a separation between personal and political.
4. In this way the essence of the ideas behind the *Bildung* conception in having become partially detached from the given educational traditions of Germany-Scandinavia, nonetheless became part of the wider philosophical-psychological discourse of the West.
5. "From the days of Socrates its meaning was systematically ambiguous, denoting as it did both 'culture of the mind' or 'civilized life' and the influences,

processes, and techniques for the making of a man. In Hellenistic times, by a subtle extension of meaning, *paideia* came to refer to the *results* of educational effort rather than the *means* for achieving the end of education—the whole or complete man, an ideal toward which one might strive but never completely attain except through life-long endeavor" (Lucas, 1972, p. 95).

6. *Paideia* comes from the Greek root words *pais*, *paidos*, the upbringing of the child. It became Latinized by Cicero to *humanitas*, to signify the education common to all human beings that was transferred by language (literacy) and culture. Here, too, it had strong ethical connotations (*areté*, virtue). By the time of the Renaissance this interpretation was recovered, although it meant originally the specialized study of the classical Greek and Latin authors of the newly revised curriculum *studia humanitatis* in the arts faculty. The original Greek root words are still found in our words pedagogy and pediatrics.
7. Interesting the reference here to "recapitulation" of cultural development, for as discussed shortly, Egan's metatheory also foregrounds a recapitulation thesis but shifts the emphasis towards a Vygotskian notion of linguistic, cultural-based cognitive tools.
8. "To the Enlightenment ideals of transparency, autonomy, and authenticity, both Marx and Freud in their different ways counterposed the actual delusions fostered by alienation, anomie, and fetish—psychological failings constituting false consciousness both of society and self" (Blake & Masschelein, 2003, p. 44). These criticisms have grafted critical theory with a predominant negativism towards society and education, even characterized as a form of *utopian pessimism*, Habermas being the exception. This charge though does not seem as appropriate for the associated "critical pedagogy" of Friere, Apple, and Giroux. Yet even this American "curriculum ideology," as Eisner (1992) argues, has remained an academic protest movement with little to no effect on U.S. school systems.
9. Such a curricular ideology suffers from two serious failings: it rarely defines the new social utopia aimed at following the hoped-for "social reconstruction" (other than suggesting vague moral values, such as "true equality" or "non-colonialist"), and it conceives of educational practice in purely *instrumentalist* terms: "it has not questioned the very concept of educational *praxis* itself but conceived it as an *instrument* for liberation or repression. Educational *praxis* still receives its meaning from the goal or end at which it should aim, here conceived as a utopia" (Blake & Masschelein, 2003, p. 50). Worse, it is thus ensnared in the very problematic common to school education being run on a technicist model: "It thus remains itself subject to the same instrumentalist logic that it deplores at the heart of the capitalist system" (p. 50).
10. One could include here also North American curricual: U.S. standards document like AAAS (1993) and Canadian *Common Framework* (Council of Ministers of Education, 1997), insofar as both are explicitly lacking an encompassing educational metatheory. The Australian statements on schooling appear similar to typical traditional content-based school curricula ("academic rationalism"), especially IB and AP structured senior science courses. The newer U.S. and Canadian policy documents, it is admitted, do attempt to broaden

the understanding of *scientific literacy* beyond just content knowledge. The Canadian document, in fact, takes the "knowledge" component as only one of four "foundation" dimensions of literacy, but as mentioned, is heavily influenced by the contexts of the STSE paradigm. A recent interpretation of this STSE policy document states that the "goals of science education have broadened to include *personal goals* such a life-long learning; ... *social goals* like developing informed citizens and a well-educated workforce; and the traditional *academic goals*" (Roscoe & Mrazek, 2005, p. 12, original italics). One clearly sees here the three main educational goals underlying the document but identified by Egan as being at odds, without any recognition on the part of the authors that fundamental problems lie at hand. It seems that the goals so stated have been extracted as general *principles* while ignoring (or being unaware of) the theoretical frameworks they are embedded in. As such, these policy documents can be taken as representative of the "second option" for literacy mentioned in Chapter 2 (also Roberts' "Vision II"), being the "inclusivist" option of DeBoer (2000).

11. "In general, pre-service teachers are taught to begin thinking of instruction by asking how a student learns, how a student can be led toward a body of knowledge, and how to evaluate what students know and are able to do. Such tasks promote a role of the teacher as 'instructor', one who is accountable for implementing formalized state-mandated curriculum. This view of curriculum as an organized framework for guiding, directing, or controlling a school's...work then determines the range of approaches available to teachers. Thus, this curriculum-as-teaching manual perspective restricts a teacher's potential ... [and] professional autonomy" (Vásquez-Levy, 2002, p. 117).
12. "It represents the blending of content and knowledge into an understanding of how particular topics, problems, or issues are organized, represented, and adapted to the diverse interests and abilities of learners, and presented for instruction. Pedagogical content knowledge is the category most likely to distinguish the understanding of the content specialist from that of the pedagogue" (Shulman, 1987, p. 8).
13. PCK itself has not been an involved area of science education research, although the field is expanding. Van Driel et al. (1998) closely aligned the idea with a teacher's *craft knowledge* and stressed its tight link to specific subject-matter topics (they examined chemical equilibrium) and cautioned against inferring to general cases of science learning. They see PCK as a rich area for further study to link research on teaching with research on learning. Abell (2007) has provided the most recent comprehensive review, combing the literature for studies on CK, PK and PCK. She quotes Geddis (1993) defining PCK as "the transformation of subject-matter knowledge into forms accessible to the students being taught" (p. 675). She admits the category has come under criticism for vagueness and some have questioned Shulman's model, adding other concepts like "pedagogical *context* knowledge or concerns" to the lexicon. Some admit to difficulties in providing proper *methods* for research to collect appropriate empirical data to confirm, disconfirm, or expand upon the term. While PCK has been mostly embraced as a useful theoretical construct, research remains "in a

formative phase, where researchers continue to define terms and methods to guide their work" (Abell, 2007, p. 1123). She provides valuable sub-topics to better define the CK, PK and PCK classes (Abell, 2007, Fig. 36-1, p. 1107) and suggests *five* components to PCK: orientations to teaching, knowledge of science learners, curriculum, instructional strategies, and assessment. What is *missing*, however, are two important factors: how NoS knowledge influences CK and PCK (deliberately ignored) and the crucial role played by an educational philosophy and metatheory. Kind (2009) in another review suggests that the "transformative PCK" category seems more effective than an "integrative PCK" that ignores subject specialist CK.

14. This is understood to be a proactive stance. Research into science teachers' PCK also indicates that the relationships are reciprocal in nature, in that CK can be re-evaluated in light of teaching experience and reflection on classroom student learning (Van Driel et al., 1998).
15. That the curriculum needs to be made problematic implies that a *philosophical* (and not just instructional) problem initially lies at hand which requires resolution. This inportant topic is too often overlooked in curriculum theory or in the science education literature; see Fensham (2004); Geddis (1993); Klafki (1995); Lijnse (2000); Schulz (2011); Vásquez-Levy (2002); Witz (2000); thereto, Aikenhead (1996) has argued the learning science involves a culturally-rooted "border-crossing" on the part of the student, to negotiate the transition from the personal "life-world" to the "school-science world."
16. This is a revision of Fensham's original chart (2004, p. 149, Fig. 10.1).
17. At the lower levels at secondary schools in Canada (grades 8 to 10) the science literacy conception according to the STSE paradigm structures the contexts for learning (based on the *Common Framework* document; Council of Ministers of Education, 1997), as explicitly stated by the BC Ministry of Education *Integrated Resource Packages* (IRPs), which determine the "Prescribed Learning Outcomes" (PLOs) to be taught.
18. Egan's (2005b) *An Imaginative Approach to Teaching* presents several useful frameworks that illustrate how this metatheory can be applied to curriculum and practice for different subject topics and themes. This book, along with the continuing research of the Imaginative Education Research Group (IERG), has brought his ideas to bear in fruitful ways in a number of diverse cultural classrooms around the globe, being applied from primary to tertiary education. My discussion will focus solely on his high-level theory in *The Educated Mind*, and the reader is referred to his other work (2005b) and encouraged to visit the *ierg*-website (www.ierg.net) for related information about theoretical applications for teacher classroom use.
19. He writes that an educational metatheory "shows how to realize in individuals a certain conception of education. Without some such conception, all the research findings in the world are educationally blind, and with such a conception, it is unclear what research findings have to offer" (Egan, 2002, p. 181).
20. This does not necessarily force a complete separation between the two disciplines but it certainly appears to further widen the gap between psychological theories and educational practice.

21. "Piaget ... is not interested in those concepts which occur 'artificially' as a result of instruction, but rather in those concepts which *develop* spontaneously or naturally as a result of normal interactions with the environment. He is interested in what happens of necessity in the development of cognition. An educational theory is concerned with making value choices among a variety of possibilities. One does not *choose* to do what is necessary ... it makes no sense to try to teach concepts which develop naturally" (Egan, 1983, p. 16, original italics).
22. Howard Gardner (1991) has also come to this realization: "The category of 'natural development' is a fiction; social and cultural factors intervene from the first and become increasingly powerful well before any formal matriculation at school.... Once the child reaches the age of six or seven, however, the influence of the culture—whether or not it is manifested in a school setting—has become so pervasive that one has difficulty envisioning what development could be like in the absence of cultural supports and constraints" (p. 105).
23. See Egan's 1997, p. 30. Bakhurst (2005) writes: "Those that invoke the slogans that the mind is 'constructed,' 'distributed,' 'relational,' 'situated,' or 'socially constituted' to maintain that culture is in some sense constitutive of mind, and that therefore the nature and content of an individual's mental life cannot be understood independently of the culture of which that individual is apart. Strong culturalism can take various, more or less radical forms ... [it] starts from the old intuition that reductionism (or eliminativism) about the mental leaves out something crucial...the missing ingredient is not primarily consciousness or phenomenology, but the sociocultural context of mind. Two intuitions lie behind this claim. [Firstly] that meaning is the medium of the mental, and meaning is ... a social phenomenon. [Secondly]...the human mind, and the forms of talk ... should be understood on the model of tools; and like all artifacts, we cannot make sense of them independently of the social processes that make them what they are" (pp. 413–414).
24. "What the imagination can grasp is enabled and constrained by the logic inherent in the various forms of knowledge and by the psychologic inherent in the process of human development. So the dynamic of this scheme is a troika of generative imagination guided and constrained by epistemological and psychological forces" (Egan, 1997, p. 189).
25. Egan has focused to a greater extent on the mental effects due to the advanced stages of literacy, especially as displayed by human culture in the relatively recent past, the last 5,000 years or so, above all regarding changes to Greek society and thought. The "romantic" stage can also be seen as forming a *transition* between mythic and theoretic, while "ironic" is that end stage where "language becomes aware of itself" and its own limitations. For the latter, Egan's archetypes are, in ancient times Socrates, and in modern times Kierkegaard and Nietzsche. He distinguishes a positive "sophisticated irony," which can manage doubt while preserving aspects of mythic, romantic, and philosophic stages, from an "alienating irony" of the typical postmodernist sort—it yields instead to radical epistemic doubt, to rejection and

dismissal of the other understandings. "The product of alienating irony is impotence; sophisticated irony is liberating and empowering. The aim of educational theory is to keep alive as much as possible of the earlier kinds of understanding in the development of irony" (Egan, 1997, p. 162).

26. Egan is careful to distance himself from the two earlier and unsuccessful versions of recapitulation: the *logical* one, which insisted a particular cultural (knowledge) content must be followed (Herbert Spencer), or the notorious *psychological* development version with its biological basis, and now discredited (once supported but later repudiated by Dewey and Piaget). See Langer (1988). Egan's reference is strictly to socio-cognitive *tools* as invented in human cultural history.
27. "Intellectual tools, or sign systems, begin, to use Vygotsky's terms, as interpsychic processes and become intrapsychic within the child" (Egan, 1997, p. 29).
28. This is exactly what Egan laments about contemporary education, along with the well-intentioned but misguided practice of using unsuitable psychological theories when transmitting "inert" knowledge, and all the while hoping to juggle three conflicting goals.
29. And the newer explanatory cures offered by studies in neuroscience on the so-called "adolescent brain," while suggesting some physiological constraints, hardly address the issue of their *development* of *mind* (Kwan & Lawson, 2000). This is once again to defer to a reductive empiricist notion of mind, possibly organismic.
30. It also formed part of the later critiques by Lawrence Cremin and Paul Hurd at the time of the 1960s scientist-based curricular reforms (Bybee & DeBoer, 1994), but to no avail.
31. "A narrative is a continuous account of a series of events or facts that shapes them into an emotionally satisfactory whole. It has in common with a story that shaping of emotion, and so words are often used synonymously, but it is different in that narratives can be less precisely tied into a tight story, less concerned with emotion, more varied, more open, more complex" (Egan, 2005b, p. 99).
32. As examples, Ptolemaic or Keplerian astronomy; Newtonian or Einsteinian physics; phlogiston or Lavoisier's chemistry; young-earth theories or evolutionary geology, etc. For those researchers who now argue for the value of stressing model-based reasoning, the particular limits and defects of the respective models of such theories become sharp. Such instances can serve as an effective teaching strategy to help students discover how knowledge can progress, and thus illuminate for them the precarious nature of how science actually advances—along a "paradigm shift" (Kuhn) or along the lines of progressing or degenerating "research programs" (Lakatos). Duschl (1990), on the other hand, has presented an intriguing alternative view, using Laudan's philosophy of science, as better suited for science educational purposes, making good epistemological use of several key historical case studies. Alas, conventional instruction rarely elucidates the *conceptual* structure of even the successful, dominating theory/paradigm (e.g. Newton, Darwin, plate tectonics, early quantum theory, Nersessian (1989), Allchin (2013),

Dolphin (2009), Niaz (2002, 2001, 2009, 2010)), and even less so does it take sufficient time, if at all, to study the fascinating historical examples of how science deals with cases of competing theories. Where historical development is mentioned, that only too often borders on myth, as many HPS researchers have long complained (Allchin, 2003).

33. One avoids the term "growth" here because it is tainted with Dewey's preconceptions and tied to a biologized view of mind.
34. But granted, this failing could also be leveled at conventional science education (at upper levels) taken as the "academic rationalism" (Eisner, 1985) program of the status quo (Hirst/Platonic knowledge-based tradition). This is evidenced not only today (Fensham, 2002; Pedretti et al., 2006), but by its very inception during the 1950s/1960s as the major curriculum reform wave (Keeves & Aikenhead, 1995; Duschl, 1990; Klopfer & Champagne, 1990; Roberts & Oestman, 1998).

CHAPTER 4

PHILOSOPHY OF SCIENCE EDUCATION, EPISTEMOLOGY AND NATURE OF SCIENCE (NOS)

Every science curriculum, regardless of its professed goals, should at least make clear to students what science is and how it is practiced. Dispelling misconceptions of the nature of science is the first step toward true science awareness. Then follows an understanding of the nature of the enterprise.

—Morris Shamos (1995, p. 224)

The reciprocal relationship of epistemology and science is of a noteworthy kind. They are dependent on each other. Epistemology without contact with science becomes an empty scheme. Science without epistemology is—insofar as it is thinkable at all—primitive and muddled.

—Albert Einstein (1949, p. 683)

Rethinking Science Education: Philosophical Perspectives, pp. 139–208

4.1. PHILOSOPHY OF SCIENCE EDUCATION, EPISTEMOLOGY, PCK, AND CONTENT KNOWLEDGE

This chapter is about putting into question the nature of curriculum *substance*, thereto, asking questions pertaining not just to the nature of scientific knowledge, but how this knowledge is represented as subject content knowledge (CK) in science education, which then informs both students' and teachers' views of their *image of science* (Englund, 1998; Matthews, 1994; Ryan & Aikenhead, 1992). Such considerations of knowledge perspectives and construction are commonly referred to in the literature as "epistemologies" of the subject, of the teacher, and of the learner, an area of active research for several decades, and whose studies often tend to be linked to research and debates on the theme of the *nature of science* (NoS). The intent here is not to survey and discuss the studies of the three respective epistemologies themselves, for this would yield an informative book in its own right,[1] but to primarily concentrate on school subject epistemology with related debates in the nature of science as discussed in science education and philosophy of science (PS). This will include referencing the valuable contributions on these themes from educators, philosophers, historians, scientists, and others involved with the history, philosophy and sociology of science (HPSS) reform movement. While it is recognized today that epistemology plays a vital role in the teaching and learning of science, its status is still undervalued in teacher and graduate education.[2]

4.1.1. Epistemology, Belief, and Epistemic Aims

That science instructors and their technical textbooks are so concerned with accurate and exhaustive transmission of canonical scientific knowledge clearly reveals the central significance of epistemology to science education. One can identify this preoccupation of academic sciences courses (a chief aim of school and college science) with the constricted and popular rendition of the customary *knowledge-aim*. Here is another area where philosophy of education (PE) discourse can provide relevance, for the knowledge-aim or truth-aim has been fundamental in the traditional view of education, including its *liberal* construal—notwithstanding significant attacks on that objective from different educational perspectives (e.g. progressivism, post-analytic, postmodernist).

> Although transmitting knowledge is not the only aim of education, it is surprisingly substantial in its ramifications. Because we can compare various educational practices to determine which ones better advance students' knowledge, the knowledge-aim offers educational guidance, justifies central

educational practices, and exposes complexities in the educational policies it supports. (Adler, 2002, p. 285)

Science teachers plainly assume their courses or textbooks provide (technical) knowledge, indeed substantially *true* knowledge—and for the most part they would be correct (e.g., propositional knowledge of "final form science," Duschl, 1990).[3] Yet being philosophically inclined means giving pause to reflect on what *basis* this can be claimed (expertise of the authors? authority of the scientific community?). HPSS-based reforms do insist, of course, that CK (of teacher or curriculum) requires expansion and corrections (e.g., historical and epistemological *context* to be properly understood and learned), as will be argued in this chapter.[4] But first stepping back and asking about justifying CK, or "what is knowledge?"[5], is to venture into both philosophy and PE territory (the right segment of the triangle in Figure 1.1). The kinds of answers to these questions have vital educational ramifications. How, for example, can one justify teaching evolutionary theory if its stake in knowledge and truth cannot be established against intelligent design claims? Or taking the "culture wars" into view, is cultural indigenous knowledge of nature *true* scientific knowledge? Are there other kinds? If so, how are they legitimated? How do we best distinguish them from science?[6]

Students, when not just assuming the authority of the textbook or teacher, occasionally wish to have explained to them the grounds for knowing, grounds that can only partially be established when "doing science" (i.e. scientific inquiry). Four possible harmful *dispositions* to knowledge students can develop from science classrooms are cynicism, dogmatism, skepticism, and relativism, and Norris (1984) rightfully asks "can all these be avoided?" Teachers require philosophical intelligence not just for telling these apart, but for awareness when they crop up during instruction and for strategies to overcome them.[7] Thankfully there already exists a tradition in PE that can assist them, which has sought to demonstrate the relevance of epistemology for education (Adler, 2002; Carr, 1998, 2009; Siegel, 1998).

The standard account of knowledge is "justified true belief" (JTB), which stipulates three conditions in order for someone to say they "know X." For instance, science educators would not be satisfied if a student stated they "know" the Earth orbits the Sun but could not provide any evidence for this proposition. In this case the student has a *true belief* (two conditions met), but without justification couldn't be said to have attained knowledge. Even if philosophers have brought forth serious challenges to JTB,[8] this doctrine of traditional epistemology still retains its value in assisting science teachers' thinking about the differences among knowledge, belief, and justifying conditions in the classroom as they arise (Southerland, Sinatra, & Matthews, 2001). It highlights the drawbacks of traditional instruction that can overstress the value of rote learning, algorithmic problem-solving, and decontextualized subject content, especially if tied to a policy of exaggerated standardized testing (Hofer & Pintrich, 1997; Mercan, 2012).

JTB can equally shed light on other cases that can occur where knowledge and belief appear conflated, such as when a student has learned content but refuses to believe it (e.g., "I understand evolution, but I don't believe it"; "I can explain the Bohr model but don't believe atoms exist"). Southerland and associates (2001) have provided an overview of the differing conceptions and occasional clashing meanings concerning how the two terms "knowledge" and "belief" are employed and understood in the separate research fields of philosophy, educational psychology, and science education. This academic difference in standpoints complicates the educational landscape but is not to be appraised here. They also raise the important pedagogical question whether science education should limit its aim to providing knowledge (or understanding) and not demand changing student beliefs (as typically required by conceptual change research). An interesting exchange of opposing views between Smith and Siegel (2004) and Cobern (2000, 2004) on the meaning and use of the terms in science education illuminates that science teachers not only need to sort out their own presuppositions to knowledge and beliefs but require sensitivity to historical and cultural dimensions of these concepts while attending to philosophical arguments.

Within the field of science education research proper, Norris (1995, 1997) has analyzed how the JTB view of knowledge finds expression in the aim of *intellectual independence*, one key content-transcendent goal articulated since Dewey and progressivism. He identifies several serious shortcomings of past and recent formulations of this goal (e.g., as found in constructivism and notions of scientific literacy). Norris notes especially the philosophical controversy surrounding the question to what extent, if any, non-experts can reason independently of experts' knowledge and community—hence, to what extent they can be justified to trust in authority and yield to scientists' judgments (and by association, their textbooks). The outcome of the dispute remains contested, but it appears some reliance is indeed unavoidable.

The degree to which intellectual independence is attainable (or not) has major ramifications for the character and educational aims of science education reform movements (like STSE, SSI, HPS, social-action). It could impose severe limitations, depending upon the stipulated objectives and overall ambition they desire to advance for the discipline, notably which independence-based goals they mistakenly assume school students can rightfully achieve.[9]

4.1.2. Epistemology, PCK, and Content Knowledge (CK)

One suggestion of this book is that a philosophy of science education (PSE) informed by Shulman's (1987) viewpoint would seek to influence science teachers' specialist CK through HPS awareness and education. With such broader and deeper HPS perspectives taken on their own disciplinary field(s) together with an acute attentiveness to where curricular materials (especially textbooks) provide either grossly over-simplified or even erroneous views of scientists or the NoS, teachers would be in a better position to present a more accurate and comprehensive view of science for their students. One can make this argument on its own grounds irrespective of educational metatheory, and it retains its validity regardless of which education system one happens to be working within, either the Anglo-American "curriculum" tradition or the Continental "Didaktik" one. Moreover, as Fensham (2004) has been quoted earlier as stating, neither the former tradition nor the latter actually "problematizes" the subject content to be learned, rather both have "held strongly to the idea that the content for school science subjects should be determined by what is accepted as lying within the content of the corresponding disciplinary science" (p. 158). Teachers do not in general question such content either, or its assumed underlying epistemology, although the HPSS reform movement has addressed these inadequacies for some time.

Even the use of an educational metatheory, it was pointed out (whether *Bildung* or Egan's), does not normally question the adequacy of the science CK as typically presented by either state-mandated policy curricular documents or textbooks written by experts in their respective disciplines (secondary or tertiary). (These are the two primary sources, along with their own university academic science training, that teachers heavily rely on for planning and instruction, and that strongly define their attitudes and beliefs about scientific knowledge, its development, and their identity.) Recall that a metatheory serves the principal purpose of allowing for content *transposition* for educational purposes but leaves basically untouched the question of content knowledge adequacy—the fact that canonical content as commonly presented is insufficient for a broader and deeper science subject understanding. As one critical commentator has stated: "The textbook as it now exists is necessary but not sufficient. The full flavour and excitement of science as a creative process cannot be experienced in a historical and philosophical vacuum" (Brackenridge, 1989, p. 80).[10] A PSE serving as well as a kind of "Didaktik analysis" could then in addition critically examine both the common CK and the growing literature of HPS-influenced research and curriculum materials for appropriate use in classroom pedagogy (many examples are provided by Allchin, 2013), thus contributing to develop a teacher's *pedagogical content knowledge* (PCK).

(To anticipate the last chapter, looking at this issue from a hermeneutic perspective, this would mean to allow the horizon of a teacher's *epistemology*—or understanding of subject content—to expand through awareness of the need of NoS insight and HPS integration—the other horizon of significance, as provided by research—as an essential component of their pedagogical content knowledge.)

Duschl (1990), in *Restructuring Science Education,* hints at PCK when he writes:

> Right or wrong, appropriate or not, a teacher makes numerous decisions on a daily basis concerning the design, delivery, and evaluation of instruction. An effective decision maker considers the learner, the learning environment, and the nature of the subject matter. The issue is a teacher's ability to make informed judgments that eventually lead to meaningful learning on the part of students. A central component of this decision-making process is a teacher's subject matter knowledge.... To achieve such educational goals, the teacher needs to acquire a special type of knowledge base. In addition to knowledge of schools, learners, and of teaching strategies [PK], teachers also need a special knowledge base of the structure of the subject. (p. 2)

He argues that such a base must include knowledge from the history and philosophy of science, requiring teachers to emphasize not only "what" is known (learning *of* science) but also "how" it is known (learning *about* science). Such a stand would mean a shift in the focus of science education, he insists, the need to overcome the "epistemological flatness" of "curriculum materials or instructional strategies that do not give a complete picture of the concepts being taught" (Duschl, 1990, p. 41).

The Swedish educator Tomas Englund (1998), drawing upon the situation in his own country, has also criticized the "structure of the discipline" as a starting point for curriculum and the common approach for science teaching:

> Whenever that happens, it becomes difficult to "problematize" the school science content; it is much more likely that it will be taken for granted. In such cases, the meaning that students are offered by their school subject—what I will call the *educational* content of school subjects—is always at risk of being dominated by a narrow view of the socialization function of education. This situation unquestionably is linked to the fact that the moral and philosophical aspects of education and socialization have been neglected. (p. 13, original italics)

Englund identifies the overbearing *instrumental rationality* characteristic of educational thinking and research as applied to the purposes of schooling as the predominant reason for this neglect, and this could equally provide another reason why philosophy in general has been accorded such a low value, and perhaps why the need for PSE has not come to the fore sooner. As argued previously, science education has always been burdened with the "socialization imperative," which at times has taken on primarily utilitarian, even extreme economic overtones (e.g., to educate for reasons of increasing the professional "pipeline," or increasing national economic status, or increasing the critical thinking of citizens for judging techno-scientific claims in society, thus to enhance democracy, etc.).

> Problematizing school subject content is a moral and philosophical endeavour that cannot be addressed by scientific-technical rationality. The central questions of philosophical inquiry in education are about the worth of knowledge and meaning offered to students. Scientific technical questions about efficiency and effectiveness are quite different questions—yet efficiency and effectiveness questions tend to be more prevalent than philosophical inquiry, on the educational research agendas of many countries. (Englund, 1998, p. 13)

He suggests that school subject content can be categorized at least three different ways according to how the *educational significance* is conceptualized (pp. 19–20):

- *Epistemic (school subject) content*: determined "essentialistically and scientistically." For science education this means the relation between a science discipline and the school subject in terms of how the knowledge is structured and how the key concepts (gravity, electricity, chemical bonds, genes, etc.) are internally inter-related to others in a formal conceptual web of *scientific meaning* (this is taken as traditionalism).
- *Contextual (awareness) content*:[11] Here he seems to mean those relations of a subject as linked to other broader contexts "such as the relationship between the individual and society, individual and nature ... dealt with, explicitly or implicitly, as educational content." The non-traditional "curriculum emphases" of Roberts (1982; like "self as explainer," STS) as to how science epistemic content is (or should be) linked to other humanistic or techno-societal concerns and issues, can be placed here.
- *Socialization content*: "includes the different meaning-creation contexts or discourses where different conceptualizations of the relation studied [contextual content] are expressed" (Englund, 1998, p. 20). He means the justification of why a subject is required to be learned and the reasons for educating the individual (*educational aims*), and these tend to accompany what is taught (overtly or not) in schools as "companion meanings." He lists "patriarchal,

scientific-rational, or democratic" (one could easily add "colonialist") and indicates how science content is provided with an ulterior *instrumental meaning* according to its socio-cultural embeddedness and group "ideological" interests.

In the first category the CK is taken as decontextualized and its "educational content" (or significance) is confined to formal knowledge acquisition, which serves as its (restricted) academic *scientific meaning*. In the second case, the educational content is widened and understood to reside in the context in which the epistemic content can be situated, providing for *contextual meaning*. In the last case, the educational significance locates the contextual meaning of CK within the milieu of broader educational philosophical goals, policy decisions, and group interests, marking its *socio-instrumental meaning* (as previously discussed in Chapter 2).

Examining the last category, one immediately identifies the broader educational goals listed earlier: "scientific-rational" can be associated with the Platonic knowledge-based project (Egan), "democratic" with the intents of Deweyan-based educational philosophy.[12] The point that Englund wants to make is that if curriculum is viewed in socio-cultural light, even the traditional "epistemic content" (his first category) of typical science classes must carry with it other associated or "companion meanings" as to the ultimate educational reasons why the material has value and must be learned. Teachers indeed usually fall back on some sort of explanation when cheeky or exasperated students ask "why are we learning this stuff?" In other words, students usually seek meaning beyond the narrow academic-scientific one. "Knowledge for knowledge sake" as pure "academic rationalism" (i.e., Eisner) will rarely suffice, although the "scientific-rational" line of justification usually includes claims about learning "truths about nature," or about "how things really work," although the former assumes a clean objectivist-realist notion of science, and the latter harbors a simple science-technology link—both of which a more sophisticated science curriculum that include HPS and STS aspects would help debunk and dispel. That is, curriculum and CK taken as "contextual" and more authentic and not just narrowly "epistemic." One final reason usually given for "epistemic content" when arguing in "scientific-rational" mode for socialization is that mastery of such knowledge serves as a gateway to higher education and professional careers (often in engineering, medicine, or pharmacy, possibly even science teaching), though the evidence has shown this applies only to a minority of students (approximately 10–15%).

As can be seen, the *companion meanings* often implied,[13] which are conveyed to students (overtly or covertly) with "epistemic content," have serious epistemological as well as practical significance—for along with the details of laws, concepts, scientific descriptions, and processes learned in classrooms "they are taught what knowledge, and what kind of knowledge, is worth knowing and whether they can master it. They are taught how to regard themselves in relation to both natural and technological devised objects and events, and with what demeanor to regard those objects and events" (Roberts & Oestman, 1998, p. ix). Along with whatever *attitudes* towards science, knowledge, science learning, and their own self-image (which in the case of physics we know is not attractive to many females) invariably associated with companion meanings, are the following regarding CK: that it is largely disconnected to the students' everyday lives or has little wider socio-technological purpose or implications; that strict adherence to learning decontextualized scientific knowledge ("epistemic content") and its evaluation using standardized and high stakes testing is somehow representative of how scientific knowledge is generated (if not a mirror of the enterprise) itself; that mastery of such formal knowledge structures is sufficient and sole grounds for professional entry; finally, that CK *logically* structured is ahistorical, objectivist, and complete—and thus *not* historical, interpretive, and tentative. The argument to focus on here is the last point, *that CK and the science curriculum common to schools is not epistemologically neutral.*

As a preliminary comment, it should be mentioned that CK need not necessarily be construed narrowly as "epistemic" with its accompanying "scientific meaning" typical for traditionalism, although as the status quo this remains widespread at the upper levels for the case of senior physics, chemistry, geology, and biology. It has been the explicit purpose of policy and curriculum developers as published in both American and Canadian "standards" documents of the 1990s to move away from this conception and tight disciplinary structure to encompass broader *contexts* for this knowledge. It needs to be pointed out that traditional CK is in truth supplied by several companion meanings formulated as three (of seven) "curriculum emphases" (Roberts, 1982): (1) to build a "solid foundation" (logic of subject topics in succeeding years), (2) to give "correct explanations" ("products"), and (3) to ensure "scientific skill development" ("processes"). These three alone, and taken together, usually serve as mutually supporting rationales for teachers when inducting their students into academic science at the upper 11 and 12 grade levels. Opposed to this view, the conception of content in U.S. standards documents was purposely taken as broader than the conventional sense, as Bybee (1998) had emphasized:

> The definition of content in the *National Science Education Standards* and the *Benchmarks for Science Literacy* is broader than "valid science" and Correct Explanations. To be sure, there is traditional content associated with physical, life, and earth science, but content is defined to include also the nature of science, technology, history of science, inquiry, and science in personal and social perspectives. (p. 163)

The Canadian *Common Framework* (Council of Ministers, 1997) taken as a "standards"-type document, also expanded the understanding of CK beyond traditional disciplinary "knowledge," which serves as only one of four "foundation" frameworks for specifying *scientific literacy* (the other three being STSE contexts, skills, and attitudes). Unfortunately these kinds of documents only tend to have force at the lower levels, and attempts to broaden the scope of CK outside of the restraints of "epistemic" for grades 11 and 12 specialty science courses simply has not happened—Denmark being an exception (Thomsen, 1998). Actual cases of attempted implementation efforts beyond strict "epistemic" in Alberta (Canada) and Australia (using STSE contexts) had even met with considerable resistance from external stakeholders (academic scientists, media, and parents; Fensham, 1998; Orpwood, 1998). As Bybee has expressed one pitfall of the reform efforts,[14] a "narrow view" of CK continues to prevail among interest groups that disallow other *interpretations* (exacerbated in absence of educational philosophy and metatheory, as argued). In effect, it should not be surprising to Englund, Bybee, and others that an impasse often arises with the two differing "educational contents" (or educational *significance*) or perspectives taken on CK. This is because the *epistemic CK* conception is customarily entrenched in a "scientific-rational" ideology (or as "academic rationalism," entailing its own companion meanings) whereas the *contextual CK* is often located within a "democratic-socialization" ideology of educational purposes and goals (Gaskell, 2002). The various stakeholders beholden to these two educational conceptions or philosophies (or "visions" of literacy; Roberts, 2007) are themselves divided as to the *chief aim* of what should count as science education (expressed in strictly *utilitarian* terms; Roberts, 1988): either to educate persons in science *for* academic science ("science for future scientists") or *for* democracy ("science for citizens"). We notice clearly once again, unfortunately, how CK is equally and unavoidably tied to disparate educational ideas or philosophies (fixed as ideologies among stakeholders) previously identified by Egan as being at fundamental odds and seeking to undermine each other.

4.2. PHILOSOPHY OF SCIENCE EDUCATION AND NATURE OF SCIENCE (NOS)

The concern in this chapter is to concentrate on the *epistemological* companion meanings, those assumptions or perspectives that lie (often hidden) behind common science curriculum as they tend to inform student and teacher epistemologies, and how an improved understanding of the NoS among both can improve science teaching and learning. It probably goes without saying that any PSE that has as its main concern the education of persons into the disciplines of science (narrowly construed)—or better phrased, an education of the scientific enterprise (widely construed)—must consider what studies into the epistemology and NoS have revealed and how this bears upon analyzing and critiquing subject CK. This means the need to canvas assorted discussions and debates pertaining to the general construal and generation of scientific knowledge undertaken in the academic disciplines of the history, sociology, and philosophy of science, including a newer array of assorted academic research called collectively "science studies" (Hodson, 2008). Such an undertaking, it is understood at the outset, cannot be an exhaustive one for this book, since these different fields of study as they have matured in the past half-century or so have accumulated a vast array of studies on different aspects of science and its diverse sub-sciences. They show a range of views and controversies, from the major theoretical positions of realism, positivism, and pragmatism in the PS (Laudan, 1990), to full blown knowledge relativism as found in certain positions held by those in the "strong program" of the sociology of science (critiqued by C. Norris, 1997; Ogborn, 1995; Slezak, 1994a, 1994b). Two invaluable resources are the detailed discussion very recently offered by Hodson (2008), and an older review by Duschl in the 1994 *Handbook.* The latter had examined studies in the history and philosophy (but *not* sociology) of science as to their diverse contributions to educational developments in science education proper during the 40-year period 1950–1990, when major transitions were taking place.

The intent here is to broaden the definition of CK as, for example, the U.S. *Standards* documents (AAAs, 1993; NRC, 1996) had originally hoped to do (and many of their HPS ideas and themes are invaluable), by arguing that although *epistemic content* is necessary it is hardly sufficient for education either *in* science or *about* science. Interest in studying the "epistemology of school science" has already produced several significant prior studies,[15] and research continues into both teacher and student epistemologies[16] but, sadly, such studies have too often shown that an improved alignment between teacher and NoS epistemologies, which several authors have recognized as of central importance, is still far from being realized (Höttecke & Silva, 2011; Matthews, 1998a).

In line with other reform efforts, the case made here should be seen as contributing to achieving those objectives as found in other international "standards" documents, with their explicit emphasis to include NoS instruction. They form part of a larger appeal for more fundamental changes in the attitude and approach to science teaching, as accentuated in the U.S. *National Science Foundation Report* (NSF, 1996) for undergraduate science education as well (Mason & Gilbert, 2004). The argument has been made that NoS inclusion enhances the educative value by raising the interdisciplinary and cultural dimensions of science courses—since it embeds scientific development in cultural and historical contexts while fostering among students a greater emotional satisfaction of curricular themes. This all contributes ultimately to their better appreciation of science and understanding of the natural world (Allchin, 2013; Matthews, 1994). Furthermore, a curricular stress on the inherent but too often neglected epistemological dimension could contribute significantly to a substantive improvement in CK understanding. The science education researcher Norman Lederman (1998), having noted that science education suffers from "subject matter without context," contends that unless NoS teaching is made explicit ("given status equal to that of traditional subject matter") the science literacy of students and citizens will hardly improve, and thereto, neither can they develop a proper scientific mindset.

Comment About NoS Terminology

We will follow Osborne, Collins, Millar, and Duschl (2003, p. 717) in distinguishing among *three* distinct albeit inter-related components: (1) nature of scientific knowledge, (2) methodologies of science, and (3) science institutions and their social practices. However, the discussion is limited predominately to the first, a confined perspective on those distinctive features of science concerning its *epistemology* and *ontology* (whose definitions were previously provided in Section 1.2). The second is usually framed as "scientific inquiry" and occasionally confused with NoS, while the last is frequently subsumed under the rubric of the "sociology of science." (I agree with Lederman that "nature-of-scientific-knowledge" and "scientific inquiry"—or more commonly "processes"—should be kept distinct, although they indeed interact and overlap.)[17] It is of course understood that both teachers and students need to develop an understanding of all three vital components, as several authors have earlier come to stress (Bauer, 1992; Driver et al., 1996; Hodson, 1985; Shamos, 1995).

Driver et al. (1996, pp. 16–23) have presented *five arguments* for NoS understanding, linked explicitly to increasing the science literacy of the public, which have now come to be widely acknowledged though not necessarily equally accepted in the science education community:

1. *utilitarian* argument: NoS is necessary for people to make sense of science and technological objects encountered in their everyday lives;
2. *democratic* argument: NoS is necessary for people to make sense of socio-scientific issues and partake of democratic decision-making;
3. *cultural* argument: NoS is necessary for people to appreciate science as a significant factor in contemporary culture;

4. *moral* argument: NoS is necessary for people to become aware of the norms and moral commitments of the scientific community (ideals of rationality, objectivity, social consensus, self-correction, etc.);
5. *science learning* argument: NoS is necessary since it supports successful learning of (epistemic) content.

Not all these arguments carry the same force, and some are surprisingly similar to those put forward earlier in the 1980s for teaching STS, especially the first two, which have been refuted by Shamos (1995). Yet because there is a shift in focus to NoS comprehension rather than requiring competence in evaluating technologies and technological impact on society (where even the experts can differ over frontier science and risk analysis, for example), the warrant for these arguments has changed and seems more reasonable (given the examples listed by the authors). That being said, the immediate interest of concern here is aligned with the third and fifth arguments.

Notwithstanding such *rationales*, which are important when arguing for HPS and NoS inclusion in classrooms and for re-writing curricular materials, the *empirical findings* of the last 50 years allow for some generalizations that can surprise, disappoint, and discourage science educators. Lederman (2007, p. 869) concludes:

1. K–12 students lack adequate NoS understanding;
2. K–12 teachers lack adequate NoS understanding;
3. Teachers' ideas of NoS are often not translated into classroom practice;
4. Teachers consider NoS a lower priority for instructional outcomes;
5. NoS conceptions are best learned through explicit instruction.

Lederman complains of inadequate measurement instruments in some research work (biased NoS notions among researchers) and the "superficial" nature of current testing strategies of NoS given the little known "mechanisms" that can contribute to changing teachers' and students' views. Indeed, the bulk of all this research work and perhaps its primary redeeming value seems to consistently confirm the same conclusions as listed above. One other discovery has been made: "The longevity of this educational objective has been surpassed only by the longevity of students' inability to articulate the meaning of the phrase 'nature of science'" (Lederman, 2007, p. 832), a conclusion Lederman had uttered ten years previously. Considering how entrenched the conventional academic paradigm has become at precollege and college levels—which will be further examined below—one is only surprised that others are surprised at these results, since "science teachers and science curricula seem rigidly bound to a tradition of communicating facts or end products of science while generally neglecting how this knowledge was constructed" (McComas et al., 1998, p. 541). The argument that science teaching should also put emphasis on "how knowledge is constructed" is itself an old one, going back to Schwab (1962, 1978), Dewey (1916/1944), and Mach (Matthews, 1991; DeBoer, 1991), and likewise one might add its longevity as argument has only been surpassed by the longevity of its neglect. The revived HPS movement has reiterated it once again.

Textbooks are not the sole culprits, but they bear much responsibility: "for most science students, a description of the NOS is relegated to a few paragraphs at the beginning of the textbook quickly glossed over in favor of the facts and concepts that cram the remainder of the book and generally fill the course. And the ideas put forth in textbooks...concerning the nature of science are almost universally incorrect, simplistic, or incomplete" (McComas et al., 1998, pp. 541–542). Taking the inherent attitudes and epistemology of this paradigm with its curriculum and how it views and structures epistemic knowledge into account, on top of how science teachers are themselves educated, then the first four items listed above fail to astonish. If anything, one could conclude that such results cry out for a fundamental shift in attitudes and thinking of what "counts" as science education, for real HPS reforms and available resources, and the absolute *necessity* of a "philosophy of science education" on the part of teachers and the discipline itself.

One more obstacle needs to be addressed, namely the charge that there is no agreement among the experts as to what NoS is and that because of this lack of consensus it is sheer folly to try to articulate NoS positions for teacher awareness and inclusion, and hence for student mastery:

> the fact of the matter is that we have no well-confirmed general picture of how science works, no theory of science worthy of general assent. We did once have a well-developed and historically influential position, that of positivism or logical empiricism, which has by now been effectively refuted. We have a number of recent theories of science which ... have hardly been tested at all.... If any extant position does provide a viable understanding of how science operates, we are far from being able to identify which it is. (Laudan et al., 1986, p. 142)

Given this confusing situation, three philosophers of science (Eflin, Glennan, & Reisch, 1999) have come forward and offered some advice to the perplexed science educator. Much of what they suggest is indeed useful, including laying out: (1) the common ground *shared* among philosophers as to what constitute the basic tenets of science;[18] (2) what is *not* shared;[19] and, more importantly, (3) the advice that educators must become more familiar with the details of the controversies of the debate among philosophers, historians, and sociologists—which is exactly, of course, what HPS advocates stress.[20] These polarized camps have made the business of science education a messy and complicated affair—it has become increasingly difficult to navigate a pedagogical course between competing views "from diehard realism to radical constructivism" (Rudolph, 2000, p. 404).

Rudolph (2002, 2000), having noted the disagreements among academics and the "vague generalizations" of typical standards policy documents, is among a few thoughtful detractors insisting science educators should reevaluate the usefulness of NoS statements and instruction.[21] He argues that no "single nature of science exists" (he allows for pluralities)—referencing the "disunity of the sciences" debate and outlook (i.e., Galison & Stump, 1996)—and even if one did, at best only a simple and "partial representation of it could ever be captured in the school experiences designed for students" (Rudolph, 2002, p. 65). But this would unavoidably involve a selection process, he admits, raising more questions about omissions and the decision-making process that is invariably linked to external socio-political factors and reasons, and thus must be further justified on these same grounds. Given that the epistemology of science, he continues, has *always* been subject to selection for socio-political reasons when constructing curriculum (he cites Dewey and Schwab as past cases), and since there is no reason to ignore this fact, therefore the task of educators is to recognize the prehistory and *first choose* the appropriate social ends they think worthy and *then* tailor their NoS selection accordingly. This is in essence an argument to subordinate NoS to the socialization context or imperative (as an educational *philosophical* position), to submit the construal of the image of science to serve public policy or other socially determined ends. A kind of argument, one should add, that falls square into the territory of a *philosophy of science education* that must be willing to engage with it. One role for a PSE *could* certainly be that of constructing an image of science from a grab-bag of selected NoS tenets commensurate with specified social ends, although I would personally dispute that project, because it wishes to subordinate education to socialization for historical reasons.

Certainly the challenge can be acknowledged, but it cannot be taken up in this current book. Nonetheless, there is some merit to his claim fronting the disunity thesis as well as the fact that the epistemology of science has been selected to serve social ends before. The position as articulated in this book is that neither education nor epistemology should be subordinated to socialization, regardless of previous cases, and that such a educational philosophical stance is found wanting. That is not only to disagree that such cases must be taken as precedent setting, rather that epistemology must be taught and learned for its own sake, and most importantly, must serve as a corrective to supplant a mistaken epistemology (and history) already existing. And one can disagree that a selection process *must always* be linked socio-politically (and not philosophically, logically, educationally, or pragmatically, say). Moreover, it seems that his position hinges chiefly on the view that NoS tenets are so fragmented and diverse that a pick-and-choose type selection is unproblematic. This *nominalist* view (shared by many sociologists of science) can be contrasted with the *essentialist* view, holding to a more singular view or template, an ideal of science that can be gleaned and articulated as a somewhat more "unified" NoS notion (Matthews, 1998a). On this dispute the jury is still out. Further, the answers to the dilemma are also relative to whether one takes a more *internalist* (epistemic) or *externalist* (sociological) perspective on the question, or possibly hazards a combination of both, as some contemporary philosophers of science like Haack (2003) and Giere (2005), or cognitive psychologists like Nersessian (2003) have articulated (also Duschl & Hamilton, 1998). Both perspectives are essential to understand scientific development and should be taken as complementary and not contrary (as Hodson, 2008, equally argues).

Most science educators notwithstanding grant his point about "partial representation," are willing to live with this, and are content as well to articulate a *common list* of essential NoS theses (see below) *where broad agreement does exist* and where their "generality" need not detract from genuine insight and understanding about the scientific enterprise (Lederman, 2007; Osborne et al., 2003; McComas & Olson, 1998). One can accept a *range* of NoS clarification, from more general to more sophisticated articulations depending upon the course, grade, and learner age-developmental stage. This spectrum would encompass the list below, as a bare minimum, but also include more elaborate *epistemological frameworks* (e.g., Duschl, 1990), up to still more erudite discussion, such as the *realism/instrumentalism controversy*.

1. *Empirical basis:* scientific knowledge is based on observations of the world.
2. *Observation and inference:* the senses and extension of senses give us information; Inferences are interpretations of observations.
3. *Methodology:* there is no one step-by-step scientific method; science relies on experimental evidence, rational arguments, peer review, and skepticism.
4. *Laws and theories:* are different kinds of scientific knowledge. Laws describe relationships of phenomena, while theories are inferred explanations. They do not progress into one another.
5. *Subjectivity:* current theories and laws influence investigations and observations. Observations are theory-laden. Personal values and agendas also influence how scientists work.
6. *Creativity:* scientific knowledge is created from human imagination and logical reasoning, and based on observations and inferences.

7. *Tentativeness:* scientific knowledge is durable yet subject to change; science historically exhibits both evolutionary and revolutionary changes.
8. *Socio-cultural embeddedness:* science is part of social and cultural traditions, and scientific ideas affect, and are affected by, this milieu.
9. *Applied science:* science and technology are different but impact each other.

Probably all can agree that these points are at a level of generality such that they can be immediately communicated to teachers and inculcated in students, although such a list is not without its critics (Irzik & Nola, 2011; Matthews, 2014). While this may be true and helpful, regrettably it is exactly this generality, alternatively, that can mask the need to expose other essential aspects of NoS. One notices the failure to explicitly mention *induction* or *inference to the best explanation* (as key reasoning tools) and the neglect of the important role that *models* play in science and in reasoning—indeed, *the need to distinguish theories from models*—and hence, an avoidance of any mention of the *status* of theories or models as realist or instrumentalist (*realist/instrumentalist debate*). Also missing is the notion of the *underdetermination* of data (Duhem-Quine thesis). These are, it seems to me, serious omissions that need correction. Such important facets need not be avoided (see Holton & Brush, 2001 physics textbook on the Duhem-Quine thesis, and especially Section 4.7 on the realist/instrumentalist debate).

4.3. KUHN, SCHWAB, AND SIEGEL ON SCIENCE EDUCATION AND TEXTBOOKS

Thomas Kuhn (1970)—the great revolutionary himself—had ironically remained very conservative on the matter of HPS integration, and argued openly for the retention of the mythical historical picture of the textbook presentation of science, its practitioners, its inquiry and its epistemic CK as typically found in the conventional "normal science education" paradigm. He stressed its value in upholding this ideal for the training of young competent "puzzle solvers" in advancing what he called "normal science" of the day. The textbook plays a fundamental and *conservative* role in this objective and helps reinforce "normal science," or the dominant (dogmatic) paradigm. Successful science as conducted by most scientists most of the time, he argued, is primarily about preserving that paradigm and maintaining the social consensus around it—and *rarely* about pushing boundaries into new territory that could lead to a conceptual revolution and a new paradigm (although this sometimes happens, and its preconditions—in opposition to Popper's views—are mainly *non-rational*; Kuhn, 1970, pp. 138–143). Kuhn even readily admits that such a textbook-centered pedagogy stifles imagination and innovation.[22] As a physicist *and* historian of science he has been among the few in his day to have shown not only an interest in science education (marginally improved today, largely due to the HPS movement; Holton, 2003), but to have commented on its vital role in the training of scientific minds: "...science education remains a relatively dogmatic initiation into a pre-established problem-solving tradition that the student is neither invited nor equipped to evaluate" (as cited in Siegel, 1978, p. 302). [23]

Siegel (1979) has criticized this educational approach described by Kuhn as dishonest, irresponsible, and cognitively restricting—and so an altogether unsatisfactory pedagogy, which I believe is correct—although Kuhn, as an insider (physicist), should be given credit for portraying how the scientist-discipline centered (traditional) paradigm (entrenched since the 1950s reform) actually seems to think it should educate—*certainly this belief holds sway in all first year undergraduate introductory courses*. Stinner (1992, 1995b) argues, moreover, that in classrooms at the upper levels Kuhn's insights degenerate into the de-contextualized problem-solving techniques now identified to be at the heart of the *problem of context* (as discussed previously).[24]

It is precisely because of the reality of this problem that, exactly contrary to what Kuhn holds, "education" understood as *training* in "problem exemplars" in order to become efficient "puzzle-solvers of normal science"—typical for both physics and chemistry classes—does *not equip* the aspiring young scientist for proper and successful scientific *research* work. Even less does it give the general student a conceptual understanding of science—in large measure because such problems are *ahistorically* formulated, highly abstract, and usually artificial.[25] Kuhn's paradox, which he recognized, that the rigid training in solving artificial problems should somehow prepare the novice practitioner for engaging the kind of authentic problems scientists encounter in research, has been resolved by studies of student learning in physics education research (PER): there is no paradox because he is simply *wrong*. Rather it is primarily—though not exclusively—through scientific *apprenticeship* (mentoring) at the graduate level that the young scientist becomes equipped for doing research while learning the norms and habits of the scientific community of which s/he is a part—a predominately socio-linguistic dimension.[26] Kuhn was certainly aware of this aspect and included it as the vital *second* component of training for "normal science" (1977, pp. 229–230; 1970, p. 47). But the general science student is clearly at a severe disadvantage, even those specialized at the baccalaureate level and not continuing on, when it comes to perceiving what science is *about* and the haphazard nature of true research work. As Matthews (1988) observed: "Karl Popper said of Kuhn's normal scientist that he has been badly taught, that he has technique without understanding" (p. 70). *Would only that such an assessment were more widespread among senior science classroom instructors*. Even if we decide to accept Kuhn's popular account of the division of science into "normal" and "revolutionary" stages as a reasonably accurate *descriptive* account (which some philosophers such as Toulmin question, preferring instead "micro-revolutions"), and equally accept his view of current academic science texts as basically puzzle-solving, paradigm-fixated training manuals, we need not accept his or others' *prescriptive* conclusions that seem to follow from them. Such as, because the average science student or would-be scientist will never be of a caliber to initiate a conceptual revolution (a la Newton, Lavoisier, Wegener, Darwin, or Einstein), our role as educators can be limited to, and quite satisfied with, training students solely for the "normal science" of (the prevailing paradigm of) the day (or as is usually the case, *of the past*). This does indeed seem to be *one pertinent feature* of current traditional/textbook pedagogy, and reinforces the pseudo-historical view, one among many, that science progresses because of a "list of successive solitary geniuses."[27] Such an unfortunate perception on the part of students easily leads to feelings of inadequacy, that they are not able to contribute to science, and reinforces their alienation from the sciences (negatively referred to as the "mystique of science" by Lemke, 1990).

A different perspective for science education is given by Schwab (1962), a well-known biology curriculum reformer. According to Siegel (1978) he appears to hold a similar picture of Kuhn's two-fold division of the scientific enterprise, what he terms "stable" and "fluid" phases, and equally views the former as dogmatically inclined, yet he nonetheless reaches quite opposite conclusions. For Schwab, the orthodox and dogmatic "stable" science is to be overcome by *critical inquirers* who consciously question and wish to push it into a "fluid" phase—quite unlike Kuhn, where instead they are (and *should be*) transfixed by the controlling paradigm of "normal science," only to be reluctantly pushed forward when shocked by irritating anomalies. Science education should then be about creating not competent puzzle-solvers but competent *fluid enquirers*, and textbooks are not about indoctrination into "normal science" but rather an initiation to question and challenge the limits of "stable" science (precisely what Kuhn argues strongly against, partly because they are not equipped to do so).

Schwab (1962) was quite correct when he emphasized that the traditional curricular focus on formal (content) knowledge at the expense of an emphasis on proper scientific processes and developing critical inquiry habits (as Dewey had emphasized) is to seriously misrepresent the *nature of science* as a "rhetoric of conclusions" (p. 24), which is how most textbook presentations can be caricatured. The scientific enterprise does indeed encompass both "stable" and "fluid" aspects, but I would not characterize these two as synonymous with Kuhn's "normal" and "revolutionary" science phases, as Siegel (1978) does. This is a fundamental misrepresentation. It is because of Kuhn's analysis of "normal" science that I believe it can lend itself to a further division into "stable" and "fluid" phases in a way that still does justice to Schwab's depiction of science—where both occur *simultaneously* (see also Elkana, 1970). Many other commentators have stressed this dual character in similar ways, contrasting "finished science" from "science-in-the-making" (Sutton, 1996), or "textbook" versus "frontier" science (Bauer, 1992), or even "public" versus "private" science (Elkana, 1970; Martin, Kass, & Brouwer, 1990). All attempt to capture the actual *two-fold nature* of (pure) science as: (1) a finished *end product* (of "declarative" or "expository" or "systematic" knowledge) as found in handbooks or textbooks (but *not* as found in professional research journals), and (2) the essence of scientific *research and inquiry*; the former being taken as static and the latter more dynamic in nature.[28]

The relevant point for science education, however, is that this "frontier" aspect is usually ignored in traditional pedagogy and thus an unduly distorted or excessively one-sided representation of science must result.[29] Furthermore, a proper *historical treatment* could spotlight the atypical yet decisive "theory-change" characteristic of science (Rogers, 1982), which both general science learners *and* aspiring scientists *need* to know (Duschl, 1990; Kalman, 2010; Niaz, 2009).

From a pedagogical perspective, as Siegel argued, there is an even more serious charge against Kuhnian-type talk of dogmatic training, as education in "normal science": *authoritarianism*. This can indeed be an unfortunate side-effect of how instructors present science, and traditional pedagogy seems particularly susceptible to this charge, where students must accept with a sort of "blind faith" the claims of the textbook or the teacher.[30] Even laboratory exercises (as *personal experience* of "scientific inquiry") can be seen as special instances of "cooked" examples to merely reinforce the claims previously presented in class—as is regrettably too often the case: a kind of *weak* indoctrination (Harris & Taylor, 1983). As Rogers (1982) has made clear, to avoid this accusation, the student's confidence in the teacher and textbook *as authority* is only justified if the grounds are made explicit as to the reasons and inquiry processes that *stand behind* scientific knowledge, and from which its authority derives. (He held that a *presentation of history* could achieve this, allowing a balance between mere content knowledge and experimental work, extending "knowing that" with "knowing how.")[31]

Duschl (1988) accused the traditional presentation of harboring a "hidden positivist epistemology," which is then inculcated into students, yet this is overdrawn. Although curricula could display aspects of such an outdated epistemology, and some do, it seems that textbooks in general show no such sophistication. It would be helpful if they did, because then one could much easier identify their implicit philosophy of science. Instead they often present an amalgam of confusing, sometimes contradictory positions with respect to PS topics.

The following represents only a small sampling of typical textbook cases. Harris and Taylor (1983) noted, for example, that the American Physical Science Study Committee (PSSC) physics (hard core traditionalist) mixes a simplistic inductivist view of methodology (which has been roundly criticized) with a naïve realist ontology, in its interpretation of models and theories for atoms and molecules (for example)—a departure from a pure positivist (instrumentalist) position. And it mixes theoretical constructs (model) and observational objects in the same way that actual working scientists do—not surprising because PSSC was developed by scientists during the 1950s reform wave. *In other words they make the cognitive slide with ease across the theoretical construct/observable object divide—a controversial move that engenders considerable debate in philosophy of science.* This observation cannot be emphasized enough—it constantly presents itself in typical textbook writings, and, so I maintain, *is the major contributing factor to students' naïve realist conceptions*. The British *Nuffield* physics also shows such inconsistencies, and its misguided empiricist-inductive "discovery-inquiry approach" is unmasked in reality as a teacher and curricula manipulated exercise (also critiqued by Hodson, 1996). Sometimes the same (historical) experiment is used to *confirm* a theory (i.e. validation; PSSC), or to introduce the theory (i.e., inductivism; *Nuffield*). In the latter case pupils can ostensibly "discover" the theory on their own (naïve inductivism). Both curricula suggest Newton's laws can easily be inferred by students as generalizations from observations with experiments using "blocks and clocks" (inductivist thesis; "blissful empiricism"), which is patently *false*. For some curricula, such as historical-based *Harvard Project Physics* (HPP), popular in the 1970s, the authors identified hypothetico-deductive explanations, which they saw as consistently Popperian. Popper, of course, is hardly a positivist.

Chemical textbooks I have examined (also physics texts) usually entail a *naïve falsificationist* stance, the view that a "crucial experiment" refutes a theory, rather than merely a hypothesis (Hebden, 1998, see also Rodriguez & Niaz, 2002).[32] Earlier, Robinson (1969) argued that from the *language patterns* of biology and chemistry textbooks he had analyzed students could be presented with alternately a realist or instrumentalist conception of theoretical entities. Munby (1976) also found this (the significance of language will be explored more in Chapter 5). Again, it is only the latter that corresponds with the classical logical-positivist or -empiricist position. Selley (1989), commenting on the revised *Nuffield* physics and chemistry texts for upper grades in the UK, noted little overt philosophical discussion, except for the occasional hypothetico-deductive and naïve-falsificationist arguments. He concludes that "most school science books are utterly unsceptical" and that "the information and explanation being presented is simply the truth" (p. 29). In the same vein as Robinson he holds that the "confident, assertive style" of the text passages could be read as implying either a naïve realist metaphysics or a positivist-type instrumentalism. Gallagher (1991, p. 123) along with Selley also finds that texts present the body of scientific knowledge as "revealed truth," with "little attention given to the nature of science, or to how the knowledge of science is formulated or validated." Science can be portrayed to varying degrees "from empirical to positivistic to Kuhnian" (p. 124).

One wonders how science teachers who are usually uninformed about PS can be expected to navigate such interpretively vague, conflicting, and even dogmatic assertions as normally found in their textbooks. Moreover, one should then hardly be surprised to find teachers' preconceptions of NoS and philosophical issues of science equally contrasted, clouded, and confused, since they tend to rely heavily on their textbooks (Gallagher, 1991; Hodson, 1993b; McComas et al., 1998). Thereto, it hardly surprises that students exhibit a range of epistemological commitments, some even holding to inconsistent and contradictory views, such as mixing "objectivist" with constructivist-relativist views of knowledge (Roth & Roychoudhury, 1994, p. 17).

More to the point, and given this typical situation in contemporary science education, the question as to *what kind* of philosophy of science would properly characterize science and should best be transmitted to teachers and students, and hence would help reform curricula and textbook writing, has certainly not been answered. To just continue with what has been done in the past, to ignore or pretend PS is irrelevant, is obviously not an option. Clearly, "history" and "philosophy" are already "inside" textbooks and the curriculum but not in a form that provides authentic views (Allchin, 2000; Niaz & Rodriguez, 2001). Then again, surely not just *any* PS will do, for there exist several, and they are contested (Curd & Cover, 1998; Ladyman, 2002; Nola & Sankey, 2000). At one extreme end, for example, Feyerabend's (1988) "anarchistic philosophy" has polemically suggested that science is not served by *any* one method—frankly polemically asserting "anything goes." (Feyerabend, being quite critical of "Western science," has also led the philosophical charge for anti-realism, and especially epistemological and cultural relativism.) But a multitude of methods does not mean *no* method, nor that all are equally valid (Nola & Sankey, 2000).[33]

Should precollege and college instructors continue to accept and be satisfied with their roles as Kuhnian-type science educators—to train students to become "uncritical, competent puzzle-solvers of normal science" with the sole professionalist *aim* in mind?

The physicist Fritz Rohrlich (1988a) had argued, contrary to Kuhn, that for an adequate understanding of science at least *four philosophical issues* about the nature of science must be addressed in science education, and likewise these are indispensable for the training of scientists.[34]

The absence of HPS in the curriculum has also been typically justified with the view that science education should *distinguish* between the "context of discovery" (or development) and the "context of justification"—a misapplied positivist principle—and privileges the latter over the former. In other words, the discipline is more concerned with "what is known" (knowledge as a finished product) than "how" or "why" it is known (knowledge as discovered and/or constructed, and theories as tentative). *The bulk of textbook organized knowledge* (including its *language*, Sutton, 1996) *and traditional instruction reflects exactly this prejudice* (Schwab and the "rhetoric of conclusions"). This state of affairs is clearly unacceptable:

> Precollege science education curricula should address not only *what* is known by science, but should also include *how* science has come to arrive at such knowledge. To teach what is known in science is to stress declarative scientific knowledge. To teach how the scientific enterprise has arrived at its knowledge claims is to develop declarative and procedural knowledge of scientific developments. Knowledge *of* scientific development as opposed to scientific knowledge, then, is knowledge of both why science believes what it does and how science has come to think that way. In the distinction between scientific knowledge as a curricular objective, and knowledge about scientific development as a curricular objective, it is that the latter is more inclusive. This is the distinction between learning science, and learning *about* science. (Duschl, Hamilton, & Grady, 1990, p. 239, emphasis added)

4.4. EPISTEMOLOGY, HPS, AND STUDENT LEARNING THEORIES

Duschl (1994) comments that HPS studies have had the general effect of influencing the conceptualization of both *curriculum design* ("what to teach")—which is our predominant focus—and *instructional practices* with their learning models ("how to teach")—which I have come to treat with a large measure of skepticism. I will briefly examine the second topic to start with.

On this latter point, one should perhaps begin on a note of caution, that identifying epistemological positions in either textbooks or of teachers is complex and challenging to carry out. Moreover, Smolicz and Nunan (1975) had warned some time ago in one of the earliest reviews examining the "philosophical and sociological foundations of school science" that philosophical investigations of science may not lead to helpful inferences where "educational directives" can be generated. "The relationships and interactions between the image of science, the philosophy of science and science education are indeed complex" (p. 103). With respect to the influence of epistemology *on learning models*, that advice has not stopped efforts of several subsequent researchers, cognitive scientists like Carey (1986) and Nersessian (1989; 1992, 1995), nor science educators like Posner, Strike, Hewson, and Gertzog (1982)—with their influential conceptual change model based on Kuhn—or even Duschl's (1990) elaborate alternative epistemological model of science learning based on Laudan, to draw out exactly those kinds of educational learning directives.

These attempts to use epistemological models of science for explaining student learning processes I claim are misguided (whether based on Kuhn or Laudan in the past, or lately on Giere, as Izquierdo-Aymerich & Aduriz-Bravo, 2003 now argue) because they confuse learning the *nature of science* (NoS) with the *nature of learning science* (NoLS), although as subjects they may indeed be related. This seems to me to confuse epistemology (as to what it can offer) with *educational metatheory* and how students come to know and make meaning. For if learning best comes about through *narrative*, then the NoLS is better explained by narrative models of learning, or possibly Egan's metatheory (acquisition and use of cultural/cognitive tools that engage the imagination) and not analogies drawn from philosophy of science cases, which in themselves are subject to much continued dispute. And if language truly plays a central role (as many are coming to believe), then Gadamer's (1960/1975) insight of how "understanding" takes place, proceeding as an "expanded horizon" carried by students' "interpretive narrative frameworks," then conceptual change can be viewed from an entirely different and perhaps more fruitful perspective. NoS theories should therefore not be taken as a model for learning science, rather as stated, solely as a way to improve and expand both teachers' and students' CK, because learning the epistemology of science should be a content goal in its own right. To disagree that epistemology should substitute for educational theory is to part company with Duschl on this point ("the marriage of psychological principles and epistemological principles," 1990, p. 6) along with the accompanying research and school of thought that sought to bless this marriage since the early 1980s.

If anything, science educators should have been more circumspect considering that in the past they had assumed an epistemology of science (inductivism) and based their pedagogy of *inquiry* upon it accordingly, which in hindsight had proven to be mistaken.[35] Another example is Duschl and Gitomer's (1991) main critique of the well-known Posner et al. (1982) model of *conceptual change* (based on a Kuhnian reading of science) with their thesis that students' constructivist learning (changes to their epistemological framework) proceeds along parallel cognitive lines of normal and revolutionary change. They argue, correctly, this Kuhnian view is too hierarchical and holistic (demanding simultaneous changes to ontological, axiological and methodological commitments) instead of "piecemeal" change—as Laudan had argued in his own PS—and which they rely on in their critique. But today neither is Laudan considered to have defined the "official" or authoritative PS (if such a project is even achievable). The Kuhnian reading had also influenced the link between cognitive psychology and science education, where "weak" and "strong" restructuring are the personal correlates of the two types of changes and where "cognitive schemas" (*schema* theory) are analogous to scientific "theories," but even here significant tensions exist between psychology and epistemology (Duschl et al., 1990). I do not wish to rehearse the criticisms of these epistemological learning theories of conceptual change here, which *have* offered educators some insights, to be sure, but which seem to have seen their day—science education today is shifting towards a Vygotskian language-influenced, socio-cultural perspective of learning (Duit & Treagust, 1998, 2003).[36]

The main problem, as argued here, in using epistemological theories is to presume not only that knowledge "restructuring" as occurs in scientific communities over time is analogous to individual student learning (which is tenuous at best)—granted, Nersessian (1989, 1992) discusses some interesting historical cases in comparison with actual scientific discovery and science learning[37]—but that epistemology can *stand in* for the purposes of an educational metatheory and philosophy when describing and explaining the learning process. It seems hardly beneficial that science education should be burdened with the duty of "getting the philosophy of science right" for the purposes of better illustrating how the epistemology of student learning should mirror the epistemology of science. Science educators could well find themselves in the uncertain and uncomfortable position of running after the latest interpretations in PS, as they once did concerning their desire to match the latest learning theory in psychology (i.e., behaviorism and Piaget):

> How to teach science is not a scientific question and neither science nor philosophy of science can give us infallible guidance on how to proceed. Scientific theory does not contain within it a means of teaching and learning science, nor does scientific method [or discovery] represent a significant means of acquiring scientific knowledge (except for those involved in scientific research).... Adoption of learning methods based on a model of science [epistemological, cognitive or otherwise] is not a logical requirement of a commitment to acquaint pupils with an understanding of the methods and procedures of science. To make such an assumption is to confuse aims with methods. Not all learning experiences should attempt to mimic scientific method. (Hodson, 1985, p. 41)

Hodson explicitly argues, in the case of student *learning* science, for the *separation* of epistemology and psychology (although he grants they can be related). We would support this viewpoint, but as further argued, also the separation of epistemology and education on this matter.

4.5. EPISTEMOLOGY AND CURRICULUM DESIGN: SCIENTISM AND NATURE OF SCIENCE

What a *philosophy of science education* should do is help uncover and analyze epistemologies of science assumed to lie behind curriculum, that is, investigate how the epistemology(ies) of science are or *are not* mirrored in the epistemology of school science. Several previous studies have already undertaken aspects of this task.

Cawthron and Rowell (1978) argued that the common epistemology of school science "generally projects an image of science which can be called empiricist-inductivist" (p. 33), as is typically displayed in the Baconian-type, step-wise linear "scientific method" starting from naked observations, leading by inference to inductive hypotheses, to general laws and objective knowledge. They hold that Popper's hypothetico-deductive, falsificationist philosophy of science has been "little studied by science teachers and educationalists" and that "no clearly defined Popperian tradition exists in school science" (p. 36). They argued that the reason for the dominance of the empiricist-inductive view (despite Popper's clear rejection of it) was due to the success of "discovery learning" at the time which appeared in the 1950s and overlapped with "science as an inquiry approach" (Hodson, 1996). This was endorsed by Schwab (1962), psychologists, and the new reforms being instituted "top-down" by state ministries and scientists in the wake of the "Sputnik shock." As they wrote:

> It all seemed to fit; the logic of knowledge and the psychology of knowledge had coalesced under the mesmeric umbrella term "discovery" and there was no very obvious reason for educators to look further than the traditional inductivist-empiricist explanation of process. (p. 38)

The irony being at the exact same time Kuhn and several others were heralding a new radical post-empiricist phase in the philosophy of science. They note though that Popper had reinforced the older view by also having emphasized *one method* for science (his alternative hypothetico-deductive one) and assuming that both theories and their empirical consequences could still be discriminated by use of a "neutral observation language." "That such a neutral language is taken for granted by school science texts is ... abundantly apparent" (Schwab, 1962, p. 38). Along with this mythical image of method have come accompanying distorted images of science—its inevitable progress linearly forwards and closer to truth (termed "convergent realism")—and of the scientist as the individual heroic explorer guided by the search for truth or "objective knowledge" (p. 42).

Hodson (1985, p. 27) likewise criticizes the legacy of *inductivism* ("long since abandoned by philosophers of science") and its role in perpetuating the myth of one dominating "scientific method." This myth, singled out and heavily criticized in a book by the chemist Bauer (1992), still tends to be widespread. Likewise, I have discovered the paradox that while the latest Canadian junior science texts often avoid this term and talk of general scientific "processes" (observing, hypothesizing, modeling, etc.) that students need to learn, too many senior specialist science texts still refer to this phrase. Hodson also comments on the inconsistency of British curricular materials of the day, some employing an inductivist outlook (*Nuffield Chemistry*) whereas others (*Nuffield Physics*) explicitly reject it (p. 35). These glaring inconsistencies in curricular materials, mentioned earlier, make the identification of an inherent *single* epistemology difficult.

When examining the critics, however, it becomes apparent that "method" is usually interpreted in science education in different ways, and the step-wise inductive method discussed by Cawthron and Rowell (1978) is at odds with Duschl's (1994, p. 444) identified *hypothetico-deductive step-wise method:* (1) select hypothesis, (2) conduct observations, (3) collect data, (4) test hypothesis, (5) reject or accept hypothesis. This certainly appears (in highly simplified form) more akin to Popper's (1963/2002) procedure of science progressing by "bold conjectures and refutations." There is also a clear resemblance here with Dewey's own description of "scientific method" as problem-solving (1916/1944, pp. 152–163), taken as problem, hypothesis generation, testing, and application (with the caveat that it should be practical oriented and situated in the student's immediate social environment). Certainly one can admit that science textbooks and classrooms often can emphasize observations and structure laboratory inquiry in an inductive-verificationist mode (akin to 1950s logical-positivist philosophy) or imply that scientific knowledge grows predominately in this way (which is fallacious)—this occurs when *theories* and finally scientific *laws* (which are erroneously thought to develop from them) are described as easily proceeding from data or "careful observations" and then "proven." Such language is indicative, certainly, but we hold Duschl to be more on the mark when he states that hypothetico-deductivism is characteristic: "science teachers would immediately recognize [this] as the standard scientific method." Teachers often ask a student to "prove your hypothesis right or wrong," and this procedure furthermore is on occasion mistakenly thought to apply to the testing of theories in general. Having noted this discrepancy, let us simply state that both *naïve* inductivism and *naïve* falsificationism are prevalent and characteristic of school science.

Hodson (1985, 1988) has argued for an "epistemologically more valid science curriculum," since teachers' inadequate understanding of the philosophy of science "leads them to project an unfavourable image of science and the activities of scientists through the hidden science curriculum" (Hodson, 1995, p. 47). The problem of "scientific method" only brings to the surface in an obvious visible way the epistemological pitfalls of a science curriculum that leaves many other aspects and attitudes concerning science "hidden," as Hodson and several writers have drawn attention to—what can be called the *implicit* philosophy of the science curriculum. Several authors have termed this *scientism*. Bauer (1992) refers to the assumption of *epistemic privilege*, the view that "science and only science is capable of generating true knowledge, and moreover that science can generate any true knowledge we may wish to have. Of course, the supposed secret of science's success was the scientific method" (p. 144). He stresses "science is not scientism." The philosopher Susan Haack (2003) describes the term as meaning "an exaggerated kind of deference towards science, an excessive readiness to accept as authoritative any claim made by the sciences, and to dismiss every kind of criticism of science or its practitioners as an anti-scientific prejudice" (pp. 17–18). Writing earlier, Nadeau and Désautels (1984) would have agreed with her general statement and Bauer's point, and see it consisting in two things:

> Scientism is thus, in our view, both the attribution to scientific activity of an exclusive claim to legitimacy and the active belief that science itself justifies its particular status, because a valid answer to any question pertaining to it can only be found through the application of its comprehensive method. (pp. 13–14)
>
> Scientism clearly runs counter to the genuine legitimacy and value of science. (p. 15)

Congruent with this exclusivist epistemic and methodological claim (usually understood in strictly empirical terms), which is clearly a distortion, Duschl (1988) in describing scientism adds the inherent pedagogical *authoritarianism*: "a view in which scientific knowledge is presented as absolute truth and as a final form" (p. 51), along with the lingering legacy of *classical empiricism* and *logical positivism* as the cloaked epistemology behind curricula. The physics education researchers Roth and Roychoudhury (1994), alternatively, identify (and derisively label) the hidden classroom and student epistemologies as "objectivism."

Nadeau and Désautels (1984) insist the "epistemological factor" must be incorporated in science teaching to counter such myth-making in science classes. "In our view, the next step must be *to ensure that science teaching at the high school level is based on the objective of instilling in the students a critical approach to scientific activity, as opposed to scientific mythology*" (p. 15, original italics). Their monograph, although dated, still presents several relevant insights and useful pedagogical devices to counter such myth-making (itemized below). (One notes that their advice given over 25 years ago to consider the "epistemological factor" has not been heeded.) The Canadian science educator Glen Aikenhead (2002a), writing more recently and with explicit reference to their monograph, calls scientism an "ideology" embedded within science curriculum that acts as a kind of Trojan horse "by concealing its values when teachers attempt to enculturate students into Western science" (p. 151).[38]

The dangers inherent to scientism as existing in both Western culture and the Western school science curriculum should not be denied or minimized; this can be better illustrated by looking at two cases germane to one main thesis of this book (the need of developing a PSE), the first with respect to the world of academia, the second with respect to science in the modern world. To the first, Gadamer's own philosophical hermeneutic project in his magnum opus *Truth and Method* (1960/1975) can be understood as a sustained argument against precisely this ideology with its presumed inductivist methodology and "epistemology of objectivism" forcing itself illegitimately upon the humanities.[39] Gadamer in fact takes explicit issue with the claim of epistemic privilege. To the second, in glancing at the contemporary relationship between science and society, several science educators (following the criticisms of sociologists of science—now referred to as "science studies," being a recent research branch of science education), when recognizing the institutional and industrialized transformation of modern science into "techno-science" (Lacey, 2008; Sula, 2009), have warned against the naïve notion of a value-free and autonomous *asocial* science as perpetuated in schools due to the continued neglect of teaching the other dimension of *current science as a techno-social practice*. Lacey (2008) even argues that the "commercialization of knowledge for profit" has brought about a new ethos for science, one he calls the "commercial-scientific ethos" (p. 311). Jenkins, too, has written:

> The commercialization and industrialization of science and its integration with contemporary technology are also important for the values of science itself. The central point here is not that all those engaged in what may be termed industrialized science are concerned directly with making saleable or otherwise useful products. It is that the industrialization of science, which has gathered such pace since the Second World War, has led to a set of social relations, priorities and values that represent a new kind of scientific activity whose malignant features are not always readily distinguished from the beneficial and benign. (Jenkins, 1992, pp. 231–232)

The society sourced "scientism" so characteristic of technoscience, that some champion while others deride, is of a different sort than the classroom sourced kind often impugned by science educators, since it consists in taking no notice of the first sort—an irony (although both lack self-awareness, the ability to critically reflect on their own epistemological grounding.). Seen in this light alone, textbooks when serving as the primary source of CK must yield a one-sided and artificial representation of the modern scientific enterprise, contributing another feature to scientism, a kind of *blind idealism*:

> school science education must surely be more sensitive than it sometimes seems to be to the industrialization of science and all that this implies for its traditional portrayal of science as the disinterested pursuit of objective truth. The scientists of school science texts, Mendeleev, Joule, Scheele, Rutherford, Maxwell, Newton and Darwin, are not just deservedly famous individuals. They represent a kind of science that has little correspondence with most of the scientific research undertaken at the end of the 20th century. (Jenkins, 1992, p. 232)

While I take note of the wider cultural implication of this ideological worldview, what I nonetheless find too frequently lacking among insightful critics, especially Aikenhead and his talk of the "Trojan horse curriculum," is the acknowledgement of the need to be just as mindful of avoiding the inculcation of beliefs and attitudes of the opposite extreme, either one of *anti-science* or one of indifferent epistemological relativism. The latter is the standpoint that all cultural knowledge claims are on par (metaphysical, religious, legendary, metanarrative, populist, or what have you), science being merely one among many and possibly not even exemplary at that (Loving, 1997; Nola & Irzik, 2005). In our age we suffer from *dangers at both ends*, where the term "scientific" holds the dubious double connotation of being honorific as well as pejorative, depending upon one's prior views and allegiance to the camps of modernist or post-modernist thought and critique, which surfaced in the late-1990s in academia and the media because of the *Sokal hoax* and "science wars" (Gross et al., 1996), as stated previously.

Considering scientism we find that

> "scientific" has become an all-purpose term of epistemic praise, meaning "strong, reliable, *good*." No wonder, then, that psychologists and sociologists and economists are sometimes so zealous in insisting on their right to the title. (Haack, 2003, p. 18)

At the other end, "science is largely or wholly a matter of interests, social negotiation, or of myth-making, the production of inscriptions or narratives; not only does it have no peculiar epistemic authority and no uniquely rational method, but it is really, like all purported 'inquiry,' just politics" (Haack, 2003, p. 21). Haack wisely points out that a defense of science must carefully navigate the opposing shoals of scientism and anti-science, also typified by what she calls the stand-off between the "old Deferentialists" (especially the older logical-positivist school, but also Popper and Lakatos) and the "new Cynics" (Quine, Feyerabend, radical sociologists and feminists, rhetoricians, semiologists, and philosophers outside PS). A useful PSE, as suggested here, would need to judiciously proceed in a similar vein with its presentation of science for educational purposes, but this cannot be taken up here (Hodson, 2008, 2009; Kelly et al., 1993; Nola & Irzik, 2005; Ogborn, 1995).

Yet we must take leave of this worthy discussion involving modern science as techno-science with its implications (where STSE reformers have been more cognizant and argued for curriculum inclusion of socio-technological impact issues [Pedretti & Nazir, 2011]), as well as the anti-science quarrel, to concentrate instead on better illustrating "scientism" in the curriculum and the theme of *nature-of-science* (NoS) more directly.

Nadeau and Désautels associate *five* myths with scientism as manifested in school science curricula and classrooms:

1. blissful empiricism (knowledge derives directly from observations);
2. credulous experimentalism (experimental "proof" determines truth);
3. naïve realism (scientific knowledge as a direct reflection of reality);
4. excessive rationalism (only science brings us closer to truth);
5. blind idealism (scientists as isolated explorers, disinterested, objective).

These same five can also be identified among McComas' (1998) catalogue of *15 myths* attributed to school science, which teachers need to be made aware of and which should be addressed in classrooms.[40] Evidence to support the claim of the popularity of these myths of scientism among students (learner epistemologies) was accumulated in a national sample study of Canadian students by Ryan and Aikenhead (1992). Using their own newly devised "views of science-technology-society" (VOSTS) test-instrument of 114 items, they surveyed the responses of over 2,000 grade 11 and 12 students (urban and rural) to determine their NoS preconceptions and compared these to the five Nadeau-Désautels descriptors listed.

They generally discovered statistically significant agreement with the scientism claim: 22% held to blissful empiricism (theories are discovered from facts), while another 40% held to a mixture of inductive discovery *and* creative invention of theory (only 4% held that theories are creative inventions of mind); 22% held to "one method," while 31% believed "evidence proves a theory true," both indicators of credulous experimentalism. Another 42% held to "one method view" but also allowed for *creativity* of scientists; 19% held to naïve realism when asked about models in science (37% held they were "close to being copies of reality"). Only with "classification schemes" did 81% of students allow for human inventive character; 64% held to the simplistic and erroneous view that hypotheses become theories that become laws (McComas *myth #1*), which can be linked with excessive rationalism. Thereto, 36% held to a view that science progresses by *disproving* previous theories (akin to Popper), while 31% held to knowledge change and growth by *reinterpretation* (a view more akin to Kuhn, according to the authors). Finally, for blind idealism, 52% held that scientists are mainly unbiased and objective, and 47% rejected the idea of consensus-making when deciding whether to accept a theory (both non-Kuhnian views). The authors also found that students confound science and technology, a myth identified by McComas (*myth #14*) but not by Nadeau-Désautels. They write: "chances are great that when students talk about science, they are probably talking about technology, specifically medical and environmental investigations" (Ryan & Aikenhead, 1992, p. 564).

Several recent studies continue to reinforce many of these findings and the claim of scientism as manifested in school science (Lederman, 2007; Lederman & Abd-El-Khalick, 1998). The intensive cross-grade study by Driver et al. (1996) of British students (9, 12, and 16 years old) also yields considerable evidence of the Nadeau-Désautels thesis and several of the McComas myths. Ryan and Aikenhead (1992) conclude: "student naivete concerning the epistemology of scientific knowledge could seriously undermine the current attempts to increase scientific literacy" (p. 572), concurring with Gallagher's previous view (1991) that the portrayed image of science for students is "both inaccurate and inappropriate" (p. 132). Driver et al. (1996) have pointed to studies that reveal how kinds of teacher epistemologies influence students' understanding of NoS, and depending on how *language* is used, students *can* be helped or hindered to develop more sophisticated understanding (see Merzyn, 1987). (The role of language will be addressed in Chapter 5.) Although teachers tend to be "eclectic" in their perspectives "they have not had opportunities themselves to reflect on and clarify their own views on the subject." Rather "the dominant picture of science lessons" that emerges "is of teachers tending to represent science as a body of facts together with a set of mechanical empirical processes" (p. 149). Their own observed research experiences in classrooms "suggest that current teaching practices are portraying a limited perspective" of NoS, while the English National Curriculum presents "a restricted epistemological perspective" (p. 143).

4.6. EPISTEMOLOGY AND PHILOSOPHY OF SCIENCE: OBJECTIVISM AND REPRESENTATION

When re-examining the Nadeau-Désautels' myths of scientism, several of these can be linked with well-known views in the older philosophy of science. Thus, "blissful empiricism" corresponds to classical inductivism, and both this myth and the next "credulous experimentalism" taken as verificationism indicate parallels with classical logical positivism. It is not altogether surprising then, that when science educators wish to denounce school science epistemology they often refer to it derisively as "positivism," which has become a catch-all phrase (see Roth & Désautels, 2002; Burbules & Linn, 1991). Ryan and Aikenhead (1992) are typical, and their view gives insight into their own interpretation of their survey instrument: "Student views that converge with Barnes, Holton, Kuhn, Snow, or Ziman are considered to represent a *worldly* perspective. Views that diverge from this contemporary literature are thought to be *naïve*. Naïve views are often identified with logical positivism" (p. 561). Unfortunately such labels are misleading and inaccurate and science educators need to be more circumspect. Scerri (2003) serves the same warning to chemistry educators. "Naïve realism" cannot be assigned to positivism, for it was *anti-realist* and instrumentalist when it came to theories and entities. (Such aspects of positivism, including its strong anti-metaphysics bent, can in fact be more aligned to epistemological relativism of current postmodernist thought, which assumes anti-realism and takes truth and knowledge to be relative to a socio-conceptual frame).

What is more disturbing is that too often science education critics do not distinguish naïve from *critical* realism—a viable and eminent school of thought in current PS. (Here the nature and status of scientific models, theories, and entities are debated and of great importance [Laudan, 1990].) Moreover, if "credulous experimentalism" is taken as instances of Popperian falsification, it must *exclude* positivism. Indeed, this myth, adding to it "excessive rationalism" and "blind idealism" together show some surprising similarities to Popper's PS. One could no doubt also find some affinities with simplified versions of Lakatos' views concerning *myths #2, #3 and #4*. Hence, it would be more accurate if science educators associated scientism more closely with several positions of the "old Deferentialists" (Haack) or what has been termed "the received view" of science (Hempel, 1966; Suppe, 1977). The philosopher of science Ladyman (2002, p. 95) lists *seven* attributes that Popper, the logical positivists, and logical empiricists all shared:

1. science is *cumulative*: progress of science is a steady growth in knowledge;

2. science is *unified*: there is a single method; sciences are *reducible* to physics;
3. logic of confirmation or falsification; such evaluations are value-free;
4. sharp distinction between observational and theoretical terms; observation and experiment provide a neutral language for the testing of theories;
5. scientific terms (language) have a precise and fixed meaning;
6. epistemological distinction between the context of discovery and justification;
7. sharp demarcation exists between scientific and other belief systems.

Clearly most myths of scientism, especially if one includes those of McComas, can be associated with one or the other of these stated attributes. More importantly, the coherence of these attributes and attitudes can function as a single "scientistic" worldview. What is also obvious is that Kuhn's and Feyerabend's philosophy of science rejects all of them (including critical realism), and insofar as the science curriculum assumes or implies or outright accepts one or several of these theses, it can be rightly accused of harboring an outdated and discarded PS. Modern PS has truly entered a post-Kuhnian phase (and there can be no doubt he symbolizes the breaking point; Nola & Sankey, 2000), albeit it remains divided and disturbed regarding the worth of his central theses. I submit a better case can be made designating the epistemology of school science as well as that of the "received view" in PS, which underlies scientism, as *objectivism*.[41] By this is meant the epistemological standpoint with its tenets at the core of Enlightenment philosophy as it developed in tandem with the advancement of science since Galileo and Descartes in the 17th century:

> The idea of a basic dichotomy between the subjective and the objective; the conception of knowledge as being a correct representation of what is objective; the conviction that human reason can completely free itself of bias, prejudice, and tradition; the ideal of a universal method by which we can first secure firm foundations of knowledge and then build the edifice of a universal science; the belief that by the power of self-reflection we can transcend our historical context and horizon and know things as they really are in themselves. (Bernstein, 1983, p. 36)[42]

Certainly modern philosophy since Descartes has wrestled enormously with the question of "how to know the real?" and delineating "what is real?" (ontological problem of *metaphysical realism*), and it is well-known that the Kantian tradition had prohibited knowledge about the "thing-in-itself" and severely curtailed the ambitions of empiricism. (Rockmore, 2004,

p. 17 notes that Kant represents a *shift* from ontological to epistemological realism, often unacknowledged in the Anglo-analytic tradition.) Still, modern epistemology as a theory of knowledge was essentially a *theory of representation* (Rorty, 1979), and various schools of thought sought to enunciate how knowledge could best be grounded, what kind of representation would be appropriate, and how it could be secured. Both rationalism and empiricism (playing out the ancient tension between intuition and observation) were greatly influenced by the developments of modern science to look for answers to metaphysical realism and the epistemological quest. The rationalist tradition (e.g., Descartes, Leibniz) not surprisingly saw in mathematics and mechanical physics a model for attaining sure knowledge (hence the primacy of "reason" and the *a priori* access to truth), while empiricism (e.g., Locke), and much later logical positivism-empiricism (also Russell and the early Wittgenstein), remained very skeptical of theoretical constructs and looked to the senses, observation, induction, and the experimental sciences and their method as a better foundation (primacy of sensory experience and the *a posteriori*). With the collapse of *foundationalism* generally—as Taylor (1987), Rorty (1979), and Gadamer (1960/1975) contend, although Rockmore (2004) admits important defenders are found today—representational realism as common to both traditions has suffered credibility along with the *correspondence theory of truth*.

A paradox is revealed by the fact that the evolution of science—itself a curious synthesis of rationalist and empiricist ideas and attitudes (no doubt contributing to its immense success)—that helped shape debates in philosophy (still occurring today in PS) has itself remained relatively immune to clashing developments in philosophy in the past 300 years, above all the quarrels in epistemology over foundationalism and the status of knowledge:

> It is at least interesting that foundationalism is not accorded attention in cognitive disciplines other than philosophy. For instance, it plays no visible role in modern science, where everything happens as if foundationalism were an irrelevant consideration. Neither Galileo nor Newton nor Einstein nor indeed any reputable scientist in modern times has ever tried, or even voiced the concern, to put science on a foundationalist epistemological basis. Yet philosophers deeply acquainted with science, such as Descartes, Kant, and Carnap, have made this effort, as if science without an epistemological foundation was somehow incomplete. (Rockmore, 2004, pp. 45–46)

This is not to say scientists have been entirely unconcerned about epistemological issues and the nature of knowledge, as the disagreements over realism and instrumentalism in the long history of science clearly show us (to be discussed in next section). Scientists have quarreled not only with philosophers and theologians but also among themselves, particularly for those individuals wearing the dual philosopher-scientist hat (e.g., Galileo, Descartes, Leibniz, Priestley, Mach, Helmholtz, Eddington, Bohr, Heisenberg, Eger, Gould, and others). Although it may be true to assert as Rockmore does that science has been in the main oblivious to the foundationalist project as taken up primarily by philosophers, it has not been unmindful when it came to ascertaining the status of scientific knowledge, and here science and philosophy exhibit a rich and reciprocal history, well-described by Cushing (1998) and Westfall (1971). And the tenets of objectivism can be recognized as the common and often hidden background to most philosophizing.

The ethos of modern science, unlike epistemological foundationalism, is purposely *fallibilist* (the other exception being pragmatism)—hence subject to correction on empirical grounds—yet this fact has not stopped "the modern debate on knowledge, perhaps even the analytic debate throughout the 20th century, [to] mainly consist of a series of variations on Platonic themes" (Rockmore, 2004, p. 29). In a similar vein Ogborn (1995, p. 19) refers to this line of thought as the "rationalist programme" and "rationalist ideology." There seems little doubt that a Platonic notion of knowledge as "direct access to things" (and as Rockmore notes, quite often ignoring the Kantian epistemological revolution) inhabits both Enlightenment and scientific philosophy—although *falsely* projected onto science—including science CK: the claim to attain immutable and certain knowledge, those facts, entities and laws assumed *not* subject to change and historically contingent, and that somehow when it comes to truth—especially about the natural world—science manages to ultimately "get things right." Such a position can be voiced in stronger and weaker forms, and most philosophers of science (inclusive of positivists as unrepentant empiricists, though restricted by them to the observation language) including Popper would argue that science "tracks truth" in some way, the weaker form only admitting that scientific theories "approximate truth." Such a position in whatever version is called *convergent realism*. One immediately identifies several myths of scientism discussed earlier here, namely "excessive rationalism" and "naïve realism"—although depending upon how the argument is framed and the issues understood, there need not necessarily be anything excessive or naïve about such views; the myths could signify degenerate and overly simplified views of legitimate objectivist positions. The Nobel Prize physicist Weinberg (1992) asserts that unless science is getting closer to truths about nature, it must remain condemned as an irrational activity.

There has always been the concern as to how *certain* scientific knowledge could be: "Are the laws and theories of science nothing more than one contingent way to describe the world, or do they yield truth?" (Cushing, 1998, p. 12). The school curriculum as asserted here exhibits an unconscious and unbroken allegiance to Platonic realism underlying its presumed objectivist notion of knowledge, taking it matter-of-fact to be *ahistorical* and beyond significant change, and hence why *epistemic content* (CK) can be laid out the way it traditionally is in science textbooks of all kinds: technical scientific knowledge (be it Snell's law of refraction, motion, chemical bonding, cell theory or cellular respiration, etc.) is assumed to "stand above things" and therefore can be abstracted out of the historical matrix, structured without context, and "known" without any hint of its developmental history—the historical context being thought irrelevant. One can call this *the fallacy of Platonic realism that justifies epistemic content knowledge*.

Duschl et al. (1990) correctly emphasize that CK and any related discussion of a particular theory of science (e.g., plate tectonic theory) is usually taught from the ahistorical *context of justification* mode (without regard to predecessor theories) instead of the historical *context of discovery* mode (which includes alternative predecessor theories and their controversies). Such an instructional *staging*, taking a Kuhnian view, is thus made strictly within the bounds of paradigmatic normal science, which is quite common and how most material is presented. Moreover, this staging assumes knowledge objectivism. Not only is such an instruction "epistemologically flat," it is usually accompanied by historical versions that are highly over-simplified and often outright inaccurate (Allchin, 2000, 2003).

If objectivism as a worldview inherent to both philosophical foundationalism and scientism is to be abandoned, what other viable alternatives now present themselves, other than skepticism and relativism? What better conceptions of what "knowledge" is, and entails, can be found (Mason, 2003)? There are some thinkers who now argue our very idea of epistemology must be forsaken while the time-worn notions of "reason" and "knowledge" are to be radically reconceptualized. Some in science education have seen an alternative in philosophical *constructivism*, but this move is highly contested (Matthews, 1998a).[43] Others have suggested this could mean a *turn* "from epistemology to hermeneutics" as Rorty (1979) and Gadamer (1960/1975) have argued—this anticipates the next chapter. There remain other philosophical and literary traditions that can be looked to and drawn upon.[44]

It is no coincidence that the fallacy of Platonic realism dominates science education, for it is widespread in academia:

> The modern version of Platonic realism is widely represented in the writings of modern philosophers and working scientists committed to a grasp of the mind-independent external world. Long after Kant, Platonic realism survives in our time in the persistent, stubborn, dogmatic, indemonstrable but widespread commitment by philosophers, philosophers of science, and scientists alike, distantly following the Cartesian form of metaphysical realism, to claims to know the mind-independent external world as it is in all its many forms. (Rockmore, 2004, p. 32)

Having assumed to have correctly grasped the "external world" as behooves objectivist prejudice—to whatever *degree*, especially assuming the objects of representational realism are largely suitably mirrored—then the play of scientific words, definitions, symbols, equations, and processes as constitutive of the bulk of curriculum can be learned for themselves in a decontexualized fashion and socio-cultural and historical vacuum. Though textbooks and reform standards-type documents certainly often parrot the "tentativeness of scientific knowledge," this seems mere lip-service compared to how textbook epistemic knowledge and classroom learning are structured.

Moreover, the reader has hopefully recognized that certain other tenets of objectivism as quoted above (by Bernstein) have clearly found a home in the science curriculum: the "subject-object dichotomy" that lies behind the transmission language model of instruction and conceptions of learning (including typical subjectivist-based cognitive learning models); the "ideal of universal method"; the notion of a "unified universal science" (ultimately reducible to physics) witnessed to by the very conception of the name of our discipline itself using the all-encompassing term "science" education; lastly, the belief of being able to "transcend the historical context"—the last point refuted by Gadamer's insights about the nature of history in philosophical hermeneutics.

We come now to another key tenet of objectivism quoted above, the crucial topic of the "correct *representation* of what is objective," that is, the representational nature of scientific laws, models, and theories. The assumption that our constructs "represent well" is taken for granted in all science courses—the *historicity* of these items masked by both the *language* employed and the *logic* of disciplinary structured edifice (Kuhn, 1970; Sutton, 1996). But this overlooks the fact that these representations (theoretical constructs) have frequently undergone a transformative process (in symbol, concept, and meaning) during development and revision in scientific communities (e.g., gene, animal, atom, electron, planet, element, mass, law of freefall, Ohm's law, and many, many others) and some remain objects of much contention in current PS (realism/instrumentalism debate). In the physical sciences they have become increasing abstract and removed from the everyday world of both the layman and the learner. Some have even been led to conclude that *representationalism* has been abandoned in science itself:

> Representation is always representation of something else. The Cartesian view of the scientific object banishes sensation in favor of geometrical schematism, which also holds in Newtonian mechanics. It later gives way to field theory, in which (when things are replaced by an aggregate of physical relations) representation takes the form of a purely symbolic relation. In the latter case, complex mathematical relations refer to or symbolize, but no longer represent, an aggregate of properties as distinguished from things.

> In the standard, or Copenhagen, interpretation of quantum mechanics, the so-called measurement problem suggests that the very idea of an independent object must be abandoned. Yet at a time when epistemological representationalism has lost its importance in science, it still remains central in philosophy. (Rockmore, 2004, p. 90)

The conclusion seems highly premature, and Giere (1999) has argued to the contrary that representation understood as a modeling task in science can proceed independent of foundationalism (see further Section 5.1.1). Quantum mechanical (QM) field theory has surely pushed the limits of traditional representation, but its epistemological interpretation with respect to realism remains very contentious, to say the least (Cushing, 1998). Yet the irony should not have escaped Rockmore that while QM appears to put into serious doubt several of the aforementioned tenets of objectivism, the physicist Weinberg, whom he correctly chides, still holds openly and proudly to his naïve realism. Nor should problems in micro-physics be taken to stand for the situation of the sciences in toto.

Without question, laws, models, and theories as sophisticated and highly *explanatory* representational (and idealization) modes of reasoning remain the pride of science and the hallmark of its success, and hence *representationalism* in its many forms needs to be treated with the utmost respect in science education, even so, if one just wishes to accept them as heuristic devices (as tools) and leave the contentious realism aspect and question to the side. To dismiss them simply as "naïve realism" (as too many critics, including Aikenhead and Roth are wont to do) is to do a disservice to both science and science education, where the latter, unfortunately, too often fails in properly accounting for the value and differences of laws, models, and theories for learners (Hodson, 1991). Let us now briefly articulate this.

For my part, I take it as self-evident that to build a scientific mind requires in large part the ability to *explain* phenomena and *justify* one's explanations. This is exactly where modeling and understanding the role of theories must come to the fore. Both *explanation* and *argumentation* are taken to be at the epistemic core of scientific reasoning and the research on the value of understanding theory change and modeling in learning science (Gilbert & Boulter, 2000; Matthews, 2007) complements the work of those who suggest that it directly reinforces critical thinking, that is, the quality of logical reasoning and argumentation discourse in the science classroom (Giere, 1991; Osborne, Erduran, & Simon, 2004), as well as better illustrating the character of historical conceptual revolutions (Duschl, 1990; Nersessian, 1989, 1992, 1995, 2003).

An understanding of scientific explanation should encompass fundamentally *four* aspects: insight into the explanatory role of laws; thereto, secondly, insight into the *conceptual structure* and *historical establishment* of a *theory* (including the empirical laws it can deduce and the rational nature of theory change, Duschl, 1990); thirdly, the kinds and functions of *models* it contains, and hence, finally, the nature of scientific *reasoning*. Scientific explanation in its widest sense is often taken to mean "theorizing," but this is commonly construed as primarily inferring hypotheses, creatively constructing models, and assessing (testing) their individual predictions (Giere, 1991 and Figure 4.1).

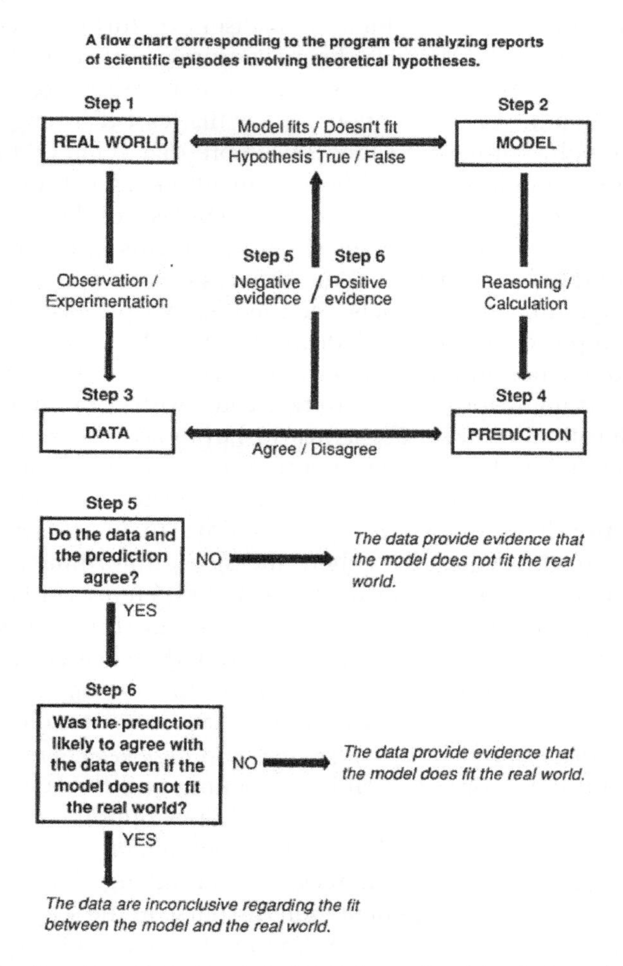

Source: Taken from Ronald N. Giere's (1991) *Understanding Scientific Reasoning* (3rd ed.) © www.cengage.com/permissions

Figure 4.1. Scientific reasoning schematic.

Though this partial picture is true and certainly helpful, it overlooks the place of prior theory and the *theory-ladenness* of data collected in the inquiry process. A theory in itself (in the physical sciences at least) is a complex conceptual creation that is (usually) comprised of a set of models, each restricted to specific domains intending to capture and abstract a given aspect of reality (e.g., Newtonian classical mechanics employs various *particle* models—free particle in uniform motion in the limit of low-velocities; uniformly circulating particle subject to a net centripetal force, etc.; Halloun, 1996).[45] As a rule theories cannot be tested directly, only indirectly through the use or application of their models. From an epistemological perspective models are said to *mediate* between theories and reality. On this view models are *subordinate* to theories, but what they lack in broader explanatory power they make up for in their ability to present immediate testable consequences. (In the normal course of teaching the subject matter of a given theory, instructors should explicate these differences.) Students then, must also come to recognize these four aspects of explanation. They can be provided with opportunities to explore theories and laws, and in appropriate laboratory and field settings can confirm *established* laws and models, occasionally even test *their own* hypotheses and models.[46] Theories though, usually remain very much outside any such classroom possibility.

On the other hand, one way assessment of theories can be undertaken is by using a *historical approach,* by choosing studies of those examples in the past where rival high-level theories have confronted each other. In those rare cases where theories clash (e.g., Ptolemaic or Keplerian astronomy; Newtonian or Einsteinian physics; phlogiston or Lavoisier's chemistry; young-earth theories or evolutionary geology, etc.), the particular limits and defects of their respective models become sharp (Dolphin, 2009; Nersessian, 1989, 1995; Niaz, 2009, 2010). Such instances can serve as an effective teaching strategy to help students discover how knowledge can progress, and thus illuminate for them the precarious nature of how science actually advances. Such ideas present a viable alternative to the reigning scientism of the curriculum and help illustrate the distinction among (and proper roles of) laws, theories, and models.

Duschl (1990) argues that because theories are the primary explanatory engines in science, a study of their history, structure, evaluation, and change should serve as the guiding cognitive framework for the disjointed and over-crowded information-based epistemic content arranged solely around subject topics (e.g., light, heat, motion, bonding, reactions, body systems, etc). "It is a common mistake of laypeople [and the media] to equate scientific theory with scientific fact" (Duschl, 1990, p. 7). He asserts that unless students begin to understand how theories are rationally debated, displaced, and confirmed by the scientific community, they will fail either to see science as a rational enterprise or to respect the nature of scientific knowledge. (Sometimes these disputes can take decades: plate tectonics about 60, and QM about 25 years [Cushing, 1998; Dolphin, 2009; Niaz, 2010]). Based on HPS research "an understanding of the growth of scientific knowledge is best obtained through an understanding of the development of scientific theories" (Duschl, 1990, pp. 7–8). Drawing substantially on several thinkers like Lakatos and Laudan, and avoiding many of Kuhn's theses, he draws up six *epistemological* models to characterized knowledge growth and theory change in science: goals-of-science hierarchy, levels of theories, argument pattern for testing theories, four criteria for evaluation, triadic network, and tripartite process. Especially useful are his four criteria (logical, empirical, historical, and sociological) whereby theories can be judged and one can distinguish among core, frontier, and fringe theories. Furthermore, he has provided several helpful examples for teachers of how epistemological frameworks can be applied to actual historical case studies (theory of periodicity of elements; motion theory; plate tectonic theory).

Duschl's presentation is not of the usual cumulative sort, and students can discover the nature of theory change without failing to grasp the overall rational nature of scientific reasoning, of argumentative discourse, and deliberation. His is among the most elaborate and useful epistemological-based science education yet developed. (It is useful because it explicates an authentic epistemology of science that teachers can incorporate; however, because it is strictly epistemological it suffers the failing of ignoring educational metatheory.) It can be taken to directly challenge and overcome scientism, especially those tenets of objectivism that present content knowledge predominately as static, highly durable, and final, as ahistorical and thematically isolated. One epistemological drawback is his failure to properly account for the nature and relation of *models* with the theories he discusses. Science educators today have shifted from solely theory explanation to the role of models and scientific argumentation (Gilbert & Boulter, 2000; Matthews, 2003b, 2007; Schulz & Sivia, 2007; Osborne et al., 2004).

In the analysis of the function of theories and models in science as useful for science education, one needs therefore, once again, to discriminate between the common "knowledge-justification" framework and the neglected "knowledge-development" framework and to reinforce the latter. (This distinction is often compared, though in a rather crude and inaccurate manner, to Kuhn's two aforementioned distinctive science stages, "normal" and "revolutionary.") Regrettably, this is usually not done, not even by those stressing modeling for science education. This is not clarified in science education generally (for instance, models for "light rays" or the "Bohr atom" are taken as historically static and as direct visual representations of reality); Kuhn, too, says little about its worth for theory change and revolutions.

On the other hand, the philosopher of science Malcolm Forster (2000) distinguishes among *three levels of theorizing* that form a hierarchy: at the top are fully developed (and paradigmatic) *theories*, which govern a science; then *models* form the middle tier, which allow for the concrete applications of the theory; and *predictive hypotheses* of the model at the lowest tier, which allow/disallow for fitting the model to the data (building on Giere, see Figure 4.1). He also argues, when considering the contentious question of the ontology of theories, that there exists a "trade-off" of *truth* against *predictive accuracy* as one moves up the levels. Scientists tend to incline towards instrumentalism with their models but towards realism with their theories.

Conventional science education, we repeat, is preoccupied with disseminating the established theories of this stable stage. This brings us foursquare to the question of the epistemic status of theories.

4.7. *PSE CASE STUDY:* THE REALISM/INSTRUMENTALISM CONTROVERSY IN PHILOSOPHY OF SCIENCE AND ITS VALUE FOR SCIENCE PEDAGOGY

The science education literature has earlier addressed aspects of the realism debate (Eflin et al., 1999; Hodson, 1991, 2008; Kelly et al., 1993; Matthews, 1994; Slezak, 1994a, 1994b), especially with regard to criticisms of radical constructivists (e.g., Roth & Desautels, 2002; Von Glasersfeld, 1989) by writers like Matthews (2000), Nola (1997), and Loving (1997). Recently science educators have seen fit to defend realism again and published substantial reviews (Cobern & Loving, 2008; Ogborn, 1995). What surprises is that the position taken there openly espouses "commonsense realism," which Rockmore (2004) would criticize as a return to a version of "direct realism" (or "intuitionalism") akin to, he argues, phenomenology and Greek thought, which is widely held as untenable today. I agree with this assessment and will not address this literature, nor concern myself with the constructivism controversy, but wish instead to seek to ascertain the degree to which *critical realism* can be defended by going directly to the realism/instrumentalism debate as discussed in philosophy of science.

Realism as a philosophical doctrine is not overtly discussed in science education curricula, however, and usually two extremes of naïve realism and instrumentalism can be identified, as mentioned (Hodson, 1991; Selley, 1989). Naïve realism tends to dominate and is displayed in two general ways: often *explicitly* (asserting genes, black holes, or tectonic plates "exist")—but not obviously (when texts misleadingly state "the electron was *discovered* by J. J. Thomson in 1897" in a "crucial experiment" yet ignore the "crucial" prehistory)—or *implicitly* when deliberately making the epistemic slide across the theoretical construct/observable divide, with models taken too literally, as cited previously. Instrumentalism occurs infrequently, depending on the nature of the subject matter, although it usually manifests itself when a strong version of empiricism ("positivism") and thereto when measurements and data are overly stressed, models and theories taken as entirely fictive, and the status of theoretical entities are undervalued. Teachers and curricula could present it, for example, when discussing chemical bonding, fields, wave/particle duality, or the quasi-historical change of atomic models, including quantum mechanics (Elkana, 1970).

The realism/antirealism debate in PS is complex, taking different forms in different branches, and can appear convoluted to those of us entering the discourse from outside that discipline, though even philosophers of science have themselves admitted "these debates are often Byzantine and confusing" to those working in the field (Eflin et al., 1999, p. 114). It will be granted that the science educator has all the more reason to proceed with caution, especially since it can be acknowledged that students "on first exposure misunderstand antirealism and instrumentalism," but that does not warrant the claim that therefore "debates about realism should be avoided, and that a naively realist view is most appropriate for science education" (Eflin et al., 1999, p. 114). Such a suggestion is energetically rejected for at least *three* reasons: the inherent naïve realism and inductivism in the curriculum and the misunderstandings with respect to NoS that necessarily follow must be countered; this is based on the argument, secondly, that the debate is integral to how science has developed and its value been interpreted (mirrored by the "intelligent design" versus evolution dispute; Brockman, 2006), and thus its neglect would seriously impede proper comprehension; lastly, within the science education research community certain detrimental antirealist views have become ingrained and need to be challenged. Currently weaker and stronger forms of realism have been expressed, but it is my view (along with others) that a *modest or critical realist* version can be articulated for science education that, on the one hand, attempts to do some justice to the intricate debates within PS and yet, on the other, can be presented in an appropriate form to teachers and students.

Matthews (1994) has stated *four general theses* of such a position. He claims further that "[it] is enough to go with, and it is incompatible with empiricism, with constructivism and particularly with idealist forms of radical constructivism" (p. 177). One can take issue with his view that it *must* be incompatible with constructivism *per se*, and his claim that the arguments against "verisimilitude" have been countered (see below). His *four theses* are: (1) theoretical terms in a science *attempt* to refer to reality (uncontested by realist and antirealists alike); (2) scientific theories are confirmable (note he correctly avoids the use of problematic terms like "verified" or "falsified"); (3) scientific progress, in at least mature sciences, is due to their being increasingly true (the doctrine of "approximate truth" or *verisimilitude*, which is strongly contested); (4) the reality that science describes is largely independent of our thoughts and minds (also termed "metaphysical realism," and also contested).

The realism/antirealism controversy is actually an old one in PS, going back to the logical positivists (instrumentalists) of the Vienna circle in the 1920s—what can be considered the beginning of the sub-discipline itself (Giere, 1999). And even earlier to French thinkers such as Duhem (instrumentalism),[47] Comte (positivist-phenomenalist), and Poincaré (conventionalism), as well as German scientists of the late 19th century such as Planck (realist) or Kirchhoff and Hertz (phenomenalists).[48] These latter held to the view that "the job of physics was to provide a complete, accurate and simple *description* of the phenomena of nature, rather than an *understanding*, either in terms of models or in terms of principles a priori necessary or self-evident to the human mind" (Cushing, 1998, p. 367, italics added).

Moreover, the historical record of scientific development (HoS) equally exhibits this controversy, going as far back as the Copernican revolution (Kuhn, 1957/1985). The Ptolemaic and Tychonic (earth-centered) theories, strongly held to be "real," nonetheless permitted the Copernican (sun-centered) theory as a useful computational (anti-realist) model for over a century (i.e., "empirical adequacy"). The subsequent theoretical successes of Kepler, Galileo, Newton, and classical mechanics in general convinced many of the metaphysical and epistemic realism supposedly inherent to scientific theorizing and methodology for over 200 years. Yet at the turn of the 19th century the debate over atomism in physics and chemistry revived anti-realism (e.g., Planck versus Mach; Toulmin, 1970). The famous Russian Mendeleev, co-creator of the Periodic Table (along with the German chemist Mayer), opposed attempts (unlike today) to reduce the periodicity and nature of physical elements to atomic structure and even dismissed radioactivity and the electron's existence. Ostwald in Germany, the most well-known chemist (and textbook writer) of his time, rejected atomism as late as 1905 (Scerri, 2007)—ironically at the exact same time as physicists were proposing atomic models with *sub-atomic* particles (e.g., Thomson and Nagaoka). Then just as the majority of scientists, especially chemists, had reluctantly come to accept the existence of atoms (and electrons), the controversy reared up again and was reshaped due to Einsteinian relativity (in 1905 and 1916) and the conceptual and methodological revolution of modern quantum mechanics (QM) in the 1920s. It is well known that realist-inclined scientists like Einstein and Schrödinger (one of the founders of QM) strenuously opposed the instrumentalist (positivist) conception of QM as would become officially endorsed in the so-called "Copenhagen interpretation" by Bohr, Heisenberg, and others (Cushing, 1998; Toulmin,

1970). The epistemic status of QM remains unresolved and continues to engender rigorous debate among scientists and philosophers alike.

Within the PS community itself the scientific realism debate, although occasionally dismissed as sterile or even "dead," has experienced a remarkable revival in the 1980s and 1990s (Boyd, 1983; Psillos, 2000). This has been due to the abiding force of anti-realist arguments put forth by such thinkers as Quine, Kuhn, and Feyerabend in particular (in the 1960s and 1970s), being lately reinforced by arguments of Bas van Fraassen (1998) and Larry Laudan (1998a), and, on the other side, the concerted efforts by several authors, such as Boyd (1983), Devitt (1997), Aronson, Harré, and Way (1995), and Psillos (1999), to challenge and rebut them.

The upshot of the preceding overview is to emphasize not only the longevity, vitality and unresolved nature of this abiding controversy, but to equally help guard against certain authors (e.g., philosophers, postmodernists, scientists, sociologists, feminists, and cultural critics), science educators, and popular science writers who would dare declare the debate either dead or decided (in one direction or the other). Thus their attempt to claim the authority of thinkers like Quine, Kuhn, Feyerabend, Van Fraassen, Latour, and so on to help buttress their own anti-realist bias, or conversely, claim Boyd, Devitt, Psillos, the objectivist tradition (and even the views of scientists themselves, like the biologist Wolpert, 1992, or physicist Weinberg, 1992) as evidence to support their exaggerated claims for scientific realism. The former instance has certainly happened with some authors arguing for *science education as social action* (Roth & Désautels, 2002) or "culturally appropriate science" (Aikenhead, 1997a; Loving, 1997; Matthews, 1994; Zembylas, 2006) or for radical constructivism. One can concur with Matthews (1998a) that prematurely tipping our hat towards either camp, although tempting, would conflict with our role as educators.[49] This does not mean we must choose to remain agnostic or indifferent on this crucial subject. It *does* mean science teachers must be, at minimum, cognizant of the arguments and historical clashes pertinent to the two opposing positions.

It should be clarified that most people are not anti-realists about "observables" (such as dinosaurs, planets, volcanoes, bacteria, cells, etc.), often referred to as "common sense realism" or "everyday realism." Usually there is little dispute here, although some writers such as creationists try to question the accepted Darwinian theoretic *explanations* of certain observables, kinds, and classes. (And there can be considerable debate about *kinds* as interpretive categories: e.g. "species," "intelligence," "gene," "memory," "race," "gender," etc.) And of course there always exists the possibility, in fact it occurs quite frequently, of differences within disciplines when empirical data is scrutinized, be they among medical researchers, archaeologists, cosmologists, paleontologists, anthropologists, or what have you. But these are not cases questioning scientific realism, which is commonly taken to mean belief in *unobservables* or theoretical entities and laws.

Because of Kuhn's influence and popularity it is perhaps appropriate to restrict our examination for the sake of brevity to his claims concerning anti-realism.[50] Kuhn aims to defend the thesis, quite contrary to what many people and most scientists hold, that science does *not progress,* or is *not progressively objective*—in other words, that modern theories do not present a truer picture about nature—*across revolutionary divides* (or "paradigm shifts," when radical conceptual change occurs as theories are discarded and replaced), because historical research "gives ground for profound doubts about the cumulative process" (Kuhn, 1970, p. 3). This has led to accusations that his analysis suffers from *epistemological relativism* (Bird, 1998; Siegel, 1987).[51] As a preliminary comment one must distinguish in Kuhn's method his *recourse to history* from his *philosophical interpretations* of his historical case studies and examples. It is mainly the latter that have engendered considerable dispute, although the former has also come under scrutiny (Brush, 2000; Pyle, 2000). Another point is that relativists can be "hoisted on their own petard" given that their methodology presupposes those very aspects of science that they question and seek to undermine.[52] In his famous 1970 *Postscript* response to his critics (with an oft-quoted passage), Kuhn reinforced his earlier (1962) theses in *Structure* about NoS:

> There is, I think, no theory-independent way to reconstruct phrases like "really there"; the notion of a match between the ontology of a theory and its "real" counterpart in nature now seems to me illusive in principle. Besides, as a historian, I am impressed with the implausibility of the view. I do not doubt, for example, that Newton's mechanics improves on Aristotle's and that Einstein's improves on Newton's *as instruments for puzzle-solving*. But I see in their succession no coherent direction of ontological development.... Though the temptation to describe that position as relativistic is understandable, the description seems to me wrong. Conversely, if the position be relativism, I cannot see that the relativist loses anything needed to account for the nature and development of the sciences. (Kuhn, 1970, pp. 206–207, italics added)[53]

Several salient points are revealed here: first, Kuhn's attempt to avoid the awkward relativist charge has proven unsuccessful, and what the relativist "loses" is precisely what is disputed. Realists argue that much is lost with respect to explanation, rationality, and objectivity in science. Secondly, the wish to shift the meaning of "progress" in science strictly towards *empirical success or adequacy*—that theories are to be evaluated solely as *instrumental* tools (a typical anti-realist move, recently bolstered by Van Fraassen). This entails that the theoretical entities and/or laws that theories contain (e.g., atom, electron, field, element, gene, species, etc.) are fictitious and *do not refer* to real objects.[54] Thirdly, Kuhn's position resembles the arguments of others, including postmodernists, sociologists, and philosophers who have made the "interpretative turn" (e.g., Rorty, 1979)—those who explicitly reject the *correspondence theory of truth*. They seek instead to substitute ontological representation with constructivism, and truth correspondence either with coherentism (or *meaning holism*) or truth pragmatism. However, these are equally disputed positions (which unfortunately cannot be addressed here).[55] Fourthly, there is exhibited a strong tendency to view theories and scientific knowledge as tightly correlated with the *mental constructs* limited to scientists (or the scientific community). It is one thing to say a scientist's perception is constrained to the limitations of his theory's perspective, its underlying ontology, and community tradition. It is quite another to insist that a world independent of scientific belief, perception, and community (theory, theory-laden data, epistemology, and sociology) is impossible to ascertain. As Nola (1988) has pointed out: "a realist can well agree with the claim that we are active makers of our theories but reject, correctly, any implication that we are active makers of what our theories are about" (p. 13). Lastly, note that Kuhn exploits the *historico-inductive argument* (a kind of inference to the best explanation, IBE) using historical evidence to disconfirm realism.

In general, Kuhn and other authors often employ *three arguments* to buttress their anti-realist claims: *anti-rationality* (that choices for paradigm change are not firmly grounded in epistemic reasons but largely determined by social factors and power, or personal idiosyncrasy, aesthetics, persuasion, etc); *incommensurability* of rival high-level theories (meaning holism, meaning variance; logical non-reduction; non-translation; indeterminacy of data); and the aforementioned *inductive historical* argument. These arguments, though related, can be separated and dealt with individually.

It was Kuhn himself who later came to differentiate the first two arguments.[56] He had been stung by the "relativist" and "anti-rationalist" charges and sought to reaffirm to his skeptics (and later *against* some sociologists of knowledge; Kuhn, 2000; Nola, 2001) that his insistence that the scientific enterprise as he had conceived it was indeed *rational* and hence in its special way "objective" because of certain universal "values" or "standards" (the older positivist view would have preferred "rules") shared among the scientific community when assessing empirical evidence and competing theories. Yet McMullin (1998) and Laudan (1998b) have conclusively shown, I believe, that Kuhn has seriously underestimated the epistemic force such values have, especially when scientists analyze perplexing (rare) cases of rival theories (and where the data remains underdetermined) and must play off *explanatory power* against *predictive accuracy*. Hence this raises "a serious question about the adequacy of an instrumental construal of the puzzle-solving metaphor," argues McMullin (1998, p. 132). He concludes: "The Kuhnian heritage is thus a curiously divided one. Kuhn wanted to maintain the rational character of theory choice in science while denying the epistemic character of the theory chosen" (p. 135). Worse for Kuhn, much worse, is that both authors have themselves used historical-based inference by turning to the historical record to shore up their arguments, the exact basis he maintained established, and would continue to establish, his theses.[57] The argument of *incommensurability* (and *meaning variance* of scientific terms over time) is complex and unresolved and cannot be appraised here.[58] Suffice to say that it presents a serious challenge to scientific realism and rationality.[59] Although scholarly consensus now holds that rationality has been recognized across revolutionary divides (as discussed), realism has *not* therefore been satisfactorily confirmed. This philosophical doctrine is provided as an *explanation* of the alleged conceptual incompatibility of successor theories to earlier ones (e.g. Einsteinian versus Newtonian mechanics), and thus indirectly of historical developments in theory (paradigm) change, and so, once again we have here an appeal to the historico-inductive argument and thus, by way of *inference to best explanation* (IBE), to anti-realism.

In short, the historical IBE argument against realism can be formulated along two different but interrelated lines. One line focuses on that aspect of incommensurability relating to *meaning variance*. It argues that because the *meaning* of central terms in a theory (like "electron," "atom," "heat," or "mass") undergo radical conceptual revisions, even *meaning loss* (while other terms are simply revised or dropped), during paradigm shifts (*revolutions*), the existence or "truth" of such terms cannot be assumed, especially since such terms allow neither for inter-theoretic translation nor reduction. Quine, Kuhn, and Feyerabend argued to differing degrees in this vein, where they interwove the interpretive meaning of these terms to the *contextual holism* of the theory structure, within which such terms were tightly embedded, along with the scientific community's socially-active construal. Hence the postulated entities derived their "reality" mainly from relations in a given and historically contingent theoretical (and social) framework, along with congruent rules for empirical utility, and less so from their conjectured connection to an extra-theoretic reality to which they were said or assumed to refer.

The second line of attack is represented by Laudan's (1998a) famous assault on "convergent realism," the widely held view that realism presents the *best argument* for the success and progress of science.[60] This is due to the fact, the proponents argue, that current theories are (empirically) successful by virtue of them being true (or at least "approximately true"), and hence, as theories have replaced one another over (historical) time they have "converged towards truth."[61] Because his critique has been called the "pessimistic meta-inductive" argument (Ladyman, 2002, p. 236), his strategy was essentially one of turning the realist's "optimistic" IBE historical argument on its head:[62] because older *successful* theories (empirically accurate ones) contained entities now shown not to refer (or exist), and because these theories are today said to be *wrong* and discarded ("ontological elimination": e.g., phlogiston in chemistry, celestial spheres and epicycles in astronomy, ether theories in physics, and many others), by the same token, he argues, our present best and successful theories (like relativity and quantum mechanics), *which are no different in kind*, in all probability will suffer the same fate in the future. Hence one cannot take their posited entities as "real," and one should not believe in their "approximate truth." Therefore the so-called "best argument" of the realists is nothing of the kind. On this view science is successful not because it "tracks truth" but on the contrary (just) increasingly develops *empirically adequate* theories—and utility does *not* argue for truth.

These two lines of argument, the attack on convergent realism and meaning variance represent, I believe, the best cases to be made for anti-realism, which must be confronted head on and rebutted by realists. Science instructors should also be cognizant of them for two important reasons: a *negative* one, in that they fly in the face of popular beliefs of science and can be easily misused by anti-science popularizers; and a *positive* one, in that they help inoculate against naïve views of realism, inductivism, and progress that dominate current beliefs, textbooks, and traditionalist pedagogy (scientism). Let me then quickly describe how as an educator one could proceed.

It is probably best to shun technical talk of "incommensurability," but *meaning variance* cannot be, it seems, ultimately avoided. Consider the case of comparing key theoretical terms like "mass," "length," "time," or "gravity" in Newton's and Einstein's theories, as commonly presented in secondary and tertiary instruction. Usually one is told that Newton's theory, while having been "succeeded" by Einstein, was "not entirely wrong," and Einstein's theory is "more correct" in having "expanded the range" outside the confines of Newtonian theory. Indeed according to many standard textbook formulations, the older theory is often shown to be a "special case" of the newer in certain limiting instances, such as, for example, in the limit of low velocities the equations for calculating "mass" in special relativity reduce to Newton's formulations. The student is given a picture of the cumulative growth of knowledge, of simple progress both towards "truth" and of greater predictive accuracy, of a smooth transition between the two theories, of meaning *invariance* of mass, perhaps even of a "crucial experiment" (Michelson-Morley) "proving" the classical *ether* and its assumptions of space and time decisively refuted.

Unfortunately all of this is a myth, and the student has been plainly miseducated. Both Kuhn (1970) and Feyerabend (1981, pp. 114–115) can be credited for having correctly drawn attention to this, especially with versions of this example (which they both cite, as does Van Fraassen), that central terms like "mass" cannot be understood outside of their defining theoretical context (holism), and that they do *not* co-refer to the same empirical object (although on the surface it *appears* that way; it will be admitted this interpretation is contentious). In each theory the terms have fundamentally different meanings, and this fact should be made explicit and not glossed over. While it is true that the two theories reduce in logical form (Rohrlich, 1988b), there is no semantic reduction whatsoever in their two differing *ontologies*, and the usual quick instructional move in equating them reveals a deep-seated misunderstanding of the nature of the two physical theories (Jammer, 1961). The ontological picture is in truth much more complicated and damaging for realists than written (Feynman, 1965; Ladyman, 2002).[63] And while the Michelson-Morley experiment did reveal inconsistencies in the guiding assumptions of classical mechanics, it historically did not play the role often attributed to it as a Popperian-type falsification instance (Brush, 2000).

It is equally difficult to see how this far-reaching conceptual revolution in physics can be understood as a case of "progress" in terms of convergent realism. Especially when one considers that Newtonian classical mechanics was upheld for centuries as the "true" picture of nature, where for Newton "gravity" denoted a *real* universal force operating between actual masses (and allegedly confirmed by Cavendish's experiment one hundred years later), whereas for Einstein the nature of gravity (in *general relativity theory*) is clearly an *illusion* created by the space-time curvature around a massive object. Force as the universal glue holding the cosmos together had been demoted to a mere fiction, replaced by the geometry of curved space-time. Both theories, moreover, made novel predictions, could explain laws (thus had wide *explanatory power*), and have great empirical success. Any of the other oft-mentioned "values" could also be applied to both (such as simplicity, fruitfulness, coherence, etc.). And scientists, usually thinking along convergent realist lines, sometimes describe Newton's as being "false" and Einstein's as "true" or at least the "more correct" theory of nature. It seems to me that cases such as this are worth their weight in gold for anti-realists like Van Fraassen, and hence his position of agnostic "constructive empiricism" appears the more reasonable one.

But this is by far not the end of the story. This mass variance example does support Kuhn and Feyerabend's interpretation that *some* key terms must undergo *meaning variance*, and the case for *weak* incommensurability. But one swallow does not make a summer. Other research involving the problem of reference of theoretical terms and historical case studies suggests that meaning and reference is indeed theory dependent, but not necessarily in the wholesale way the theory-laden thesis and coherentism demand.[64]

Arabatzis (2006) has carefully studied not only the historical record concerning the term "electron" and its varying *representations* (and models) in the crucial era between 1890 and 1925 but also examined the philosophical presuppositions of historiographical writing. (Most people may be surprised to learn that the *notion* of the electron is by no means an obvious one and engenders some debate in HoS and PS circles, despite its indispensable and indisputable role behind so much modern technology, including electricity, communication, and older computer and TV cathode ray devices.) He has discovered, quite surprisingly, not only that *representations* in physics tend to take on a life of their own—that they exhibit a certain autonomy—but more importantly that a "core of meaning survived changes in theoretical perspective" (p. 262).[65] What has been revealed therefore with studies in the HoS, whose significance has moved to the fore in philosophical debates, is that the *historical-inductive argument* cuts both ways, and cannot just be used as a weapon in the anti-realist arsenal.[66] It is necessary albeit not sufficient for realism that "the core of properties attributed to the entity will survive changes in high-level theory" (p. 258).

How should an educator then approach the problem of meaning variance in curriculum? Recall that the instructor must already be cognizant of meaning variance because of how ideas are learned and understood within a conceptual web of meaning, and that these are tied to a given scientific theory's ontology (Nersessian, 1989). Recall as well that students have difficulty with learning isolated concepts and especially ontologies too far removed from their everyday experience. Educators must now be additionally cognizant that when the curriculum happens to shift across a divide from older to newer theories, or between theoretical frameworks (e.g., when expected to teach about special relativity or quantum mechanics or evolution), they must verbalize and clarify that some formerly familiar terms have now suffered meaning losses, although the degree to which they suffer, whether radical or gradual, can be left open, and admitted as uncertain due to academic dispute. In other words, scientific representations, and allied with them their characteristic models (electron, atom, gene, chemical bond, etc.), can *retain a core of meaning* (tied directly to measurable features), even as they undergo sometimes drastic revisions over time when theory ontologies change.

Coming now to *the anti-realist argument against convergent realism*, realists have tried several strategies to counter it. Some, like Devitt (1997), think Laudan's historical examples are overstated and can be (easily) met;[67] others deny him that our best current ("mature") theories are no different in kind, or that Laudan's conception of "empirical success" is too flexible and does not accommodate cases of novel predictions for confirming or adjudicating among theories (Ladyman, 2002, pp. 238–243). Hence "it is arguable that contemporary science has a degree of unification and coherence, as well as mathematical sophistication, that is quite absent in many of [the] theories Laudan cites" (p. 238). It seems this argument, however, is only partially valid for some of his examples. Laudan has correctly countered that the theory of the electromagnetic ether was a "mature theory," and the argument is clearly invalid for the case of classical mechanics. Newtonian gravitational theory (a "mature science") clearly displayed all such aspects and is now considered in most lights as *false* (although it is still used). Others, such as Psillos (1999), take him more seriously and meet him on his own ground, by seeking to undermine the historical argument by restricting the number of falsified theories and entities that form his inductive base, and/or showing *continuities* with respect to those aspects of abandoned theories that contributed to the success of newer ones.[68] Hence although "truth" *per se* may not be salvable, notions of convergence in terms of "approximate truth" can be.

Notions of "progress" in terms of "approximate truth," however, are notoriously difficult to explicate, and philosophers have raised several problems with them. While most realists concede that Popper's earlier "verisimilitude" version is unsalvageable, Psillos (2000), Boyd (1983), and others (Newton-Smith, Putnam, Matthews) maintain some such concept is mandatory to sustain realism, which is probably a correct assessment. Yet Laudan (1998a, pp. 1124–1125) has charged that the notion is too ill-defined to serve the purposes to which it is put and further, this lack of clarity undermines the realist's cause. Aronson et al. (1995) attempt to rescue the idea with the epistemic notion of *similarity* with respect to models and theories as type-hierarchies of natural kinds. Giere (2004), in a similar vein but on a more restricted path, argues that a sufficient case can be made (while evading "truth-likeness") employing the notion of "similarity" related to how *theoretical models* approximate (and can be tested in) the domains in which they are applied.[69] There is currently much discussion going on in this line of research and it bears directly on science education in very practical ways. This includes how students can learn science through *model-based reasoning* that can help facilitate conceptual change (Halloun, 1996, 2004; Matthews, 2007; Schulz & Sivia, 2007), as addressed in some cognitive science studies (Nersessian, 1995, 1998, p. 163). Hence there could be a double payout for education insofar as model-based reasoning would help students learn (and reason) as well as illuminate scientific epistemology (by seeing how science "tracks truth").

> Use of the terms "model" and "theory" within the science curriculum should, therefore, be an indication of the "degree of certainty" with which we hold a

> particular view. It is quite common in school science to have a realist theory (for explanation) and an instrumentalist model (for prediction) for the same phenomena. Nor is it unknown to have alternative, conflicting instrumental models for different aspects of the same phenomena (e.g., wave and particle models of light). What is confusing for children is that the role and status of theories and models are not defined. (Hodson, 1991, p. 24)

In conclusion, one can reasonably maintain that modern science is a successful and rational enterprise with reliable—but never absolutely certain—knowledge and whose theories, entities, and laws are well confirmed (Bird, 1998; Haack, 2003).[70] What then of realism and "truth"? Although both philosophical doctrines associated with realism and anti-realism have their respective weaknesses and problems, the *empirical advancement* of science—even in ways strict instrumentalists would accept—appears contingent upon the *strong belief* among scientists (see Weinberg, 1992) that their theories are approximately true and convergent (which most anti-realists but not positivists can assent to). Clearly there is a difference between the "correctness" and the "utility" (or productivity) of a *belief* (DiSessa, 1993; Elby & Hammer, 2001). Alternatively, scientists can be, and often are, quite skeptical about the truth-value of their constructed *models* (Forster, 2000). Moreover, the epistemic estimate of the truth-value of scientists' theories is only really contentious with strict forms of empiricism (which bear their own problems), notably with *theoretical* constructs or entities (unobservables), and this primarily raises questions about their possible fictive makeup only for those few (of the many) sciences focused on explaining phenomena associated with the micro- and macro-realms, like particle physics, molecular biology, quantum and structural chemistry, or astrophysics and cosmology.[71]

> If discarded theoretical entities and mechanisms—crystalline spheres, phlogiston, humors, etc.—are a problem for realists, then the long litany of such entities that have been effectively revealed and scrutinized is a bigger problem for empiricists. One-time theoretical entities—molecules, electrons, genes, chromosomes, molten cores, the planet Neptune—have a habit of turning into respectable observable entities. The empiricist doctrine is predicated upon a distinction between observation terms and theoretical terms that the advance of science and technology renders untenable. (Matthews, 1994, p. 178)

Be that as it may, the controversy about convergent realism remains unresolved and continues to be fleshed out according to the views, arguments, and proclivities of the contestants, and one must give careful ear to how terms like "success" and "progress" are given various, even contradictory meanings: anti-realists and relativists (Kuhn and others) assert that contrary to the traditional understanding of science (as continuous, but non-linear and cumulative), it does "progress" but only as a more adept instrument for problem-solving and empirical "success" and not towards "truth." It is discontinuous and non-cumulative. Others, such as Nersessian (1998), maintain that progress during conceptual change is continuous but non-cumulative, while realists like Brown (2005) hold it to be continuous, but non-linear (and cumulative), and pragmatists like Laudan (1990) insist science is progressive but non-cumulative. All these authors have availed themselves of historical case studies and examples to defend their views. *Educators can therefore no longer assume for NoS a non-problematic continuous and cumulative character.*

How should an educator then approach the problem of "convergent realism" in the curriculum? As with the previous case of discussing meaning variance, educators can mention that while scientists may *believe* their theories converge on truth, whether or not they do–and *how* they do—is a matter of much dispute, and that the historical record shows not a smooth progression but rather a continuous but non-cumulative growth in our theory-based knowledge and views of nature. Yet neither has "progress" been so discontinuous or disjointed that it has not allowed for a significant improvement in either the empirical accuracy of our theories or our overall understanding. In effect, during change of theories some "knowledge" is lost, some revised, while new insights are gained and new predictions and advances made. Alternatively, though current ontologies appear to "match" nature better than previous ones (as in "best-fit" scrutiny) there is no guarantee that such a "match" will not be superseded, and hence our *understanding* could be radically altered in the future. This is the crux of the slogan that all "theoretical" knowledge is *tentative*. In spite of this, this awareness should not be allowed to cloud the very *certain* and *non-tentative* character of established "factual" knowledge—in other words, the "facts" (very rarely are these discarded). One thinks here of the existence of the *cosmic constants* and their peculiar values that cannot be accounted for: Planck's quantum h; electric charge e; and Gravitational constant G; or for that matter, the circulation of the blood, or that most massive objects near the earth's surface accelerate at a constant average rate of 9.8 m/s^2.[72] *That* objects fall at a certain rate is undisputed and not affected by changes in high-level theories, though the explanations as to *why* and *how* they fall depend on the theoretical framework, which *can* change (e.g. *interpretations* of the meaning of "gravity" in classical mechanics [= force] versus Einstein [= fiction]).

The issue as identified here for those in science education is not just to clarify "the role and status of theories and models," as important as that is, but likewise *one of taking seriously the nature of scientific revolutions and hence to clarify them* as well (as Duschl, 1990, and Niaz, 2002, 2009 had sought to do). Thus the problems of assessing how strong or tenuous are the continuities (meaning variance), or how deep the discontinuities (ontological elimination) across the revolutionary divide (of paradigm shifts). This important feature of scientific development can no longer be ignored or glossed over in curricula and instruction, when and where revolutions appropriately arise with CK. This sadly has taken place, and continues to take place not only because of pseudo-historical narratives but also due to the prevailing *philosophical bias* that in science, *because it is* understood to be (assumed to be) continuous and cumulative, such atypical upheavals can be avoided, at worst, or given lip service, at best. (No doubt the two have reinforced each other.)

The *HoS case of the Copernican revolution* (Kuhn, 1957/1985) is an excellent example where much has been neglected. Most physics textbooks, for example, completely ignore the pre-history in astronomy, especially the Ptolemaic model and its epicycles. Tycho's mixed helio-geocentric model is ignored, too; at most he is mentioned in passing because of his empirical data of Mars as a precursor to Kepler's laws. Students are not informed of this momentous historical event and how scientific ideas challenged many commonsensical, physical, and religious beliefs of the time—even how the radical new concept of *infinity* arose and transformed the scientific outlook with its implications for physics, philosophy, astronomy, and mathematics is simply passed over in silence (Morris, 1997). Much can be done to reform science pedagogy on this major revolution using HPS ideas: here key philosophical issues can be addressed, and it happens to be one case that I myself have used. It gives rich material for the student to learn about: *underdetermination of data* for rival theories (Ptolemy versus Copernicus); the use of *models* to "fit data" for phenomena in specific domains along with reasoning using IBE (i.e. Kepler's analysis of the three models of Ptolemy, Copernicus, and Tycho); the *meaning variance* of key terms (like "planet," "star," and "motion"); and the controversy about the merits of their relative *realist or instrumentalist* interpretations. They even learn that abandoned theories (Ptolemy) harbor models that *retain* their instrumental value (e.g., the earth-centered celestial sphere is still used for GPS navigation and by NASA). It could include discussions on how theories impact upon the cultural and religious views of society at the time, and more importantly spotlight how with Kepler (and later Newton) explanatory power *outweighed* mere empirical adequacy and predictive success (McMullin versus Kuhn) and finally helped ground the theory without question on a *realist* base. Once again the advantage of a *critical realist* position vis-à-vis instrumentalism can be acknowledged:

> Critical realists can be realist about some theories (those they believe to be true) and instrumentalist about others, which they find useful but not true (i.e., theoretical models). Instrumentalists, however, are always instrumen-

> talist and have no need to distinguish between theory and model. From the critical realist position, it is not illogical to retain a falsified theory in an instrumental capacity, provided that its status is acknowledged. The fact that it is useful does not mean that it is true. It may be that within a restricted domain of application a falsified theory is more useful than a true one because it is simpler to use. Science often approaches a realist theory by way of tentative instrumentalist models *and* may retain a structure in an instrumentalist capacity (i.e., as a convenient model) after its realist value has eroded. (Hodson, 1991, pp. 23–24)

4.8. SUMMARY

The commonplace academic curriculum structures epistemic CK of the school and college curriculum in an ahistorical (often a pseudo-historical) and decontextualized fashion ("scientific meaning") that basically cuts the student off from the ability to construct meaningful learning ("contextual meaning") and gain a deeper insight not only of the concepts being employed but especially the nature and development of scientific knowledge. Empirical findings have shown that such a pedagogy overwhelmingly presents an obsolete image of science because it lags decades behind the newer studies in history, philosophy, and sociology of science. Research has uncovered that among both teachers and students fundamental misconceptions exist (i.e. their personal epistemologies) about the NoS, especially pertaining to their understanding of its epistemology and methodology. There is little awareness of the precarious nature of cutting edge scientific research and the role of human imagination and creativity in knowledge production, nor attentiveness to the vital difference between the evolutionary and revolutionary developments of scientific concepts and theories. Science education in general fails to adequately explicate the distinctive character of laws, models, and theories at the core of explanation.

Much of the confusion can be laid at the door of the nature of specialist textbooks and the mistaken "scientism" inherent to school science epistemology as well as the failure of helping teachers develop a PSE as a second order reflective capacity. Such a philosophy as part of a teacher's PCK would seek to broaden the view of content knowledge and incorporate HPS insight to critically examine curricular materials and helping transform instruction.

Scientism was recognized to encompass five tenets and linked with other myths typically presented in precollege and college science pedagogy. It was also associated with objectivism as part of the wider epistemological foundationalist project of the Western Enlightenment, although foundationalism itself has been widely renounced, and science itself was never established on a foundationalist basis. "Textbook science" is preoccupied with promulgating the reigning theory of the day with little to no concern about how and why that theory had become established. Presented historical accounts are often mythical, systematically distorting and disguising the historicity of concepts and theories as they had originally developed in location and time. This forces education into both a *moral* and an *epistemological* dilemma. The student is asked to accept this knowledge on the basis of the authority of the teacher and the text (with whatever slim evidence is presented therein), without really knowing the grounds for the legitimacy of that authority—a type of pedagogical dogmatism. HPS reform ideas argue for better inclusion of NoS with a revised epistemology (including emphasizing scientific reasoning), and while academic dissension exists over NoS, a minimum of common ground can be articulated at school and college levels. A range of NoS discussion can be articulated from a minimal base or "list" to more sophisticated views dependent upon grade, course, and the age-developmental stage of the learner. Finally, a case study examined the realist-instrumentalism debate in PS, counting Kuhn's views as an example of how a PSE could provide clarity for educators and contribute to overcoming naïve realism while developing a critical realist outlook.

NOTES

1. Abd-El-Khalick and Lederman (2000); Abell (2007); Désautels and Larochelle (1998); Lederman (2007), and Wandersee et al. (1994) have previously covered considerable ground.
2. "Epistemology should play a more important role both in contemporary research in science and in the teaching of science. If we as educators emphasized the need to change students' epistemologies more in our teaching, then there may not only be more success in the understanding of the subject by our students, but also be dividends in the future in more directed research by future graduate students" (Kalman, 2009; Kelly et al., 2012, p. 286).
3. This has also been referred to in the research literature as the "disciplinary view of knowledge" in contrast to "personal learner epistemology" and "social practice views of epistemology" (Kelly et al., 2012). The latter defers to science studies research and how knowledge is attained and justified through discourse practices within epistemic cultures. What is significant is that "within this perspective, knowledge is seen as competent action in a situation rather than as a correct, static representation of the world" (Knorr Cetina, 1999; Kelly et al., 2012, p. 286). What is not being acknowledged is that the two stated perspectives are themselves beholden to two different epistemological philosophies, namely pragmatism and objectivism. While science education has traditionally been in the thrall of the second and is now expected to shift to the first, it could better take advantage of the respective benefits of each.

4. Even when basic science "subject matter" is taught, it is always accompanied by some context that may operate covertly (e.g. preparatory, socio-utility, etc.). Such contexts have been called "meta-lessons" (Schwab), "curriculum emphases" (Roberts, 1988), and "companion meanings" (Roberts & Oestman, 1998).
5. Also, what kind of science knowledge is of *most* worth? (A key question of prioritizing subject content).
6. A very informative discussion on such questions, including examining beliefs, learning, knowledge, and critical inquiry pertaining to the aims of science education can be found in Nola and Irzik (2005). The comments that follow can be considered supplemental to their work.
7. Certainly the relatively recent research studies to enhance *scientific argumentation* in the classroom also aim towards resolution of the issues and questions raised here, but are not of present concern.
8. These will not be discussed here; instead refer to Siegel (2010) and Norris (1997).
9. Kuhn (1970) was skeptical about what science education could achieve in terms of developing independent thought and argued instead the conservative view of reinforcing the conventional paradigm—in part because this furthered "progress," and in part because students had no competence to do otherwise. Schwab held a different view, and Siegel (1978) has contrasted the two positions, as will be discussed.
10. Egan's metatheory, however, although neutral on this topic, does not exclude the use of HPS integration, as argued last chapter—nor, for that matter, does it exclude concerns of STS reform issues—as long as these are integrated in such a way to further students' cognitive-emotive tools as "romantic" or "philosophic" frames of understanding. As mentioned, his distinctive "narrative approach" lends itself easily to historical "story-telling" in science class, as well as incorporating respective epistemological features. In other words, his argument for proper transposition can incorporate content *broadening*. This point could in principle also apply to the *Bildung* metatheory, but because it has been traditionally conceptualized more along exclusive humanistic lines (Eisner's "rational humanism"), often in contrast to the conception and methods of the natural sciences and the way they are generally thought needed to be taught, it has tended to be conceptually and institutionally isolated from science education. In Germany at the gymnasial (upper secondary) level the contrast is manifested in two different kinds of schools, *humanistische* (classics, literature, and language focus) and *naturwissenschaftliche Gymnasien* (science school focus), where the *Bildung* paradigm is usually associated with the former only. Here the bifurcation of Western culture has been rigidly institutionalized already in a nation that developed the concept.
11. I have renamed his category labeled "knowledge content," which is confusing and misleading. His intention is interpreted to surface contextual factors, and he appears to take "knowledge" in the widest sense to imply an understanding where "epistemic content" is related to its contexts. One

could perhaps have labeled it alternatively "knowing content contexts," but this is awkward.

12. Englund (1998) himself references Dewey, and it becomes clear that this kind of socialization is the one he not only prefers but—along with the rest of the authors in the book and like-minded champions of the STS reform movement earlier—insists should be the overall educational rationale for science teaching and CK orientation. He would therefore take the main task of his PSE, as I see it, to consist in "reestablishing the philosophical aspect of educational and didactic research" (p. 23) along these lines and towards this end: "Thus the educational philosophy that has focused on the relationship between democracy and education is placed center stage" (p. 24).
13. This term is borrowed from Roberts and Oestman (1998): "We argue, broadly, that the 'socialization' of students is not only a matter of their deportment, attitudes, conduct and so forth, as the term tends to be used, but is also very significantly associated with the meanings provided by their educational experiences. Science textbooks, teachers, and classrooms teach a lot more than the scientific meanings of concepts.... Most of the extras are taught implicitly, often by what is *not* stated. Students are taught about power and authority, for example.... All of these extras we call 'companion meanings'... and they function as both context and subtext for the more obvious subject matter meanings in school subjects" (p. ix).
14. Other pitfalls he mentions are the "rush to closure" and "losing sight of constraints." The former refers to how textbook publishers, local school districts, and teachers could embrace the "standards" as policy in either a superficial way or pick-and-choose aspects in order to keep the "epistemic" content and traditional curriculum largely intact. This appears to have been the case with recent implementation efforts in California (Bianchini & Kelly, 2003). The second refers to constraints of the educational system such as incurred costs to districts (professional development; resources), school culture, and teacher attitudes.
15. As examples, Hodson (1985); Cawthron and Rowell (1978); Smolicz and Nunan (1975); and Elkana (1970).
16. Including Abd-El-Khalick and Lederman (2000); Désautels and Larochelle (1998); Driver et al. (1996); Meichstry (1993); and Ryan and Aikenhead (1992).
17. "Science processes are activities relating to collecting and analyzing data, and drawing conclusions. For example, observing and inferring are scientific processes. More complex than individual processes, scientific inquiry involves various science processes used in a cyclical manner. On the other hand, NOS refers to the epistemological underpinnings of the activities of science and characteristics of the resulting knowledge" (Lederman, 2007, p. 835). See Schwartz et al. (2004) for a discussion on scientific inquiry.
18. They are: (1) The main purpose of science is to acquire knowledge of the physical world; (2) An underlying order exists, which science seeks to describe in a maximally simple and comprehensive manner; (3) Science is dynamic, changing, and tentative; (4) There is no single scientific method.

19. (1) There exists considerable dissensus over how the generation of scientific knowledge depends on theoretical commitments and social and historical factors; (2) There is dissensus over the "truth" value of scientific theories (what they claim exist in the world) independent of the scientist and/or scientific community (i.e., social constructivism versus realism versus anti-realism).
20. Current science education, however, at the undergraduate level (the background of most science educators) and pedagogical training of science educators in education faculties have both tended to ignore this, as mentioned before. Nor has this awareness penetrated the institutions responsible for training scientists or science teachers, both of whom chiefly obtain their understanding of science from formal, textbook-dominated lessons and "cookbook" laboratory exercises.
21. Rudolph (2000) has suggested that the nature of science debate has only tended to expose more heat than light, and hence the science educator should side-step issues attempting to pin-down what a *universalist*, covering methodology or aims may be, for a *particularist* view of diverse methods and goals specific and unique to each discipline. This should include sidestepping the common philosophical questions of science's ultimate *aims (axiology)*, where there continues to be little agreement ("Does science aim at truth or merely aim at empirical adequacy?"), for a narrow focus on context and practice (and also *truth*) as found within the specific discipline and group of practitioners. Although this advice may indeed remove NoS out of the vague context of standards-type documents, it does little to help the educator elaborate discipline-specific NoS and the questions of truth that inevitably arise there (e.g., in physics).
22. He wrote that in general "scientific training is not well designed to produce the man who will easily discover a fresh approach" (Kuhn, 1970, p. 166). In other words, it cannot help to generate new creative ideas that will contribute to resolving "puzzles" within a domineering paradigm or even possibly initiating scientific *revolutions* (or "shifts in paradigms")—the two different ways that Kuhn understood "progress" could occur in science (though he insisted "revolutions" were rare events in history of science).
23. See here also Kuhn (1977, p. 229) for an earlier, almost same statement.
24. He emphasizes that "textbooks imprison science teachers in a belief that the instructional sequence of assign, recite, and test is guaranteed to produce knowledge" (Stinner, 2001, p. 324). He adds: "For the majority of students, I maintain, *doing problems is to memorize 'scientific facts' and practice algorithms*" (Stinner, 1992, p. 4, origina italics).
25. "Without wishing to defend the excessive lengths to which this type of education has occasionally been carried, one cannot help but notice that in general it has been immensely effective. Of course, it is a narrow and rigid education, probably more so than any other except perhaps in orthodox theology. But for normal-scientific work, for puzzle-solving within the tradition that the textbooks define, the scientist is almost perfectly equipped" (Kuhn, 1970, pp. 165–166).

26. "Although Kuhn is right in pointing out that there is an 'essential tension' between innovation and tradition, what he failed to appreciate fully is that it is mitigated by the flexibility in the apprenticeship learning component of training practitioners. Cognitive science research indicates that there are no paradoxes with respect to the traditional pedagogical method. The textbook-type science education has *not* been successful in producing practitioners. Very few students learn the subject sufficiently well even to provide explanations and predictions of simple physics phenomena, never mind to go on to graduate school and become practitioners" (Nersessian, 2003, p. 189, original italics).
27. An evaluative review of standard physics textbooks in the 1970s/1980s (Lehrman, 1982) had shown that almost all are poor when presenting historical and philosophical dimensions of science. Their value for technological applications was generally better, but even this was uneven. Since then improvement at least when it comes to technology has been better (see text by Cutnell and Johnson, 1998), but it entirely ignores HoS. One notable exception for HPS has been Hecht (1994). But the text by Holton and Brush (1972/2001) represents the most comprehensive attempt to *reshape* a college physics textbook using HPS themes.
28. These discussions always refer to "pure" or "basic" and *not* "applied" science. It is interesting that Kuhn admitted that for the latter (and for inventors) his "normal/revolutionary" distinction may not hold, nor his emphasis on textbooks as puzzle-solving initiations into normal science (Kuhn, 1977, pp. 237–239).
29. Laboratory work, often construed as helping students acquire technical skills, scientific reasoning, and a glimpse of research work but which mainly is comprised of "cookbook" type experiments combined with a simplified neo-Baconian inductivism, can hardly be judged as a mirror of actual research, according to Hodson (1996) and others. Hence the renewed call for "authentic science inquiry" (Schwartz et al., 2004).
30. Kuhn's (1970) insights on the *nature of science education* admits as much, especially concerning how readers of textbooks can be misled into thinking that applications of a theory as given in exercises *confirm evidence* of the theory—a typical pedagogical ploy! "But science students accept theories on the authority of teacher and text, not because of evidence. What alternatives have they, or what competence? The application given in texts are not there as evidence but because learning them is part of learning the paradigm..." (p. 80)
31. Again, in agreement with my view that it allows for "knowing why": "Unless the enabling and justificatory connection of process with knowledge-claim is made clear to the pupil, he can strictly have little rational ground for confidence in the educational enterprise; in no other way can the conditions upon which his knowledge-claim rest, give him 'the right to be sure'" (Roger, 1982, p. 3). Duschl (1990) would agree and has provided just such justificatory historical-epistemological frameworks for education. Rogers insists though that laboratory-based inquiry must be more genuine. He also freely admits that all "knowing that" (expository) and "knowing how" (experimental) aspects of science are conducted within a *conceptual framework*

determining the kind of questions asked, what counts as evidence, and the nature of inquiry itself (as with Kuhnian "paradigms"). Further discussion on epistemology follows in Section 5.1.1.

32. Popper's falsificationist thesis (that no theory can ever be proven true, or *probably* true—contra logical empiricism—but can in principle be *falsified*) had been critiqued by many philosophers of science, including Kuhn and Feyerabend already in the 1960s (Lakatos & Musgrave, 1970). Today most consider his thesis "dead." Much earlier, at the turn of the century, the French historian Pierre Duhem had explicitly attacked the view of the "crucial experiment" as an apparent decisive instance when rival theories in physics are compared (Duhem, 1954/1998). His ideas today form part of the so-called "Duhem-Quine thesis," which argues, in part, that theories can only be tested as a whole, as a combined set of hypotheses, laws, and background assumptions (*holistic thesis*). Hence the empirical data always remains inconclusive; that is, neither truth *nor* falsity of a theory can be determined by evidence because auxiliary and *ad hoc* hypotheses can always be imagined (or invented) to "save" the theory (*underdetermination thesis*). Curd and Cover (1998) admit that "probably no set of doctrines has had a greater influence on modern philosophy of science than those included under the designation of the Duhem-Quine thesis," although one must be aware that "an astonishing variety of doctrines fall under [that] umbrella" (p. 255).
33. Nor are all knowledge claims *scientific* knowledge claims, nor yet, are all such *cultural* knowledge claims when conflicting with Western science equally valid and acceptable (Matthews, 1994, 1998a, 2009a). Such important critical socio-philosophical issues arising from postmodernist critiques, from "cultural" and "social studies" of science, from the "science wars" debate—and taken together slowly starting to impact the values and identity of science education—cannot be addressed and critically appraised here.
34. These include: (1) the invention-discovery (or theoretical-empirical) distinction; (2) the justification (criteria) of accepting a theory; (3) the question of the status of scientific theories (instrumentalism; levels of realism); (4) the role of epistemology.
35. "Curriculum developers confused the teaching of science *as* inquiry (i.e., a curriculum emphasis on the processes of science) and the teaching of science *by* inquiry (using the processes of science to learn science). It was assumed that the attainment of certain attitudes, the fostering of interest in science, the acquisition of laboratory skills, the learning of scientific knowledge, and the understanding of science were all to be approached through the methodology of science, which was, in general, seen in inductive terms" (Hodson, 1988, p. 22).
36. What can be said is that such models are proven inadequate for several reasons, they rely on individual, cognitive mental changes while ignoring socio-linguistic factors, and conceptual change strategies based on such models have not been very effective. The mechanisms of theory change have also been elusive to identify. (See Strike & Posner, 1992, for a revision of their ideas.)

37. Duschl (1994, p. 447) sees her research and that of others in a similar vein, which takes the position that there exists "a structure that can be applied to scientific discovery," as part of the new "cognitive twist" in PS, where interpretive procedures from cognitive science are used to *interpret* PS.
38. Aikenhead (1997a) has devoted his efforts to analyzing how school science education as part of a Western enculturation process when embodying scientistic epistemology can alienate students, especially First Nation youth who may not share the same values, "ways of knowing" and seeing the world. He has offered an unusual and constructive perspective for science educators, to consider their task and the curriculum as mediating a "cultural border crossing" for their students, between the worlds of academic science and their own home cultures (Aikenhead, 1996, 1998). This dovetails with the views here insofar as I would stress the *language* component as a key feature of the cultural disconnect for students (stated in Chapter 5). My view does not dovetail insofar as Aikenhead tends to emphasize cultural relativism.
39. See Gadamer's "Introduction," also the "Foreword to the Second Edition." Bernstein (1983) writes: "His entire philosophic project can be characterized as an apologia for humanistic learning. Gadamer, throughout his long career, has sought to show that the humanistic tradition, properly understood, is an essential corrective to the scientism and obsession with instrumental technical thinking that is dominant today" (p. 180).
40. The first corresponds to *myth #5*: "evidence accumulated carefully will result in sure knowledge" (McComas, 1998, p. 58); the second to *myth #4*: "a general and universal scientific method exists" (p. 57) and *myth #6*: "science and its methods provide absolute proof" (pp. 59–60); the third somewhat to *myth #13*: "science models represent reality" (pp. 66–67), although this is not strictly true. Whereas McComas fronts the idea of models here, Nadeau-Désautels takes models primarily as instrumental and sees in their use a strategy to overcome naïve realism; the fourth approximately to *myth #2:* "scientific laws and other such ideas are absolute" (pp. 55–56); the last to *myth #9*: "scientists are particularly objective" (pp. 62–64) and *myth #15*: "science is a solitary pursuit" (p. 68).
41. Not to be confused with the scientific attitude to strive to be objective in argument and research.
42. He provides another description: "the basic conviction that there is or must be some permanent, ahistorical matrix or framework to which we can ultimately appeal in determining the nature of rationality, knowledge, truth, reality, goodness, or rightness.... Objectivism is closely related to foundationalism and the search for an Archimedean point. The objectivist maintains that unless we can ground philosophy, knowledge, or language in a rigorous manner we cannot avoid radical skepticism" (Bernstein, 1983, p. 8).
43. See for example von Glasersfeld (1989): "The word 'knowledge' refers to a commodity that is radically different from the objective representation of an observer-independent world which the mainstream Western philosophical tradition has been looking for. Instead 'knowledge' refers to conceptual structures that epistemic agents, given the range of their present existence within their tradition of thought and language, consider *viable*" (p. 124). It is

curious that both von Glasersfeld and the later Kuhn (Bird, 2002) developed Kantian constructivist positions, yet Kant is very much in keeping with Western objectivism.

44. "One is the emergence of a weaker, anthropological, culturally based, historically relative views of knowledge in Wilhelm von Humboldt, Herder, Fichte, Hegel, Marx, Dilthey, Gadamer, and other writers. The other is the rise of classical American pragmatism. Pierce, and those influenced by him, including James and Dewey, share his anti-Cartesian concern to formulate a nonfoundationalist view of knowledge which, with the possible exception of James in his radical empiricist phase, gives up familiar claims to know the mind-independent real as it is" (Rockmore, 2004, p. 27).
45. "A theory may be thought of as a family of models. Different models are derived from a theory using different idealizations, different simplifying assumptions, and different auxiliary hypotheses. Many different models can be derived from a single theory. For instance, if we assume that there are six planets, which are small point masses, then we get one Newtonian model of the solar system. But if we assume that there are 7 planets, or if we model the Earth as bulging at equator, then we get different Newtonian models of the solar system" (Forster, 2000, p. 236).
46. With the growth of computer-based simulations in classrooms today, students now have the opportunity to run modeling scenarios for some subject topics, especially in physics and chemistry.
47. One must be careful here about attributing positions to complex thinkers. Duhem did indeed hold the line against atomism and other theories positing *theoretical entities* at the turn of the 19th century (as many others did as well, including the famous chemist Ostwald, who only converted towards the end of his life). On the other hand, he presented at least three good arguments against the instrumental (anti-realist) view of theories as *mere calculating devices*: 1) actual scientific practice and aims; 2) novel predictions; 3) the theoretician's search for unity/unification to explain phenomena. And his adherence to science as a "natural classification of laws" can be considered as a form of "structural realism" (including Poincaré, and Worrall today). Hence Psillos (1999) places Duhem "between realism and instrumentalism" (p. 37).
48. Matthews (1994, Ch. 8) traces the conflict even further back, to Platonic "realism" and Aristotelean "empiricism." He presents other insightful historical examples of the recurring simultaneous clashes between the two opposed philosophies: Copernican/Ptolemaic realism and Osiander's empiricism; Newton's realism and Berkeley's empiricism; and Planck's realism and Mach's empiricism.
49. This certainly is the case with Roth and Lucas (1997). In examining their research work they can quite fairly be accused of indoctrinating their students into an anti-realist/radical constructivist ideology.
50. There is considerable overlap with certain Feyerabendian theses such as *incommensurability*. Arabatzis (2006) argues that Feyerabend, unlike Kuhn, developed the more sophisticated philosophical arguments and hence restricts his attention to him when it comes to analyzing *meaning variance*, for

example. Bird (2002) agrees with this general assessment of Kuhn and so explains his lack of overall relevance for philosophers. Yet notwithstanding these views, Kuhn continues to exert considerable influence across disciplinary fields. And it appears that philosophers of science have rather too quickly written him off, says Laudan (1990). As Forster (2000) himself admits, to be anti-Kuhn is taken as being a card-carrying philosopher of science.

51. Laudan (1990) has explicated *six significant theses* comprising full-blown epistemological relativism: (1) non-cumulativity, (2) theory-ladenness and underdeterminancy of data, (3) holism, (4) shifting standards of success, (5) incommensurability, and (6) interests and social determinants of belief.
52. They insist on using the "objective" historical record to show the inherent subjectivity of science, the indeterminacy of the empirical data, and the inadequacy of induction. Yet their method is objective in intent; it assumes the adequacy of the data in question (historical cases) and is inductive in approach.
53. Kuhn had not moved from this viewpoint in his later writings (see his 2000, p. 242).
54. There are in-between positions. Nancy Cartwright in her important *How the Laws of Physics Lie* (1983) accepts the existence of entities but maintains anti-realism for laws. She qualifies this by distinguishing between *phenomenological laws* (i.e., as in applied physics and engineering) and *fundamental laws* (the typical generic and famous theories in physics and chemistry as found in textbooks). She maintains that while the former are fairly accurate at describing how bodies behave (are "true"), they nonetheless remain restricted in scope and hence trade off explanatory power for truth. The latter suffer the opposite defect in having wide explanatory power but lack truth because they cannot be used in specific real case instances. Note her position is somewhat a reversal of the earlier mentioned *phenomenological* views of science.
55. Briefly, the contextual or holistic theory of meaning holds that the meaning of any one term is dependent upon its internal relation of all other terms (or beliefs), which hang together in a kind of *semantic web*. To change any one term's meaning is to send a "ripple effect" throughout the entire web, substantially affecting the meaning of all terms. Thus the meanings of terms are connected to the surrounding *context*. Strong versions of both correspondence and coherentism have serious defects, and their weaker forms are usually easier to defend. Quine, Kuhn, and Feyerabend all adhere to strong versions, as do relativists. Realists are divided as to whether or not realism entails a prior commitment to truth correspondence (Psillos, 2000). Devitt (1997), for example, argues that while realism and truth should be addressed as distinct problems, realism in order to be viable requires some form of correspondence.
56. "By separating the issues of comparability [i.e. rationality] and commensurability [i.e., anti-realism], he believes he can retain a more or less traditional view in regard to the former while adopting an instrumental one in regard to the latter. The radical challenge [in his book *Structure*] is directed not at rationality but at realism" (McMullin, 1998, p. 132).

57. "Neither the history of science nor Kuhn's philosophical arguments show that scientific revolutions cannot be resolved by rational arguments based on evidence and shared rules. By treating paradigms as indivisible wholes and failing to appreciate the ways in which rules and aims can be rationally debated, Kuhn has seriously underestimated the role of reason in paradigm debates.... Thus, according to McMullin and Laudan, Kuhn's thesis that subjective, psychological, and rhetorical factors must play a *leading role* in all scientific revolutions is ill supported and false" (Curd & Cover, 1998, p. 239; my italics). See Pyle (2000) for a critique of the Kuhnian historical interpretation of the chemical revolution of Lavoisier.
58. Forster (2000, p. 245) agrees. The *incommensurability thesis* developed along two separate lines: one was defined by Quine and based upon considerations in the philosophy of language; the other was defined independently (but almost simultaneously) by Kuhn and Feyerabend and based upon considerations in the philosophy of science. Kuhn, too, under criticism, had later come to modify his own thesis and aligned it with Quine's *untranslatability* thesis, though with differences (Kuhn, 2000). Sankey (1994) has provided a comprehensive and critical account of the two versions of the thesis. See also Bird (2002) for a negative and Brown (2005) for a more constructive assessment.
59. "The problems of rivalry, content comparison, and progress are the most important issues raised by [the] thesis. They represent a challenge to the rationalist seeking to understand theory-choice as informed by a critical appraisal of genuinely alternative theories. They are a challenge to the realist inclined to view theory-change as resulting in an increase of truth about the world" (Sankey, 1994, p. 4). With regard to Feyerabend's version we read that his "conception of meaning and his skepticism toward the ontological implications of scientific theories presented a serious challenge for scientific realism" (Arabatzis, 2006, p. 244). Bird (2002), however, maintains that the thesis no longer presents the danger it once did and has since lost its "philosophical significance" for most philosophers of science (p. 444). Sankey (1994) holds a similar view: "If we like, the word may be retained as a loose name for the cluster of related problems having to do with theory change" (p. 221). Arabatzis, alternatively, considers the challenge posed by the meaning variance aspect of the thesis as quite serious and one that *must* be met by realists.
60. This has come to be referred to as the "no miracles argument" for realism. Putnam helped establish the slogan when he argued in 1975 that "[it] is the only philosophy of science that does not make the success of science a miracle" (quoted in Psillos, 2000, p. 715).
61. Most scientists take this traditionalist view as self-evident. A typical example here is the embryologist Wolpert (1992). His case is unusual, however, insofar as he is intimate with the writings of philosophers of science like Kuhn and the relativist arguments of some sociologists. Moreover, he presents many historical case studies to buttress his conclusion: "But it is precisely in this respect that science...is special: for the history of science *is* one of progress, of increased understanding.... And in the last fifty years the progress in,

for example, understanding biology at the molecular level has been astonishing. Science is progressing in that the truth is being approached, closer and closer, but perhaps never with certainty. But very close approximation can be a great achievement and is infinitely better than error or ignorance" (p. 100).

62. Kuhn, as cited earlier, had maintained that scientists and people in general tended to this popular outlook because of (mis)*education* and the falsified histories in textbooks. Laudan in effect reinforces this.
63. Classical mechanics comprises at least *three* distinct and incompatible ontologies within its historical formulation. First, the original theory of forces acting at-a-distance (Newton); secondly, the field-theoretic approach, which understands gravity as a *field (potential)* occupying all points in space, acts locally and is analogous to electromagnetic field theory (also related to modern gauge field theories and, to some extent, to Einstein's radically transformed tensor-field theory in general relativity). Thirdly, the "analytic mechanics" approach developed by Euler, Lagrange, and Hamilton in the 18th and 19th centuries, based on the "principle of least action" and the energies of the system (of masses). (Einstein's theory thus presents a fourth ontology.) One assumes these three classical ontologies represent different formulations of the *same* theory only at the cost of holding to an anti-realist (positivist) view since all three are *empirically* equivalent, although not *evidentially* equivalent. Such an equivalence is commonly and explicitly presented in tertiary physics education, although they "refer to different entities and posit different explanatory frameworks" (Ladyman, 2002, p. 255). This dilemma raises the question as to which ontology "fits" better, or if nature has a preferred ontology, the discovery of which is the task of "our best theories" to map. Unfortunately, all three classical ontologies have been abandoned as classically formulated, while analogous aspects were transferred in QM. Feynman (1965, pp. 50–55) makes much the same point when comparing the three different "interpretational schemes" (p. 54). "They are different because they are completely unequivalent when you are trying to guess new laws" (p. 53). "We have not decided between the last two yet [field method and energy principle].... The best law, as at present understood, is really a combination of the two in which we use minimum principles plus local laws. At present we believe the laws of physics have to have the local character and also the minimum principle, but we do not really know" (p. 54).
64. This does not mean we can go back to the earlier causal theory of reference-fixing by such *semantic* realists as Kripke and Putnam, although philosophers like Bird (2002) often enlist them against Kuhn and company. "Both views err in opposite directions; incommensurabilists permit wholesale referential change, while those who hold to a bare causal theory permit little, if any referential change. Moreover both views conflict with standards accounts of the historical stories told of scientific revolutions. In these accounts, some terms retain their reference, others change their reference, new items are introduced to refer to new entities and old-terms previously thought to refer are declared non-referring" (Nola, 1980, p. 527). Nola had reached this conclusion by studying the historical cases of the revolutions associated with

the theory of electrical attraction ("effluvia") and the theory of combustion ("phlogiston"). Arabatzis (2006) presents several weaknesses in the arguments used by Putnam against the well-known Feyerabendian theses.

65. "Throughout that period nobody doubted that the electron was a universal constituent of matter, with a certain mass and charge, and that it was the agent of radiation.... This aspect of the construction of the concept of the electron supports, albeit not conclusively, a realist attitude towards the electron" (Arabatzis, 2006, p. 262). "Any future theory about ... electrons will have to incorporate, but perhaps also *reinterpret*, these 'well-known causal properties' of electrons" (p. 257, my italics).
66. Arabatzis (2006) concludes "historical case studies can play a dual role, either negative or positive, in disputes over scientific realism. They can either undermine realist intuitions or neutralize the antirealist implications of meaning change. However, this neutralization does not lead automatically to a realist position" (p. 260).
67. Matthews (1994, p. 177) even cites Devitt to the effect that the problems raised about verisimilitude "can and have been answered." I could not disagree more. The arguments Devitt presents on pp. 161–174 cannot be considered as decisive refutations in my view, especially his reliance on *partial reference* when comparing the Newtonian and Einsteinian discontinuity (p. 165) or his critique against incommensurability (p. 168). He certainly does not argue from historical evidence, rather from philosophical premises.
68. "If the relevant realist arguments are sound, then the fact that our current best theories may well be replaced by others does not, necessarily, undermine scientific realism. All that it shows is that a) we cannot get at truth all at once; and b) our inferences from empirical support to approximate truth should be more refined and cautious in that they should commit us only to the theoretical constituents that do enjoy evidential support and contribute to the empirical success of theories" (Psillos, 2000, p. 721). The examples of abandoned theories, however, are much more serious an instance against realism than cases where previous theories have been "absorbed" into other theories, with *some* entities either redefined or dropped (e.g., Newton into Einstein). Such cases are primarily about *meaning variance* and although difficult, as discussed, they are not about *ontological elimination*, such as phlogiston theory, which even Devitt (1997, p. 161) admits presents a strong case. This is a major difference in kind. Neither Ladyman (2002, pp. 244–252) nor I have been convinced that Psillos (1999) has effectively met the test posed by Laudan's historical counter-examples of the caloric theory of heat or the electromagnetic ether theory of light. Both of these were mature and enjoyed novel predictive success. "Even if there are only one or two such cases, the realist's claim that approximate truth explains predictive success will no longer serve to establish realism" (Ladyman, 2002, p. 244), precisely because the *theoretical entity and the entire ontology* has been dumped. Here the "relevant realist arguments" as presently structured do not appear to be sound.
69. "Giere's view is that to the extent to which the model and the real system are similar, we may say that the model provides a better or worse approximation

to the real system. He suggests that the notion of similarity between models and real systems provides the resource for understanding approximation in science and avoids 'the bastard semantic relationship' of approximate truth" (Psillos, 1999, p. 273). Psillos, while strongly supporting Giere's approach, nonetheless maintains that Giere's abandonment of "truth-likeness" is premature. He argues for the retention and defense of this concept, even—and especially—in light of Laudan's familiar attack on it, and that Giere can hardly avoid some use of this notion. Giere's revised conception of the *aim* of science is given in (2005): "The aim of science is not 'a literally true story of what the world is like,' but merely the production of models similar to limited aspects of the world in ways determined by the scientific context. Some of these aspects may be of the type traditionally designated as 'theoretical,' such as the atomic structure of the air" (p. 157).

70. "A remarkable fact about modern science is that as the number of phenomena which science has investigated has grown, the number of theories needed to explain them has decreased. And those theories have been deeper and more general, and, correspondingly, more integrated when it comes to explaining the phenomena.... From the point of view of the constructive empiricist, who thinks our theories are empirically adequate to the phenomena but are most probably false, this fact should not be merely remarkable but really quite extraordinary ... the ability of a theory to integrate with other theories and its ability to produce novel and unexpected true predictions constitute evidence for its truth that goes beyond observational success" (Bird, 1998, p. 150). Haack would agree with this view but is more fallibilist.
71. This likewise raises the contested thesis as to the conjectured inter-theoretic reduction of the sciences, either within or across disciplines (e.g., biology to chemistry to quantum physics), which some philosophers of biology and chemistry increasingly question (and cannot be addressed here). They deny that such reductive assertions are ultimately required to yield predictive success and that there exist *autonomous levels* of explanation (refer to Kitcher, 1998, against inter-biological reduction, and Scerri, 2001, against the oft-assumed chemistry reduction to quantum physics). Weinberg (1992) remains a champion of reductionism.
72. One should also include in this list true *discoveries* of entities like x-rays, electrons, and neutrinos, and especially the accumulation of *empirical laws*, like Kepler's three laws of orbits, laws of thermodynamics, and so on. For the latter even Kuhn (1977) himself explicitly admits the importance to acknowledge the non-tentative and cumulative nature of this *empirical base*: "As science develops, they may be refined, but the original versions remain approximations to their successors, and their force is therefore either obvious or readily recaptured. Laws, in short, to the extent that they are purely empirical, enter science as net additions to knowledge and are never thereafter entirely displaced" (p. 19). Unfortunately this important admission is little known among his many supporters and relativist-inclined defenders.

CHAPTER 5

PHILOSOPHY OF SCIENCE EDUCATION AND NATURE OF LANGUAGE

> Each science, as a science, has in advance projected a field of objects such that to know them is to govern them. We find quite another situation when we consider man's relationship to the world as a whole, as it is expressed in language.... For to live in a linguistic world, as one does as a member of a linguistic community, does not mean that one is placed in an environment as animals are. We cannot see a linguistic world from above in this way, for there is no point of view outside the experience of the world in language from which it could become an object.... *Being that can be understood is language*.... Thus hermeneutics is ... a *universal aspect of philosophy*, and not just the methodological basis of the so-called human sciences.
>
> —Gadamer (1960/1975, pp. 449, 471)

Rethinking Science Education: Philosophical Perspectives, pp. 209–257

5.1. INTRODUCTION—THE SHIFT FROM EPISTEMOLOGY TO ONTOLOGY: THE "INTERPRETATIVE TURN" AND HERMENEUTICS

5.1.1. Science Education and Language-Focused Research

In the discussion thus far three fundamental theses for improving science education have been raised and defended: the value of Egan's metatheory as an overarching conception for the educational endeavor (also as *practical planning* framework) of teaching and learning science subjects ("science literacy"), including its value for informing the educational philosophical component of teachers' PCK; secondly, the value of surfacing epistemology and NoS for broadening both a teacher's CK and a student's understanding of science subject development; thirdly, the merit of PSE for the research field, teacher identity, and professionalism. The first addresses the "why," "when," and "how" to teach from a predominant *educational theoretical* perspective, arising from considerations in the PE. The second addresses the "what" to teach (and encompassing the "why") by enabling a critical perspective on curriculum and textbooks arising from considerations in the philosophy, history, and sociology of the sciences (PS). The third provides a synthetic perspective to create awareness in order to allow the other two to even "get off the ground" by influencing teacher attitudes, beliefs, and frameworks for curriculum planning, design, and implementation (Figure 1.1). As emphasized previously, the common *formal content* of science discipline subjects must be taken as problematic by the teacher when curriculum planning and teaching for two reasons: it requires an adequate *transposition* for learning purposes (*Bildung*, Egan), and it requires an adequate curriculum *broadening* to include NoS for more inclusive and accurate science comprehension purposes (HPS). The duty of a PSE, it has been put forward, is to offer critical and reflective platforms for enabling both of these to take place in general, while Egan and HPS were suggested as specific ways to address them.

It can be admitted that these two tasks are certainly challenging and comprehensive enough for science teachers to consider when unit or lesson planning in their classrooms, and for training programs in education faculties to accomplish with aspiring teachers starting out afresh and wishing to enter the profession. These two tasks *taken alone,* so it is argued, would go a considerable distance in improving contemporary science education, as some studies that have focused exclusively on NoS inclusion have already shown. But there is one more significant component, however, that has come to prominence relatively recently in science education research circles that must equally be considered when discussing science education reform: the *nature of language* (NoL) and its complex and often hidden role in science learning and NoS understanding. This additional component therefore represents another dimension of science education that a "philosophy of science education" must deliberate upon and scrutinize. To clarify at the outset, however, the contribution to NoL discourse addressed here will avoid the "philosophy of language" debates typical to Anglo-analytic philosophy, and avoid in general a comparison of language philosophies in the two traditions (Anglo-American and Continental-hermeneutic; Medina, 2005; brief discussion below). The task to be undertaken is to pursue the recent language discourse as found in the research literature pertaining to learning science in classrooms, but chiefly from a hermeneutic language perspective and bearing in mind the critiques of the Anglo-analytic tradition by Rorty (1979), Wittgenstein (1958), and Gadamer (1960/1975).

The central importance of language for science teaching and learning has now become widely recognized within the research community, and since the 1990s the literature on this subject has grown significantly (Carlsen, 2007; Fensham, 2004; Sutton, 1998; Wellington & Osborne, 2001). Wellington and Osborne (2001), drawing on 30 years of research, have provided a very useful book for classroom teachers by illustrating the importance of literacy when learning the "language of science," including a discussion on how the language *of science education* in classrooms differs from science technical language. Kelly (2007) has provided a comprehensive review of the role of spoken and written discourse in science classrooms, building upon the earlier ground-breaking work of Lemke's *Talking Science* (1990).[1] Yore and Treagust (2006) comment on the work coming out of an international conference that sought to explore the diverse research links between language and science literacy. Anderson (2007) has identified the language-focused "socio-cultural" research program as one of *three* major traditions on science *learning* that have now come to distinguish the diverse kinds of research in science education (along with "conceptual change" and "critical" traditions). Researchers within this third tradition, as can be expected, have emphasized different aspects of the role that language can play in both academic science and classroom communities, and Carlsen (2007) traces these perspectives to several leading schools of thought, notably the influence of Vygotsky, social semiotics, and studies of situated cognition within communities of practice.

The predominance, however, given to Vygotsky and neo-Vygotskian perspectives (Moll, 1990; Mortimer & Scott, 2000; Wertsch, 1985) on teaching and learning within the socio-cultural tradition—especially with prevailing socio-constructivist views of learning (Duit & Treagust, 1998), where Rowlands (2000) argues it sometimes misconstrues Vygotsky—along with a resurgence of interest in Deweyan pragmatism (though offering many compelling insights), have almost entirely obscured the substantial *hermeneutic tradition*. This tradition is quite eminent in Continental philosophy and the humanities. Indeed, one can generally observe that there has been very little written by way of comparison in science education literature that has taken this other philosophical and humanistic tradition into serious consideration (some exceptions being Bevilacgua & Giannetto, 1995; Donnelly, 2001; Eger, 1992; Hesse, 1980). It has certainly largely ignored arguments of those authors, such as Rorty's influential *Philosophy and the Mirror of Nature* (1979), who, though coming from the Anglo-analytic philosophical tradition, have nonetheless sought to more closely align the pragmatic school with hermeneutics.[2] This chapter seeks to offer a corrective to balance out these two prevailing tendencies within science education, principally by critically re-examining the Deweyan progressivist perspective on language by identifying its serious shortcomings—which, as argued here, several authors have glossed over or simply failed to recognize—and contrasting this with the *philosophical hermeneutics* of the important 20th century German philosopher Hans-Georg Gadamer (1900–2002).

The late physicist and philosopher Martin Eger, in a series of original papers (1992, 1993a, 1993b), had insightfully shown the relevance of Gadamer's "philosophy of the humanities," notably for science education with regard to the *interpretation* of nature, but especially of science *texts*. Hermeneutics, an age-old scholarly discipline, ties understanding to the ability to achieve personal meaning when interpreting any text (utilizing the "hermeneutical circle" method). The significance of his ideas lies in offering an alternative approach to viewing science learning and knowing, drawing science education away from psychological and cognitive science perspectives and towards philosophy and the humanities (Bontekoe, 1996; Donnelly, 2001; Gallagher, 1992). Today Eger's and Gadamer's ideas are finding useful expression in some research work (Borda, 2007; Kalman, 2011). Both explicitly shift the emphasis away from epistemology towards *ontology*, away from "knowing" in the objectivist sense to interpreting, meaning, and being. This shift, or "interpretative turn" (Hiley et al., 1991), has not been entirely endorsed as regards questions surrounding the NoL, ontology, and the relationship between epistemology and hermeneutics. The next sub-section provides science educators with a brief but updated outlook regarding these major topics.

5.1.2. Epistemology and/Versus Hermeneutics

Any discussion involving philosophical hermeneutics recognizes two current states of affairs, namely the ongoing unresolved dispute over the self-conception of philosophy, and secondly, the so-called "interpretative turn" from epistemology *to* hermeneutics.

To the first, one identifies that the modern Anglo-analytic philosophical tradition has fractured into two differing schools of thought as to what the nature and role of modern philosophy *is* and can accomplish (represented by the opposing views of Dummett and Rorty; Bernstein, 1983), as discussed in Chapter 1. This opposition is reflected as well in contrasting perspectives on language theory—which Charles Taylor has characterized as the *designative* and *expressive* traditions (Medina, 2005, p. 39; discussed below). That said, authors like Bernstein, Rorty, and Taylor nonetheless all comment on the convergence of thinking in both the Anglo-American and Continental traditions that rejects *foundationalism*, or the former project of grounding philosophy, knowledge, and language ("objectivism"), as Descartes, Kant, Russell, and the early Wittgenstein sought but failed to do.

With the current preoccupation of repudiating this formerly eminent epistemological tradition,[3] the task of "overcoming epistemology" has come to mean different things to different thinkers (Baynes et al., 1987). Dewey and Bentley (1949), for instance, sought to overcome subject/object dualism with his pragmatic focus on "transaction," the active/practical behavior taking place between the knower and known. Taylor (1987) correctly views both Quine and Rorty as abandoning foundationalism (with the former attempting to "naturalize" epistemology), while he solely targets overcoming the conception of knowledge as *representation* that lay behind the ambition of the foundationalist project since Descartes:[4] "If I had to sum up this understanding in a single formula, it would be that knowledge is to be seen as correct representation of an independent reality. In its original form it saw knowledge as the inner depiction of an outer reality" (p. 466).[5] One notes that representation plays a significant role in science and science education, and Giere (1999) argues, in contrast, for its continued importance in science independent of foundationalism. Indeed, some philosophers and science educators have argued for a "fallibilist epistemology" as a viable alternative to opposing foundationalist and radical constructivist views of knowledge and belief (Siegel, 2001, 2010; Southerland et al., 2001). The collection of papers in Carr (1998) intends to help guide curriculum policy beyond "rational foundationalism" and "promiscuous postmodernism." The discussions in these works can contribute to advancing teachers' epistemological conceptions and deliberations, whether concerning science, curriculum, or student learning.

The second aspect, as stated, acknowledges an "interpretative turn" to have taken place not only in philosophy (due initially to Heidegger, 1977) but in the natural and social sciences as well (inclusive of language theory)—though granted, still subject to much dispute—that also seeks to move "beyond objectivism and relativism" (according to Bernstein, 1983).[6] Such a move can be considered a shift in the philosophical emphasis entirely "from epistemology to hermeneutics," as both Rorty and Gadamer have claimed[7]; certainly it can be admitted that the relation between the two modes of inquiry is contentious and that differing conceptions of language inform both.

Furthermore, although there are many similarities in Rorty's and Gadamer's positions, there exist important differences as well as to the nature and task of epistemology and hermeneutics, which is instructive. For example, while Rorty would agree that Anglo-analytic philosophy of language has slowly come to abandon the notion of language as correct "picture of the world,"[8] he would disagree with Gadamer's universalist perspective of philosophical hermeneutics (with its inherent view of language as the *medium* of all understanding). Both agree that hermeneutics is not to be considered a successor to epistemology, rather that it involves an entirely different approach to comprehend the world—indeed Rorty construes it as a kind of "paradigm shift" (one that is holistic, historicist, and pragmatic). While Rorty makes a sharp distinction between the two but sees them as complementary and mutually supportive (epistemology for "normal discourse" and hermeneutics for "abnormal"), Gadamer views them rather as antagonists: hermeneutics as the universal condition of understanding (and hence of *being; Dasein*)[9] but epistemology as a failed *epistéme*-based, historico-philosophical venture whose time has come and gone. The project has died and should be buried.

Rorty correctly stresses that Gadamer had emphasized *Bildung* as historical enculturation (hence the crucial role of education) as a proper goal of hermeneutics—construed as an open project of how understanding takes place through interpretation and dialogue, a form of *inter-subjectivity*. This is seen in contrast to "knowledge" possession and obsession of isolated, individual cognition (the foundationalist project), but he would not consent that such "understanding" entails knowledge. Rorty is clear that "knowledge" is fallible and constrained to the "normal discourse" of a particular (historical) socio-cultural paradigm (explicitly referencing Kuhn's ideas).[10] But taking such a position on a *standard* of knowledge, one can argue, alternatively, must implicate Rorty's outlook as committed to the epistemic assumptions of Cartesian foundationalism.[11]

There is certainly more that can be surveyed here in the debate about the shift "from epistemology to hermeneutics." Siegel (2010), for instance, takes issue with Taylor's arguments for "overcoming" epistemology, while Suchting (1995) criticizes many of the "lessons" supposedly drawn from hermeneutics. Several very important questions exist that still need addressing, such as, if the common division between *explanation* and *understanding* is abandoned—which has long been accepted as *the* major difference between the natural and social sciences (Mason, 2003)—and "interpretation" comes to characterize all human inquiry, does or should a "contrast class" exist in opposition to it? Thereto, how can or should one demarcate the lines between the humanities and the different sciences? Moreover, how does one adjudicate between better and worse interpretations? Is hermeneutics[12] really an alternative paradigm to epistemology (as Gadamer and Rorty insist), or another, albeit extraordinary, version of epistemology itself, just not of the classical foundationalist sort (as Rockmore, 2004 and Westphal, 1999 contend)?[13]

There are fundamental issues and concerns identified here that a PSE would equally need to consider and evaluate, which have necessarily arisen in the dispute between the advocates of epistemology, hermeneutics, and their different perspectives on language, knowing, and understanding. These issues, however, cannot be pursued in the present work.

5.1.3. Comparing Egan and Gadamer

Before examining the contrasting views of language offered by Dewey and Gadamer and what this implies for science education, a brief comparison of Egan with Gadamer may be called for, since Egan has occupied such a central place in this book. Such a comparison will bring to light selected *limitations* to his metatheory. Some similarities immediately come to mind: both start with the question of *meaning* and the chief role this must play for understanding (it also performs as learning *outcome*); thereto, secondly, both have made "understanding" a central concern and the main plank in their philosophy; thirdly, both accept the *historicity* of language and the fact that this conditions human experience, learning, and knowing. Consider as well the obvious similarity between Gadamer's definition of hermeneutics and Egan's purpose for education, for I had previously quoted him as stating (p. 128, this volume):

> We can easily forget that learning symbols in which knowledge is encoded is no guarantee at all of knowing.... The primary trick in bringing knowledge to life from the codes in which we store it is through the emotions that gave it life in the first place in some other mind. Knowledge, again, is part of living human tissue; books and libraries contain only desiccated codes. *The business of education is enabling new minds to bring old knowledge to new life and meaning.* (Egan, 2005b, pp. 96–97, italics added)

Whereas Gadamer writes:

> The best definition for hermeneutics: to let what is alienated by the character of the written word or by the character of being distantiated by cultural or historical distances speak again. This is hermeneutics: to let what seems to be far and alienated speak again. (as cited in Bernstein, 1983, p. 250, footnote 36)

Insofar as hermeneutics can be considered an *educational* task (and Gadamer certainly allows for this possibility) both thinkers' perspectives intersect. As for historicity, while Egan's metatheory is premised upon the recapitulation of cultural-linguistic stages as evolved over time in human anthropology, it fails to acknowledge Gadamer's insight and stated principle of "effective-historical consciousness" or otherwise called "history of effect" (Gadamer, 1974, p. 299; *Wirkungsgeschichtliches Bewußtsein*). This principle admits that we are inescapably in a "temporal distance" in relation to the past that we cannot overcome; rather as historically-situated social beings we cannot avoid thinking within and out from a given historico-cultural tradition, whose perspective is itself transient and changing and which affects our very being and perception of past and current events, usually in unconscious ways (a major source of our *pre*judices). Gadamer puts it nicely: "In fact history does not belong to us; we belong to it" (in Gadamer, 1975, p. 278). In other words, such a *condition* of our existence cannot allow for any sort of true historical objectivity, and I think it is fair to interpret that Egan's recapitulation thesis assumes some sort of cultural-linguistic objectivism. One could argue that this rejoinder can put into question the "truth value" of the entire metatheory; alternatively, it doesn't necessarily follow that it is wrong, only that our current "best evidence" (according to present historico-cultural and anthropological understanding, so Egan would argue) speaks in its favor, but that it nonetheless remains fallible (in essence Egan offers a kind of "inference to the best explanation" argument). But I am of the opinion this doesn't detract from its current merit as an educational metatheory. What does detract is (as mentioned last chapter) that a *criterion* is missing for judging the CK of science curricular material (especially textbooks), commonly taken as ahistorical unchanging, objective fact—in line with the typical Enlightenment conception of knowledge objectivism.

Linked to the previous objection one could also make the argument that Egan remains in the orbit of Enlightenment thought insofar as he exhibits those aspects of the philosophical and psychological tradition that seek to fashion the view of the child according to developmental stages that are presumed to be known in advance and can be formulated by theory (e.g., whereas Dewey would look to psychology, science education had first looked to Piaget and later to cognitive science, Egan looks to Vygotskian tools developed socio-culturally). It should be noted that hermeneutics takes an entirely different view and does not specify any sort of conception or image of the developing child; instead it takes its lead from Heidegger's existential notion of persons as "being" (*Dasein*) that are "thrown-into-the-world." From this *ontological standpoint* they reach out and search for meaning and understanding, one which they can only achieve through the "hermeneutic circle of interpretation" as concerned and communal "languaged" beings. This is in effect an *ontological* shift away from epistemology, and one could argue that here, too, Egan remains part of the Western epistemological tradition—at least insofar as one can presume "to know" *when* and *what* stages the child will/must pass through and that this can specified as an educational program of development. (This is indicated as well by the fact that Egan talks of "mind" and not "being" of the learner.) The advantage such an ontological perspective offers is that it provides insight and a mechanism for how individual learners proceed to learn anything and actually "come to understand." The one serious disadvantage is that it does not afford the educator a framework for structuring curriculum according to the age-developmental phases of the growing child, allowing for decision-making about "when" and "why" to teach—only the "how to learn." This should not surprise, as hermeneutics is not meant to function as an *educational* metatheory exclusively, on the contrary as now seems clear, "only" for clarifying the learning process—but as such I submit it provides more powerful ideas for the educator than the constructivist and cognitive science schools.

As stated above, it was Martin Eger (1993a, pp. 11–19) who first provided an astute exposition of this "ontological turn" for science educational purposes, especially with respect to students trying to interpret the *language* of a science *text*.[14] He explicates Gadamer's notion of the "fusion of horizons" (of meaning) that takes place between the text and the learner (the language horizon of the text on the one hand, and of the student's language horizon/perspective, or *life-world*, on the other).[15] His papers are referenced in subsequent sections.

5.2. *PSE CASE STUDY*: SCIENCE EDUCATION, DEWEY, GADAMER, AND LANGUAGE

The next sections explore the role and *nature of language* in science education with the specific focus of contrasting the differing perspectives on language as found in two seminal 20th century thinkers, John Dewey and Hans-Georg Gadamer. The broad influence of Dewey's ideas on science education is generally acknowledged (DeBoer, 1991; Fensham, 2004), but, as will be argued here, the linguistic assumptions behind his progressivist educational philosophy remain largely unexplored and problematic while they continue at the same time to resonate and detrimentally affect science learning. What is suggested is that these assumptions should be made explicit for science educators while at the same time providing for the exposure to a different conception of language, one provided in the *philosophical hermeneutics* of Gadamer (1976/2008; 1960/1975). This is one that embeds instead the well-known dialogical component of language within its *ontological-historical dimensions*, one that stresses that all understanding is essentially interpretation and that sees learning itself as an interpretive event within the "hermeneutic circle" of understanding (Bontekoe, 1996).

Although such ideas may sound radically new and even obscure to most science educators, they actually constitute part of a worldwide scholarly conversation that has taken place for over 25 years pondering the nature of several different domains: language (Medina, 2005), education (Eger, 1992; Gallagher, 1992), science (Heelan, 1991; Hesse, 1980; Ihde, 1998), and philosophy (Bernstein, 1983; Hiley et al., 1991; Rorty, 1979; Taylor, 1987). This overall discussion has come in general to deliberate upon and value the significance of Gadamer and hermeneutics for these diverse fields. It is argued that science educators can directly profit from an exposure to that ongoing and cross-disciplinary discourse, and this analysis serves the further purpose of illustrating another important role that PSE can fulfill, namely *the critical appraisal of language philosophies as related to science teaching and learning.*

In what follows, I begin by reviewing Dewey's legacy in science education along with some past and current critiques of "traditional" science education, pointing out that these critiques, too, carry out an understanding of language that is akin to Dewey's. Next, in the fourth and fifth sub-sections, I detail the linguistic assumptions and shortcomings of Dewey's progressivism. In the sixth and seventh sub-sections, the Gadamerian alternative to Dewey's "progressive linguistics" is introduced, and I detail some of the work in science education that has already begun along hermeneutic lines. What is suggested is a truly progressive understanding of science education that breaks with "progressive linguistics." (This new progressivism can be called "more Deweyan than Dewey.")

As an important preliminary comment it can be admitted at the outset that Dewey in his many books had not completely clarified his views on language, let alone develop what can be considered an explicit, full blown "theory of language." Yet I believe if one carefully canvasses his works, certain assumptions and standpoints can be gleaned from them and articulated. There are some critics who may disagree with the analysis here undertaken and the views expressed, and may suppose instead that his ultimate view on language had been presented in his last book (which he co-wrote with Bentley), entitled *Knowing and the Known* (1949), although this opinion is not shared. While one can acknowledge that he does appear in that last work to indicate a conception of language as dialogic, or *language-in-use*, or at least it can be said that he seemed to have moved more in this direction later in life—as to what that could *mean*—this may not have been his exclusive understanding, nor may the later meaning have been the dominant one in the majority of his prior works. In any event, his other writings indicate a notion of language as will be here indicated and critiqued—at minimum, he may not have been consistent in his views.

5.3. THE DEWEYAN LEGACY IN SCIENCE EDUCATION

In 1916, John Dewey offered a critique of science education that, it can be admitted, still rings considerably true today. In his widely read and still fashionable book *Democracy and Education*, he wrote the following about science students:

> The necessary consequence is an isolation of science from significant experience. The pupil learns symbols without the key to their meaning. He acquires a technical body of information without ability to trace its connections with the objects and operations with which he is familiar—often he acquires simply a peculiar vocabulary. (Dewey, 1916/1944, p. 220)

Dewey correctly noted the inability of many science students of his day to translate their classroom learning into what he called "experience" (Dewey, 1938). And reading Dewey's words today, we are reminded that the status of critique aimed at science education has changed surprisingly little in the last ninety years. Many of Dewey's observations on science education could well be contemporary. It is still the case that science education (certainly in physics and chemistry education) is too often plagued by abstract symbolic and rote memory learning (whether in regard to the text and vocabulary employed, or through use of algorithmic problem-solving), *rather than* meaningful engagement and deeper conceptual understanding (Mestre, 1991; Nahkleh, 1992; Roberts & Oestman, 1998). Science instruction and curricula worldwide (which, sadly, have changed little over many decades, irrespective of several reform "waves"; Osborne & Dillon, 2008; Wallace & Louden, 1998; Van den Akker, 1998) are still characterized, certainly at the upper levels, by the restricted, specialized focus on the mastery of decontextualized and ahistorical "technical" knowledge, as analyzed in the last chapter. Aikenhead (1997b) has referred to this ongoing and domineering goal of science education (carrying its own defining "science literacy" preconception) as "technical pre-professional training" (TPT)—which only serves, at best, the small minority of students aiming at science-based careers.[16]

Most importantly, the *language* employed in classrooms is still dominated by the "transmission" mode instead of an "interpretive" one (Lemke, 1990; Mestre, 1991; Roth, McRobbie, Lucas, & Boutonne, 1997; Sutton, 1996), which in itself belies a "positivist" view of knowledge—an assumed objectivist epistemology. This transmission model, moreover, is reinforced by the teacher's overt reliance on *textbooks* (and their language), which drives both "what is taught and how it is taught" (Stinner, 1995b, p. 275)—no doubt a major factor in Dewey's complaint about students acquiring only "technical knowledge" and "peculiar vocabulary." And thus, science education is often *not* progressive in the Deweyan sense of progressivism. Science education, looking back in time from our vantage, has seemed distinctly unable, either by lack of effort or by lack of disciplinary wherewithal, to employ the tenants of progressivism as Dewey laid them out.

From the words of Dewey quoted above, a few should be underscored: "vocabulary," "symbols," "isolation," "experience," and "meaning." While these words have no intrinsic bearing on science education, it can be noted that they are loaded with assumptions about the way that language gets used in science education to impede deep learning on the part of the student. So if "vocabulary" and "symbols" stay at the level of abstraction, if they stay "isolated" in the mind of the student, then the student will not be able to gain "experience" of the scientific notion under study, nor will he or she be able to make "meaning" of it. What I want to underscore here is that language is intimately connected to the progressive critique *of* traditionalist science education. For Dewey, there is a direct correlation between the way that *language* gets used during the process of learning science, and the way students do or do not make meaning of the science under study. Dewey's observations, then, may or may not be so contemporary, depending upon whether recent critiques do or do not share the *linguistic* assumptions of Dewey's progressive educational project.

At first glance there would certainly seem to be considerable merit in his critiques of traditional science pedagogy, even sustained as they were by his linguistic assumptions (to be examined below in Section 5.4). It is no doubt because of these sorts of insights that Dewey has exerted a sizeable influence on science curricular change, more so in North America during the pre-World War II era. His main charges are: (1) that the structure of a discipline was a poor place to begin to learn a subject (whereas it should be properly aimed for at the *end* of studies); (2) that much classroom knowledge remained "inert" and useless in students' minds; and (3) that science education should be about connecting curriculum to students' personal meaning while enabling them to acquire the kind of scientific reasoning skills ostensibly employed by scientists themselves (Dewey, 1916/1944, 1938. 1945). Indeed, it is not only because of his insightful criticisms but especially his new conceptualization of the nature of schooling (as enculturation into democratic thinking and living) linked to the central role of science education within it, that he can still be seen, rightly or wrongly, as the only major philosopher of education whose ideas have significantly shaped aspects of science pedagogy not only in the past but that still hold sway in some fashion in the present, in the English-speaking world (Fensham, 2004; Kruckeberg, 2006; Shamos, 1995; Wong et al., 2001).

Although his contributions to science education reform are numerous, his principal influence came in two separate reform "waves," the first during his lifetime, in the so-called "progressive era" (roughly 1917 to 1957) of American education, and the second time (though rather less directly) with the "new progressivism" and "science literacy" emphasis of the 1970s and 1980s, including the science-technology-society (STS) reform movement (DeBoer, 1991). In both of these instances his legacy was felt in those curriculum designs, which not only called for, but actually managed a *partial shift* in the key goals of science education. The shift of emphasis was away from one focused predominately on mastery of science subject disciplines and textbook-based instruction (the customary "knowledge" aim) to instead the new *double emphasis* placed upon mastery of the alleged "science method" (as a problem-solving, socio-linguistic-based tool), as well as what has come to be known as "science for social relevance" (Bybee & DeBoer, 1994). This included a stress on more "student-centered instruction," which meant for Dewey that the subject material should be chosen from the pupil's immediate social environment and experience, that it should also exhibit some kind of application, and that *practical* (not just numerical) problem-solving would be involved within this personal field. Such innovative ideas for his time, needless to say, have today become part of the common conversational landscape and contributed to positive changes on the part of science teachers' attitudes and conceptions of science teaching and learning.

And yet these shifts did not go unchallenged by more traditional-minded science educators (who continued to lay stress instead on knowledge attained through science discipline structures, as many still do today), not even in his time. And the "social relevance" curricular theme—certainly reinterpreted and expanded since then (DeBoer, 1991)—has not been universally endorsed (and where endorsed only to limited degrees), being plagued by various kinds of substantive interpretive, value-based and practical problems. In particular, under the guise of "social context" in the STS movement, considerable disputes erupted during the so-called "curricular wars" of the 1980s surrounding the question of what the fundamental goals of science education should be, including its own *self-conception* as a discipline—whether it was to be (putting the issue in simplified terms) more concerned with discipline-centered knowledge (as *ends*) or instead "knowledge-in-use" (as *means*) for social purposes (Bybee & DeBoer, 1994, pp. 378–380; Good et al., 1985; Yager, 1984).[17] Unfortunately, important aspects of these disputes remain unresolved to this day within the worldwide science education community, especially with respect to the contending conceptions of "science literacy" (for example, how and if it should be tied to *citizenship*) as a defining goal and hence what the most suitable orientation of the curriculum should be.[18]

It was certainly Dewey's unique and particular stress that a supposed "method" was to be taken as both *means* and *ends*, as a means to how knowledge is acquired, but also as a goal of instruction, for he maintained the essence and power of science lay more so in its method of inquiry than as an accumulated body of "technical" knowledge (Bybee & DeBoer, 1994, p. 371; Dewey, 1945). And because he believed that such a method could be cultivated in students as an attitude, as a generic way of thinking and problem-solving (through the *correct use of language as a representational and social tool*)—what he termed "scientific habits of mind"—it could be applied to any subject matter. Therefore, he assumed, it would help them solve real-life problems and prepare for democratic citizenship in the emerging, ever-changing industrialized society of his day.

His sense of science education then was also intimately tied to his reformulated purpose for schools, now well-known, namely, "to ensure that students secured knowledge that would assist them in coping with the problems they faced as participants in society" (Briscoe, 1990, p. 21). Hence the paramount stress on the *utility* of knowledge, "how things worked and how to do things" instead of "knowledge about things." In this perspective, science education was to play a central and leading role in such a revised curriculum, necessarily forcing a kind of subordinate status on the other school subjects. As could be expected, such a prioritizing of subjects and aims naturally lead to resistance on the part of science and especially non-science teachers. Moreover, the chief focus on the utility of knowledge and schooling for the *predominant* purpose of fostering "critical citizenship"—especially for those topical reform movements that have taken this view entirely on board for science education, like STSE and others aligned with it (e.g. "socio-political activism")—harbors considerable problems in it own right. Namely, that it tends to diminish knowledge-in-itself as an intellectual pursuit—also the aesthetic, creative side of science—as well as other values of science education, such as for personal self-development and advancing cultural literacy, among others (DeBoer, 2000; Hadzigeorgiou, 2008).

Be that as it may, it is not my immediate purpose to critically appraise these influential ideas of Dewey here (which continue to be pursued in various aforementioned guises); others have already done so: especially concerning those ideas feeding contemporary progressivist notions that still interpret and seek to instrumentalise education as a prevailing *socializing project* (Apple, 1992; Egan, 2002; Jenkins, 1997; Levinson, 2010; Nyberg & Egan, 1981); or those criticisms bearing on the reappraisal of his various views on "science method" will not be rehearsed, taken as "thinking habits," also "process" and experimental inquiry[19]—whether related to the questionable legitimacy of his views on scientific methodology (as a type of hypothetico-deductive reasoning, itself today under scrutiny in PS and science education, as discussed)—or especially how his views on "logical attitudes" and scientific "habits of mind" bear upon current discussions on "critical thinking," which today have come to heavily criticize his original notion of generic and transferable thinking skills (Bailin, 2002; Hodson, 1996; Kuhn, 1993). Although the several factors listed can help account for the stalling of the Deweyan project over time (each contributing in its own way), the immediate focus here will be on his conception of language. It is his linguistic conception, as will be argued, that lies at the crux of what has proven to be most unworkable for what might otherwise have been a more successful program of progressive science education.

5.4. JOHN DEWEY'S LINGUISTIC ASSUMPTIONS

To truly understand the progressivism of Dewey, then, one must understand his overall orientation toward language itself. And to be clear about this orientation, one should note at least *four tenets* of what I am calling Dewey's "progressivist linguistics," his four assumptions about language. These are that language is representational, it is instrumental, it can be separated into two elements, and it exists on a separate ontological plane.

First, language is *representational*. Language symbolizes ideas and things, and in doing so, allows humans to communicate without much unnecessary toil. If people didn't have the use of language, they would need to go to great lengths, using gestures or even replicating the thing to be represented, in order to communicate. But since language is available to *reflect reality*, we can communicate without much effort. Of course, this representational aspect of language comes with one primary need: that language be used to represent *well*. Since language is a mirroring of the world, we must be careful to clean the mirror and make it shine. The better that language reflects, the better language does its job.

In *How We Think*, Dewey (1910/1991) describes the role of linguistic symbols in the thinking process as follows:

> [Symbols] are symbols only by virtue of what they suggest and represent, i.e., meanings. They stand for these meanings to any individual only when he has had experience of some situation to which these meanings are actually relevant. Words can detach and preserve a meaning only when the meaning has been first involved in our own direct intercourse with the thing. To attempt to give a meaning through a word alone without any dealings with a thing is to deprive the word of intelligible signification; against this attempt, a tendency only too prevalent in education, reformers have protested. (p. 176)

Language is a symbol system that, on its own, unforced by human will, does not lend itself to meaning. Language remains in the form of the uninterpreted symbol if not connected to experience. Language remains sedentary as a symbol system until memory and experience are brought to bear through the use of language. Unless people use language for thinking, it will remain "only" a symbol.

Second, language is *instrumental*. Language helps us to do two things, to think and to communicate. To understand, first, how language helps us to think, we can quote Dewey (1910/1991):

> To say that language is necessary for thinking is to say that signs are necessary. Thought deals not with bare things, but with their *meanings*, their suggestions; and meanings, in order to be apprehended, must be embodied in sensible and particular existences. Without meaning, things are nothing but blind stimuli or chance sources of pleasure and pain. (p. 171)

If language did not exist, then a person could only think for him or herself if that person could address every thought by interacting with its *concrete* existence. This would be such a cumbersome situation that the thinker would never get much thinking done. But the thinker can, instead, use language as a short-cut to thinking, a short-cut that need not deal with concrete existence. Language is like a road map of a city. It allows us to deal in a conceptual way with all those roads we haven't been on. Language, in other words, is a tool for thinking. It is a good, versatile, light (in fact weightless) tool. Language is also a good tool for sending meaning from one person to another. When language sends the thought of one person to the waiting mind of another, language has helped one person to understand another. Language is a tool not unlike a *conduit*; its structure allows transfer from one thinker to another.

Dewey's often-cited description of language as the "tool of tools" once again brings home this instrumental understanding of language:

> But at every point appliances and application, utensils and uses, are bound up with directions, suggestions and records made possible by speech; what has been said about the role of tools is subject to a condition supplied by language, the tool of tools. (Dewey, 1981, p. 134)

Third, language can be separated into its *intersubjective and cognitive elements*. Dewey maintains that language allows individuals to be in contact with one another, and that it also helps us to think. For Dewey, language is most often used for the former rather than the latter. People's communication differs from thinking insofar as communication does not *trouble* the nature of what is being communicated, whereas thinking always troubles the nature of the symbol into meaning-making. It is easier to communicate than it is to think, and so people most often settle for communication. Dewey (1910/1991) notes,

> The primary motive for language is to influence (through the expression of desire, emotion, and thought) the activity of others; its secondary use is to enter into more intimate sociable relations with them; its employment as a conscious vehicle of thought and knowledge is a tertiary, and relatively late, formation. (p. 179)

So language can influence, it can establish relations, or it can embrace thought. The first of these two are oriented toward intersubjectivity *per se*, while the last has a cognitive aspect as well. The use of language for relation is more ubiquitous than the use of language for thinking because the former's origins are so ancient. That language's "primary motive" is to *establish relation* intimates a linguistic anthropological prejudice that follows directly from Dewey's Darwinian understanding of primate development and Spencerian view of evolution (Egan, 2002). Before human beings even acquired language, they were social beings. The tool of language came later, and it was used first of all to foster sociability. Only later, apparently, did it take on a more cognitive function.

This Deweyan distinction—between influencing an other, entering social relations with an other, and transmission of information to an other—actually fine-tunes the aforementioned *instrumental* and *symbolic* descriptions of language. As an instrument, as the "tool of tools," language serves as a means to two possible ends. On the one hand, language may be used to nudge, or poke, or move another person. It may be used to establish bonds. On the other, it may be used as a vehicle (a conduit, as stated previously) for moving ideas, thoughts, and meaning from one person to the next. So the tool of language may be a tool pure and simple, or it may be a tool with meaning inside. It may be a hammer, or it may be a Trojan horse. Both are tools, but one has contents that are bound to inform and surprise. As a symbol, too, language may be primarily intersubjective, or it may be more informative. Most linguistic symbols, at least under this Deweyan description, have meanings that enable humans to convey social demands and social desires to one another. Yes, linguistic symbols stand for human meanings and are therefore more complex than rudimentary gestures, nevertheless, most of them lead to relations rather than thinking itself. It is only at the more self-aware, contemplative end of the spectrum of human interaction where symbols become thick with rational intentions.

Fourth, language is on a *different ontological plane* than other states of being. This last Deweyan commitment to language is best understood as the common denominator of the first three assumptions mentioned above. When language is symbolic, when it is instrumental, and when it serves the higher purposes of interaction and thinking, it has such attributes and functions because it is fundamentally different than non-linguistic life. Language is different from other sorts of experience. It symbolizes things without being those things. It is used like a tool toward certain ends without being a part of those ends. It fosters human interaction but it is not human interaction *per se*. It is used to promote thinking without being the thinking that it promotes.[20] This profound difference between language and other aspects of life is clearly demonstrated by the secondary roles that Dewey assigns to language. Language, under its Deweyan description, brings on, or sometimes stalls, but is certainly not the same as, human experience.

Dewey is certainly not unusual with regard to this presumed ontological difference. John Stewart, who calls this perspective of ontological difference a "two worlds" commitment, has noted the following:

> the two worlds claim is most basic. As reviews of the history of linguistics demonstrate in detail, this claim embodies the ontology first established in Platonic and Aristotelian formulations of the nature of language. The basic distinction between linguistic and nonlinguistic worlds was articulated explicitly in the influential Aristotelian formula that Heidegger cited: "Spoken words are the symbols of mental experience and written words are the symbols of spoken words." This became the medieval claim linking *aliquid* and *aliquo*, which was developed into John Locke's claim that words are "signs" that signify "ideas," and the connection in Wittgenstein's *Tractatus Logico-Philosophicus* between "propositions" and the "objects of thought" that they "picture." (Stewart, 1996, p. 16)

For the purposes of this chapter, there is no need to follow up on these historical echoes of language's ontological difference other than to say that Dewey was in good company.[21] As will be pointed out later, though, by joining this longstanding commitment to two worlds, Dewey derails his own progressive project. The two world commitment is both central to Dewey's other linguistic assumptions, and central to the unsustainability of progressive education in general, and progressive science education in particular.[22]

It should be pointed out, though, that the linguistic company that Dewey keeps based on our analysis, locates him primarily within one of two great semantic traditions in the history of philosophy, that the philosopher Charles Taylor has termed *designative*. This major tradition can be contrasted with the alternative *expressive* tradition, which he traces back to Herder and Humboldt in German Romanticism (early 19th century), and has continued in time to be further elaborated by authors such as Husserl, Heidegger, and Gadamer, which in turn, is linked to the "dialogic perspectives" of language by such diverse modern thinkers as Wittgenstein, Bakhtin, Foucault, Habermas, and Rorty (Medina, 2005, pp. 39–46).[23] As will be elucidated below, these two traditions, broadly conceived, mark out a fundamental difference in the conception of what language is and does, seen either primarily as "symbol and tool" and hence as contrasted with real "lived experience," or instead viewed as the medium of the "life-world" experience of human understanding itself.

5.5. DEWEY'S LINGUISTIC PROGRESSIVISM

Given the above synopsis of Dewey's linguistic assumptions, I will now offer an analysis on how these linguistic assumptions bear on progressive pedagogy. Using the four-part metric offered above, it is possible to interpret Dewey's progressivism in a linguistic way. This interpretation will map Dewey's linguistic assumptions onto his progressive, educational preferences, as well as onto his biases against traditional education.

Language is Representational

Dewey links the representational nature of language to progressivism by noting that there are *three* possible aims of education when it comes to representation: to focus on language itself ("pure symbol"), to focus mainly on the object of representation ("pure experience"), and to focus both on the object and on language. Certainly, a mere verbalism, or "verbal methods," are the choices of the traditional methods Dewey rails against (1910/1991, p. 178). And Dewey is also not interested in the uncontrollable nature of the direct experience. Progressivism opts for the third way, insisting that words *and* objects go hand-in-hand within educative experience. For Dewey, words and objects are as natural a pair as the map and its city. They must be experienced with each other if the country student is to be educated about the city. This is precisely why Dewey does not abandon education *per se* in turn for direct experience. For him, the symbol and the thing, together, are a necessity for thinking and learning.

But language can remain at the *passive level of symbol* when not pushed into thought and meaning: This is well explained by Dewey (1916/1944) in the following passage:

> The outcome is written large in the history of education. Pupils begin their study of science with texts in which the subject is organized into topics according to the order of the specialist. Technical concepts, with their definitions, are introduced at the outset. Laws are introduced at a very early stage, with at best a few indications of the way in which they were arrived at. The pupils learn a "science" instead of learning the scientific way of treating the familiar material of ordinary experience (p. 220)

When language stays at the level of symbol and abstraction, then students (and often teachers) become satisfied with mere "verbalism," with the façade of true thinking and knowing. Such instances of verbalism were rife in the science classrooms against which Dewey rails, and progressivism, on the contrary, champions experiences that are *real* experiences rather than *simulacra* of experiences. Dewey is thus not happy with the pedagogical implications of language's representational life. Progressive education would use language as a symbol system only as a means to another *end*—that end being an educational *experience* rather than educational verbalism.

Here one can think of all the books in a well-stocked library. In the library resides language. The traditional educator will give books to his or her students, but the learning will only be "book learning." The traditional student will not make personal meaning out of those tenacious, abstract symbols. This traditional student can be contrasted to the progressive student who will experience words as alive and meaningful. For the traditional student, the books remain passive symbols of someone else's authority, while for the progressive student, these same books will come to life and break out of their symbolic constraints.

Language is an Instrument or a Tool

This assumption coincides with the communal nature of progressive education. As an instrument for thinking, the word is never mine alone. It has always been initiated by someone else, and I am indebted to that someone else for making the sign useful to me. That the sign allows me to think means that someone else has helped me to think. Language serves to link together people who shared the experience of its use. It creates bonds for those who have used the same words. Language is a communicative endeavor. It facilitates interaction, communion, dialogue, and in fact all the necessary person-to-person interactions that are so necessary to a fruitful reconstruction of democratic habits. Education, like democracy, "is primarily a mode of associated living, of conjoint communicated experience. The extension in space of the number of individuals who participate in an interest so that each has to refer his own action to that of others" (Dewey, 1916/1944, p. 87). Education, like democracy, uses language for its communal *ends*.

For Dewey, the tool of language is used poorly in traditional education. It is misused because the traditional educator gives edicts and pronouncements rather than encouraging participation and discussion. The traditional educator does not use words to bring students together, and this happens primarily because the traditional educator considers the *autonomous growth of rational minds* the most important goal of education. When Dewey rails against traditional education by stating that the "funnel" metaphor of traditionalists must be replaced with the "pump" metaphor of progressivists, he reminds us that traditional education sees students as disparate "vessels" to be filled, one by one. Language, according to Dewey, *can* serve this purpose. It can act as a conduit for the transmission of meaning from individual to individual. However, if language is used in this way, then it will not fulfill the *social aims* of progressive education.

Language Has Intersubjective and Cognitive Elements

Dewey recognizes in language two elements that roughly parallel the distinction between progressive and traditional education. Language serves to bring people together, and language serves to transfer meaning from one person to another. While progressive education is concerned with the former function of language, that of bringing people together, traditional education is concerned with the latter function, that of transferring meaning from teacher to student. Here is where the two primary goals of progressive pedagogy, the goal of social amelioration and experiential learning, can be seen most starkly as two corollaries of a particular linguistic orientation. For Dewey, the intersubjective role of language is to be lauded because language brings people together for a greater social good. The cognitive role of language, on the other hand, is something to be tolerated, at best, and more preferably eschewed. If enhanced thinking is a goal of education, and certainly progressive education does not shy away from thought itself, then such thinking is better done through *experience* than through language. While progressive education embraces the social agenda of language, it eschews its cognitive elements by advocating experience over talking.

Once again, traditional education commits a linguistic *faux pas*, embracing the wrong end of the linguistic spectrum. If language functions on a continuum from sheer intersubjectivity, on one hand, to sheer autonomous cognition, on the other, then it is the linguistic mistake of traditional education to emphasize the wrong end of this continuum.

Language is on a Different Ontological Plane

The commitment to two worlds, one where language abides and another where "real" experience happens, is, as I intimated earlier, the bedrock of Dewey's progressive linguistics. It is the common denominator of language as symbolic and social/cognitive tool. This commitment to two worlds is, first of all, a way of describing the difference between something that helps to achieve a goal, and the goal itself. It differentiates betweens tools and non-tools. Second, this commitment distinguishes between a symbol, on one hand, and the thing such a symbol stands for, on the other. Third, this commitment illustrates the two ends of a linguistic spectrum. On one end, the intersubjective end, language causes something "real" to happen. On the other end, the cognitive end, language causes "mere" thinking (as opposed to "real" doing). Thus the commitment to two worlds helps to establish the various ways in which language falls short of the more serious business of "real" experience.

This bedrock, two-world, understanding of language helps to supply what might be considered *the* guiding metaphor for progressive education. Progressive education, after all, is an alternative to traditional education. It is an alternative to an older form of education that is presumed to be somewhat *less* than experiential, somewhat less than socially ameliorating. This new alternative is to traditional education what language is to the "real" world. Thus when Dewey describes the "linguistic" methods of traditional education, he invokes a two-world framework that was, and still is, quite a common understanding of language versus "real" life.

Shortcomings of Progressive Linguistics

While progressive linguistics provides Dewey with a concrete, linguistic metaphor for the difference between progressivism and traditionalism, it is important to understand that, ultimately, such an understanding of language *undoes* progressivism at the same time that it provides its bedrock. Dewey's theory of language is so tight that his only options are to either *escape from language more or less completely*, or, *make the same linguistic blunders that he accuses traditional education of making*. This can be explained with further recourse to the linguistic themes elaborated above.

With regard to *instrumentalism*, it should not be forgotten that it is in the nature of a tool to be used for a purpose, and then to be cast aside. When one uses a hammer, for example, one uses it for the purpose of driving in a nail. When that hammer has served its purpose, and when the nail has been driven, then the hammer is of no more use. It is thus in the nature of a tool that the tool should outlive its usefulness. In this regard, I repeat, it should not be forgotten that Dewey considers language the "tool of tools." And, indeed, when Dewey's progressivism is put under close linguistic scrutiny, one finds that the educational use of language is put in the same *tenuous* position as other tools. Language itself is a tool that serves other ends, ends that, once achieved, signal that language has served its role and is no longer needed. Language serves as a tool for sociability and language serves as a tool for thinking, but ultimately, language is no longer needed once it has served these ends.

Insofar as education itself should be considered a *linguistic* endeavor, Dewey's instrumental understanding of language paints him into a corner. Language, as it appears in curriculum documents, and in the words that are exchanged between students and teachers, is something to be experienced for a while as a means to further ends. The very life of education—its curriculum, its pedagogy, its student interactions—are linguistic practices to be experienced for only as long as necessary in order to master them. After such a time as these linguistic tools have been successfully used, it is time to discard them and get on with the experience for which linguistic interaction prepares one. The school provides linguistic tools to be used and then discarded when they are no longer necessary.

All of this would be fine if progressivism promoted a version of education similar to the one promoted by traditional accounts, a version where education is preparatory to the rest of life. But Dewey in fact promotes *just the opposite*. Perhaps most famous of Dewey's criticisms of traditional education is that it is preparatory rather than participatory. Contrasting his own progressivism with traditional ideas, Dewey (1916/1944) states in *Democracy and Education,*

> The first contrast is with the idea that education is a process of preparation or getting ready. What is to be prepared for is, of course, the responsibilities and privileges of adult life. Children are not regarded as social members in full and regular standing. They are looked upon as candidates; they are placed on the waiting list. (p. 54)

Dewey is in a tight spot here. He promotes language as the tool of tools. And following the progressive program, education is a venue where this tool of tools is used for community-building and for thinking. But if language is instrumental in this way, then progressive education itself, insofar as it must be an endeavor *in language*, is no less preparatory than traditional education.

Dewey's understanding of the *symbolic role of language* is no less pernicious in undercutting the progressive project. In this regard, it is instructive to look at his symbolic understanding of curriculum. In *The Child and the Curriculum,* Dewey (1901) describes curriculum as a map, an ordered set of instructions symbolizing the combined wisdom generated by thinkers of the past. The "map" of curriculum serves to show students the quickest way to negotiate "knowledge" handed down to them.

> The map is not a substitute for a personal experience. The map does not take the place of an actual journey. The logically formulated material of a science or branch of learning, of a study, is no substitute for the having of individual experiences. The mathematical formula for a falling body does not take the place of personal contact an immediate individual experience

> with the falling thing. But the map, a summary, an arranged and orderly view of previous experiences, serves as a guide to future experience. (p. 20)

For Dewey, curriculum serves as symbol for "the having of individual experiences." And this symbolic relationship of curriculum to experience, is none other than, is equivalent to, the symbolic relationship of language to "real" life. Dewey (1916/ 1944) has succinctly stated this equivalence in his influential book *Democracy and Education*:

> It has been mentioned, incidentally, that scientific statements, or logical form, implies the use of signs or symbols. The statement applies, of course, to all use of language. But in the vernacular, the mind proceeds directly from the symbol to the thing signified. (p. 222)

Here, in his own description of curriculum-as-linguistic-symbol, Dewey says something somewhat contradictory about progressive education. While the *natural* tendency, the vernacular tendency, of language use is to "proceed directly from the symbol to the thing signified," the *educational* tendency of progressive education is to create curriculum that, while facilitating the transmission of vast amounts of accrued knowledge, requires a symbolic detour on the part of the student, a detour that is neither natural nor of the vernacular. It is once again difficult to see how progressive education succeeds at its own calling to be *of* experience rather than *preparatory to* experience—this, because of Dewey's own account of the linguistic/symbolic role of curriculum.

Progressivism is similarly hamstrung by its commitment to language as both *intersubjective and cognitive*. As previously mentioned, Dewey (1910/1991) most appreciates language when it serves as a tool for human interaction. He most appreciates that "the primary motive for language is to influence (through the expression of desire, emotion, and thought) the activity of others" (p. 179). Following Dewey's philosophical anthropology of language, language's role in thinking and cognitive expression is more of an afterthought, and thus less true to the human condition of sociability. To emphasize, once again, this appreciation for language's social aspects, these words from *Experience and Nature* (1981) are instructive:

> The heart of language is not "expression" of something antecedent, much less expression of antecedent thought. It is communication; the establishment of cooperation in an activity in which there are partners, and in which the activity of each is modified and regulated [coordinated] by partnership. (p. 141)

At first glance, this bias of Dewey's against language that it is an "expression of antecedent thought" is in keeping with his bias against traditional education as a practice of transmission rather than growth. But as it turns out, this linguistic bias of Dewey's is also a fair condemnation of progressive education too. If expression of antecedent thought is, as Dewey maintains, a derivative linguistic activity, it is instructive to go back once again to the linguistic/symbolic role of curriculum. For curriculum, too, no matter how well "psychologized," is fundamentally a matter of expressing antecedent thought (Dewey, 1901). As Dewey prioritizes the intersubjective elements of language over its cognitive uses, one must wonder if education itself, whether it be progressive or traditional, isn't being de-prioritized at the same time. On one hand, Dewey explains the educational benefit of using language's cognitive elements in experiential ways, emphasizing how curriculum should be taken out of its abstract state and psychologized by students. On the other hand, he maintains that language's expressive, cognitive functions are not at all at the "heart" of language, are not its true orientation. So progressive education, because it must deal with concepts and ideas, with curriculum and symbols, continues, out of necessity, to have dealings with the less admirable elements of language. Thus progressive education, following Dewey's own account of language, uses the tool of language *in the wrong way*.

As indicated earlier, the *two-world commitment* can be considered the bedrock of Dewey's linguistic metaphorics, pitting the progressive education of experience against the traditional education of verbalism. Statements like this, about the tool of language, are to be expected in Dewey's (1916/1944) work:

> The emphasis in school upon this particular tool has, however, its dangers—dangers which are not theoretical but exhibited in practice. Why is it, in spite of the fact that teaching by pouring in, learning by a passive absorption, are universally condemned, that they are still so entrenched in practice? That education is not an affair of "telling" and being told, but an active and constructive process, is a principle almost as generally violated in practice as conceded in theory. Is not this deplorable situation due to the fact that the doctrine is itself merely told? It is preached; it is lectured; it is written about. (p. 38)

Suffice it to say, the two-world commitment is no less of a bedrock when it comes to the corner into which Dewey paints himself due to his linguistic commitments. In fact, Dewey quite obviously here uses the language side of the two-world metaphor to critique traditional education when progressive education itself can no more do without language than its traditional counterpart. Even as Dewey offers the above criticism, his rhetoric belies his argument: "Is not this deplorable situation due to the fact that the doctrine is itself merely told?" In what is meant to be a poignant critique of the progressives-gone-verbal, it should not go unnoticed to the reader of

Democracy and Education that he or she is, in fact, *a reader*, and that Dewey himself has just used words to *tell* us how the doctrine is merely told. This is not to say that Dewey must himself be a stand-in for progressivism, or that when Dewey speaks, progressivism has also made the error of speaking. It is rather to point out the more general difficulty, if not impossibility, of "getting out of language" and into ("real-life") experience. The two-world commitment supplies an easy metaphor to imply that such a thing can be accomplished, and that it can be accomplished by progressive education. But Dewey's own conception of the symbolic/linguistic nature of progressive curriculum quickly subverts this easy metaphor.

5.6. HANS-GEORG GADAMER ON THE NATURE OF LANGUAGE

To repeat, it is my contention that John Dewey's project of making science education progressive has been hamstrung by his symbolic, tool-oriented conception of language, an understanding that clings to the two-world commitment that *language is on a different ontological plane than other sorts of experience*. And thus, the failure of the progressivist project to reform science education, I maintain, can in large part be attributed to *a continued adherence to language as tool and symbol, as social but not cognitive, as part of a world that is not altogether "real."* In order to create the conditions for a truly "progressive" science education, it is not enough to follow Dewey's lead. One must also, paradoxically, *break* with Dewey. In order to make this break with Dewey and his particular linguistics, one could turn to the language theory of Hans-Georg Gadamer. To illustrate Gadamer's hermeneutic understanding of language, it is useful to revisit the four linguistic assumptions previously attributed to Dewey, and to offer a Gadamerian alternative to each.

First of all, language is *not* solely representational.[24] While it is a commonplace assumption in most Western conceptions of language to take language as primarily symbolic or representational, and while Dewey's conception of language typifies this sort of assumption, Gadamer helps to elucidate language's other-than-representational qualities. He notes:

> But the metaphor of a mirror is not fully adequate to the phenomenon of language, for in the last analysis language is not simply a mirror. What we perceive in it is not merely a "reflection" of our own and all being; it is the living out of what it is with us—not only in the concrete interrelationships of work and politics but in all the other relationships and dependencies that comprise our world.
>
> Language, then is not the finally found anonymous subject of all social-historical processes and action, which presents the whole of its activities as

> objectivations to our observing gaze; rather, it is by itself the game of interpretation that we all are engaged in every day. (Gadamer, 1976/2008, p. 32)

There are at least two parts to the claim that "language is not simply a mirror." The first is that there are many elements of language, indeed, the preponderance of its elements, that, while they certainly contribute to communication, do not symbolize *anything*. Communication theorist John Stewart explains this as follows:

> [In conversation] ... interlocutors are as engaged in negotiating their respective identities as they are in making assertions. Questions are at least as important as answers, and pause, stress, rhythm, facial expression, proximity, gesture, movement, and various unmarked features of vocal intonation contribute significantly to conversational outcomes. (Stewart, 1996, p. 23)

While it cannot be denied that some part of language is used to denote certain physical and mental concepts, much more of what language does consists in the more intangible negotiation of human ways of *being*, the "other relationships and dependencies that comprise our world" (Gadamer, 1976/2008, p. 32), that one cannot precisely point to in a physical way, or explain through propositional logic.

Another part to Gadamer's claim is that language is always already in the *mode of interpretation* even when it might seem to be only or "simply" symbolic. Take a statement like, "I heard a bear just outside the door last night." At first glance, this might seem to be the description of a physical state. It can be taken to be representational. However, such a statement, whether it is encountered in conversation or in text, cannot be taken in its human meaning unless it is subject to interpretation. Such a statement takes its listener or its reader *with it*. The statement forces into play a chain of hermeneutic dependencies. It brings out presumptions, associations, feelings, deductions, even out-and-out guesses, as to its meaning and significance. As Gadamer insists, the human subject, upon hearing or reading such a statement, is not at leisure to do just anything with such a statement. He or she must contextualize it, relate to it, fill it with meaning and dependencies. Likewise, the statement's originator, whether that person be a speaker or a writer, is never in a position to intend such a statement exactly as he or she chooses. The originator must always relinquish interpretive rights to some extent since the statement is as much dependent upon addressee as it is upon addressor. It is in this sense that the statement is in play "in the game of interpretation that we are all engaged in every day" (Gadamer, 1976/2008, p. 32).

A second contrast to Dewey's linguistic assumptions is Gadamer's insistence that language is *not* a tool. To quote Gadamer (1976/2008) at length:

> Language is by no means simply an instrument, a tool. For it is in the nature of the tool that we master its use, which is to say we take it in hand and lay it aside when it has done its service. That is not the same as when we take the words of language, lying ready in the mouth, and with their use let them sink back into the general store of words over which we dispose. Such an analogy is false because we never find ourselves as consciousness over against the world and, as it wore [sic], grasp after a tool of understanding in a wordless condition. (p. 62)

Whereas for Dewey language is the "tool of tools," for Gadamer the tool analogy neglects the lived debt that human beings always already owe to a linguistically situated consciousness. Once again, one can highlight a couple of aspects of Gadamer's linguistic critique, this time as it applies to the tool analogy. Words are the possession of everyone rather than the possession of specific individuals. Insofar as words are passed from mouth to mouth, they are changed and inflected in ways that tools can never be. While a tool can be flexible in its applications, its use never actually depends on the aggregation of individuals who have used it in the past. A tool remains the same tool no matter who has used it previously. *Language, on the other hand, has uses that have been historically and socially inflected.* The use of a word or phrase depends upon both the recent and not-so-recent use of that word or phrase by other people. A word or phrase used in one situation, used in one way, may or may not be able to be used for the same purposes, or with the same meaning, in another situation. Words "lying ready in the mouth" of one person may be significantly dependent upon the mouths of others wherein those same words were lying ready previously.

Moreover, unlike tools, words and phrases can never be used to fill a void where there were no words and phrases before. While a certain wrench may be the *only thing* suitable for turning a certain pipe, there are no words or phrases without which communication would simply stop. Words and phrases are not used to fill communicative voids because any would-be communicative void is itself steeped in a communicative event that is always already linguistically imbued. Not saying is a part of saying. Thus not saying and saying are in a relationship that is quite different from the relationship of being without a tool and being with one. Being without a tool is just that—it is a lack. Being without language is not possible in the same way as language never lacks; it rather turns upon a *different* meaning or a *different* interpretation whose silence is part of language itself.

Third, language cannot, as Dewey claims, be separated into inter-subjective and cognitive elements. Gadamer insists that all language is fundamentally the *language of dialogue*:

> language has its true being only in dialogue, in coming to an understanding. This is not to be understood as if that were the purpose of language.

> Coming to an understanding is not a mere action, a purposeful activity, a setting up of signs through which I transmit my will to others. Coming to an understanding as such, rather, does not need any tools, in the proper sense of the word. It is a life process in which a community of life is lived out ... for language is by nature the language of conversation; it fully realizes itself only in the process of coming to an understanding. (Gadamer, 1960/1975, p. 446)

In language, there is in fact no "purposeful activity," no "setting up of signs" that can be separated from the to-and-fro of conversation. If there were such a purposeful activity alone, then language would not "fully realize itself." Here one finds a stark contrast between Gadamer's hermeneutics and Dewey's linguistic anthropology. While Dewey indicates that language has, over the course of human development, taken on a second life as a purposeful activity, a setting up of signs, Gadamer maintains that language has not achieved its status *as language* if it remains symbolic and purposeful. *Contra* Dewey, the cognitive aspects of language are not an afterthought of language, a derivative element of communication. They are instead only one step *on the way* to language.

Fourth, language is part of reality itself rather than being on a different ontological plane. As Gadamer maintains in his debate with Jürgen Habermas,

> there is no societal reality, with all its concrete forces, that does not bring itself to representation in a consciousness that is linguistically articulated. Reality does not happen "behind the back" of language; it happens rather behind the backs of those who live in the subjective opinion that they have understood "the world" (or can no longer understand it); that is, reality happens precisely *within* language. (Gadamer, 1976/2008, p. 35)

These words by Gadamer not only elucidate what it means for language to be on the same ontological plane as reality. They also intimate the interpretive folly of those who would claim to get "outside" of language. As might be expected from discussions above, this one-world commitment is itself a common denominator for Gadamer's alternative understanding of language. For when language is not distinct from other matters in the world, then it cannot only take on the "lighter" consequences of standing in for something more real. Nor can it be used as a linguistic means to some real end. Nor can it exist as something merely cognitive sometimes, something primarily intersubjective at other times. Language cannot, in fact, be identified *as* anything in particular because it is impossible to extract oneself from language in order to identify its defining limits. Language works as the rest of the world works by being a part of that world.

5.7. THE RELEVANCY OF GADAMER'S LANGUAGE THEORY TO SCIENCE EDUCATION

Gadamer's views have been introduced here for the primary purpose of contrasting his conception of language with that of Dewey and progressivism, not only because Dewey has played such a seminal role in science education in the past (and to lesser extent in the present; Fensham, 2004), but also because his linguistic assumptions continue to implicitly inform science teaching and practice in different ways. And while it is true that some earlier research (in the 1970s) had pointed out "how the assumptions of teachers about knowledge and language in learning could be placed at one point or other" along a continuum from *transmission* to *interpretation* (Wellington & Osborne, 2001, p. 25), one can readily acknowledge that the transmission model is ubiquitous (Mestre, 1991). There can be little doubt that even in most upper level science classrooms one often observes the nature of a discourse that assumes and exhibits those characteristics of language here described as representational/symbolic as well as instrumentalist in orientation while ignoring or misunderstanding language as *dialogic* and *interpretive* (also where it pre-forms the life-world of learners; Tobin, McRobbie, & Anderson, 1997)—all factors that, when taken together, usually create considerable barriers for student comprehension not only for learning targeted concepts (like heat, force, or angular momentum) but especially for a broader *nature of science conception* (Duschl, 1990; Kelly, 2007; Lemke, 1990; Sutton, 1996).

An In-Class Example From Physics Education

For the sake of brevity only one example is chosen, being the study by Roth et al. (1997) on physics teacher demonstrations. This case illustrates several pedagogical points why students typically fail to learn abstract and mathematical, symbolic-framed concepts in senior classes, in this case *angular momentum*. This study seems to be indicative. A sampling, such as the following from the work of Roth et al., illuminates some common tendencies and problems that prevail in science classrooms today during the presentation of formal content knowledge, especially with respect to the role language plays—or is assumed to play—during demonstration interactions, including oral discourse and written text.[25] The authors were able to identify *six dimensions* of classroom dynamics that inhibited students from learning what the teacher intended during his demonstrations.[26] However, I will focus on only *four* of these in order to accentuate what is assumed and what is ignored by the teacher about the nature of language that tends to short-circuit the learning process. The sharp difference between an assumed transmission model of language versus a hermeneutic-interpretive one, I contend, is the major factor that underscores the learning problems at issue in at least four of the dimensions they identify. Below is only a vignette of one typical classroom discourse, taken from the study:

> *[The teacher] picks up a bicycle wheel and sits down on his turning stool, which is hidden from view for all but the students in the first row ... He invites students to observe ... He rapidly spins the wheel with its axis vertical—that is, parallel to the axis of the turning stool. This is associated with an almost unnoticeable opposite spin in his body. [He] comments, "This chair isn't very good. I'll try that again." This time, the chair makes about a one-eighth turn. "Did you just see it? Look again, look at my body mainly. What was my angular momentum just now? Zero. I'm isolated, sitting in this awkward-looking position. When I spin it, what do you notice?"*
>
> *[Student] calls out, "Opposite to the wheel."*
>
> *[Teacher] "Yes I'm going the opposite way to the wheel. When we are looking at these vectors, to start with, L was zero, wasn't it? That's my angular momentum. It's made up of two things: my angular momentum and the wheel's both 0, to kick off with." [He] walks to the chalkboard and writes: L = 0 = L me + L wheel. He continues. "The angular momentum is the vector and has direction. This is how we measure the direction of angular momentum. You see, when I spin that, when it spins, if I put my fingers in the direction of the spin, my thumb comes out the axle."* (Roth et al., 1997, p. 509-510)

In general, it can be inferred from the authors' descriptions and analysis that the teacher makes unwitting use of a conduit-tool model during "show and tell," supposing a classical transmission ("Morse code") model of knowledge acquisition (sender-receiver) and assuming its sufficiency to clarify the abstract concept of *angular momentum.*[27] He also switches with ease, as scientific experts typically do, between multi-modal representations of the concept using the specialist language when explaining his active physical demonstrations (rotating on a swivel chair while spinning a counter-wheel balance).[28]

What is revealed, however, is that student understanding and misunderstanding of the event is determined by several language-based factors that have been overlooked, that are inter-related—the so-called *dimensions* identified by the authors ("a number of influences that mediated students' descriptive and explanatory discourse," Roth et al., 1997, p. 520)—and that can be re-categorized as follows:

1. *Language of a theoretical framework:* without a background interpretive theory, students cannot distinguish "signal from noise" during observations. In other words, the language of prior theory structures what is seen and experienced. The scientist and teacher are

clearly at a distinct advantage over the novice in this regard, having appropriated scientific theory and "seeing" already.

2. *Interpretive interference:* other factors such as previously acquired "knowledge bits" and preconceptions (themselves immersed in several in-and out-of school discourses), as well as images of other demonstrations, largely contributed to their *interpreting* what they saw being at odds with the teacher's views (p. 509, "interference of discourses learned in other contexts of the physics course" and "interference from other demonstrations and images that had some surface resemblance");
3. *Confusion of disciplinary, theory-based representations linked to textbooks and teacher talk:* multi-modal symbol systems, required to represent and model physical objects—which have now come to characterize modern science, especially technical canonical language—are best acquired through use and by active participation in discourse communities, which typical school students have little opportunity to explore. "Students learn to manipulate symbolic structures without referential content and are not provided with opportunities to integrate those symbolic structures that can be used alternatively to describe the same system." (Roth et al., 1997, p. 527); In that absence students, unlike experts, find it very difficult to *combine* unlike representations referring to different aspects of a phenomenon (physical experience, mathematical symbols, graphics, hand gestures for "right hand rule," etc.), and which can indeed be taken *to stand for objects in different (linguistic) worlds*. Hence,
4. *Minimal opportunities for probing student science talk:* because students generally were given limited chances to genuinely *question and dialogue* during the event about the observed phenomena, which could have helped surface their own understandings and preconceptions, they fail to correctly associate phenomena in ways compatible with the accepted scientific canon. In other words, they generally lack the awareness to develop the ability to compare and contrast their discourse with scientific discourse. "Students therefore had no means or opportunities to assess in which way their talk was inappropriate because, from a language perspective of knowing, competence in talking science requires participation in scientific discourse" (Roth et al., 1997, p. 527). These insights have been depicted before by Lemke (1990) and others, and while it can be admitted that the constructivist movement in science education has sought to rectify this imbalance (Llewellyn, 2005), classrooms that are immersed in a traditionalist culture primarily framed by teacher-talk, formal knowledge acquisition and dominated by lecture and high-stakes testing, still suffer this charge (Tobin et al., 1997).

These points illustrate, I contend, the four Gadamerian perspectives on language stated above: that for the students in their attempt at understanding science, language performs more than just a representative or tool role, rather it is the *medium of an interpretive event*; moreover, it cannot be easily separated into intersubjective and cognitive elements, and it does *not* appear to function on two different ontological planes.

At first glance one could argue that aspects of this example could serve in fact to illustrate and *reinforce* Dewey's criticisms of traditional (science) education: while the instructor does assume a conduit model of language, as a "tool" it is used poorly (solely) to transmit information into "students' heads"; this serves, moreover, as an end-unto-itself without any application ("pure knowledge") where the teacher as authority figure (along with text) dominates the conversation (by controlled *triad* of "question-response-evaluation" or "I-R-E"; Lemke, 1990) and where active discussion and participation in the event (through action and genuine dialogue) is minimized. Dewey may even have pointed out that it shows how language remains at the level of "verbalism" (word or symbol disconnected with the object), and that learning had not occurred (could not occur) precisely because the cognitive element had dominated the intersubjective, thus proving his assumption of the separation of elements.[29]

Some of these comments appear to have force (and one can grant that on the surface such an interpretive progressivist lens can be fruitfully employed to criticize aspects of conventional classrooms), but let us look a little more closely. Notice how the instructor *has* chosen an object from the "personal experience" of students (a demand of progressivism), for he sits and tries out various movements on a swivel chair, and yet they fail to learn what he demonstrates. So, in effect he *has* provided what Dewey always insisted on, to connect the abstract "map" with "the city" as it were, in this case, the representational scientific language with real life application.[30]

It is certainly true that Dewey (1938) insisted that a kind of hypothetico-deductive problem-solving approach ("method") with active peer engagement *and talk* should exemplify *all* school experiences if they are to be "truly" educative, and that science education would ideally serve as the paradigmatic model to illustrate this (p. 81). Hence, in returning to our example, to further justly meet Dewey's criteria the students should have been allowed to not only swivel around in chairs (direct experience) but have transformed this opportunity, using the teacher's guidance, into a problem-solving activity on its own and thus, through the social use of language somehow manage to co-construct with *that* experience and external guidance the formal knowledge and language of science in the process. This is, as a matter of fact, similar in expectation to what "guided inquiry" approaches and some constructivist views of learning and knowledge have continued to suggest, although such views have now been heavily criticized from different perspectives.[31]

Hermeneutics, Epistemology, and the Role of Language in Classrooms

The buried supposition appears to be, certainly for Dewey and others, that the everyday language-world of students can be led to, or perhaps somehow enabled to co-create, scientific language providing the correct linguistic intersubjectivity, personal experiences and manipulative elements are established. Only then will the abstract, symbolic language of science link up with student meaning and allow "symbol to join up with thing"; only in this way will the language as instrument be used wisely, not solely as conduit for knowledge as *ends*, but as tool for social purposes (citizenship), when it subsumes the cognitive element under the intersubjective one—that is, the "tool" used to enhance thinking will be brought under the social-communal value of knowledge-building through problem-solving, and so extend experience for furthering thought and democratic life. But this supposition, regrettably, is not borne out, as several decades of conceptual change and constructivist-based research on alternative conceptions has shown—students' often erroneous, substitute views and interpretations of phenomena are remarkably resistive to strategies seeking a switch to canonical constructs (Duit & Treagust, 2003; Duschl & Hamilton, 1998). This can be partially attributed to the fact that their "scientific" ideas are not held in isolation, independent of the conceptual-semantic net of their language and life-world (Nersessian, 1989). They will rarely be "teased out" and replaced solely through intersubjective talk, no matter how well-intentioned or appropriate the direct experience and problem-solving milieu—but this is not to say such factors are entirely unimportant.

Vygotsky (1986) of course, had much earlier alluded to the different natures and sources of the two kinds of languages in classrooms, the "scientific" and the learner's "spontaneous/everyday" discourse, and the important interaction of the teacher-learner in the *zone of proximal development,* where the culture-bound and technical scientific language can begin to be internalized. Moreover, and bearing this in mind, our example then also illustrates that the Deweyan assumption of the ontological two-world commitment does not hold—recall that this is the bedrock of his other three assumptions. The language-worlds of students and of academic science represent if not incommensurate linguistic worlds (which hermeneutics does not allow), then nonetheless "separate worlds" that must be brought together in an altogether different kind of way or situation in order for student understanding to take place, one we call the *hermeneutic experience*, and one that has been appropriately described by Eger (1992). While it is true that the Vygotskian perspective has been useful as an explanatory lens in science education, and while we maintain that it helps further undermine the aforesaid Deweyan assumptions of language,[32] nonetheless a Gadamerian perspective presents yet another alternative view of language and how students come to learn.[33]

As a general rule, the nature of language and how it is used in classrooms can be roughly categorized as an interplay between *three* essentially different discourses, involving those of the scientific community (including textbook and canonical knowledge formulations), the teacher and the student. This can equally be viewed, appropriating instead Wittgenstein's (1958) terminology, as the case whereby each is situated in their own respective "language game." The three-way interaction of these discourses (or "three-language problem"),[34] as can be expected, can serve to either improve or, on the other hand as is only too frequent, undermine scientific literacy (Figure 5.1). That is, it serves to increase or decrease interest in science (affective dimension), as well as validate or misrepresent the NoS in the minds of students (cognitive dimension), depending upon how science is understood and how scientific knowledge is framed. In most cases, what is often revealed through the expression of these discourses (whether in verbal or written form) in classrooms—which are themselves intimately tied to their own underlying epistemologies (of the textbook, teacher and student)—is a dogmatic and philosophically out-dated portrayal of science (Lemke, 1990; Matthews, 1994, 2014).

According to the socio-cultural research tradition, conceptual conflicts that students encounter originate not primarily due to conflicting *cognitive* mental models (according to prevailing conceptual change theory) but more so in conflicting multiple discourses, each linked to its own community of practice, and in which the cognitive models are necessarily embedded and find their expression (Anderson, 2007; Klaassen & Lijnse, 1996). It is precisely such insights and the conceptualizing of the classroom as a linguistic field of contending multi-discourses that corresponds closer to a Gadamarian outlook on language, as argued here, than a conventional progressivist one.[35] With the growth of the socio-cultural research tradition in science education, as stated before, has come the realization that the model of language as used both in science and science education is not only or primarily one of useful *conduit and representation* (or symbol and tool), but equally and perhaps more so one of *discourse and interpretation* (Gregory, 1988; Kelly, 2007; Sutton, 1998). We concentrate here on the latter and ask: "What is the role of interpretation in the understanding of science?" (Eger, 1993a, p. 11). It of course belongs to the essence of the discipline of hermeneutics that the notion of interpretation stands at its core, for it has developed historically as the scholarly pursuit of understanding the meaning of ancient foreign language texts via theories of interpretation.[36] We are already accustomed to the view that one needs scholars to interpret ancient manuscripts, and that today different interpretations exist for religious texts and laws, or novels, poems, plays and movies. What is not so common is to view either education or the scientific enterprise exclusively in this light.

Specialist Science Language		**Language linked to Everyday**		**Science Education Language**	
Language special to the subject		*Language special to the subject but which is linked to student's everyday or home language*		*Language not common to the student's world outside school.*	
Terms, ideas and forms of language unique to the subject which teachers are aware of as a potential problem and therefore present and explain to their students explicitly		Terms, words and forms of language which may not be deliberately presented because teacher is unaware problems are caused by familiar words being used in more specific ways		Terms, words and forms of language used by teachers which pupils would not normally hear, see, or use except in the world of science classrooms	
e.g., trachea	Kilojoule unit	e.g., work	unbalanced	e.g., ideally	
amoeba/ cytoplasm	amplitude	energy/ excite/ radiation/ heat	positive/ neutral negative/	random	categories
neuron/ subconscious	atom/ molecule	field/ current/ resistance	salt/ acid/ mole	principle	characteristic of....
photosynthesis/ osmosis	fusion/ fission	power /excess	reproduce/ animal/ family	crucial	distinguish between ...
equilibrium	torque	cycle/ period	yield/ equate	relevant	adjacent
saturated	respiration	substance	static	source	
refraction	Homeostasis	impulse	test	factors	
cathode ray	entropy	law	prove/ valid	determines	
electrolysis	enthalpy	theory	battery	theoretically.	
Ion / ionization	carbohydrates	weight	cell	proportional to	
Isotope	transmutation	friction	uniform	relative to	
emulsion	quadratic	star	reaction/ rate	complex	
metabolism		consumer / property	error / uncertainty		

Source: Adapted from Wellington and Osborne (2001); Yore and Treagust (2006).

Figure 5.1. The three language problem.

One invaluable answer is provided by Sutton (1996). He is mentioned here chiefly because his analysis contains vital aspects of a hermeneutic view of language but without the acknowledgement of this perspective. His analysis of how the language of a scientific idea changes from its initial formulation to how it later becomes rephrased, codified and depersonalized through the different stages of publication—passing from discovery to research paper, handbook and finally textbook—uncovers the often neglected aspect of the development of scientific concepts themselves, in other words, the *historicity* of scientific language and theory. (Gadamer foregrounds the importance of the historicity of human consciousness, thinking and language.) He has argued that students' beliefs about science are *shaped by their beliefs and use of language*, and has illustrated how these are in turn shaped through textbook "talk" and classroom conversation. He has described how students encounter two possible and different conceptions of language in school science, contrasting language as a *labeling system* (the dominating one) with language as an *interpretive system* (for making sense of and sharing experience through open discourse; also characteristic of research science "at-the-edge"). The latter he ascribes to a "persuasion" view, while the former, in assuming an unproblematic labeling system, corresponds to a "transmission" (Morse code) view of communication, instruction and knowledge—also to our previously described "symbol" and "tool" conceptions (Figure 5.2). I note—in agreement with Roth et al. (1997) and our previous example—as such, it implicitly takes on board a form of epistemological objectivism and naïve realist stance:

> From a classical perspective of knowing and learning, [the teacher] has done many things appropriately. A classical perspective of knowledge treats it in modular form. Words (concepts) have meaning and refer to ontologically real objects. Looking at real objects and events provides a direct view of the concepts. Through observations, individual students are enabled to see the underlying structure. The teacher has only to provide the correct labels (we note that this is also a central part of other teaching strategies such as the learning cycle). (p. 528)

It is not only in teacher demonstrations, but as Sutton argues more broadly because of the pervasiveness of the "labeling" aspect of language and its mode of use (in textbooks and classroom discourse generally, when privileging the *transmission* model) that it directly contributes to misrepresenting the development of scientific ideas and process of research inquiry. "With such a limited sense of what language is for, and lack of experience in actively using it, [students] carry too simple an idea of science as fact-gathering and of language as fact-labelling, and they can become increasingly disadvantaged as learners" (Sutton, 1996, p. 13). Science facts, concepts, and laws are then mistakenly understood ("common sense view of science") as forming the *starting point* for curriculum and teaching (to be memorized, used in equations, or inductively "rediscovered" in lab-based activities), rather than being revealed as the *end point* of an arduous historical and critical inquiry approach: originating in human *imagination* and discovery, remaining tentative, requiring confirmation and correction in professional discourse communities, and only much later re-interpreted and structured according to the complex conceptual web of the specialist language of an established discipline.

View #1: *"Common sense" view: Textbook influenced epistemology and Language*

*Science taken as "Finished Product" and Pseudo-Historical Myth

FACTS obtained by observation and experimentation	→	*Summaries* of scientific knowledge about the world (laws and theories)
(simplified *inductivism* as method)		(theories as 'discovered')
"Factual language"; Words a labels		Memorize 'facts' and processes and confirm by algorithmic problem-solving

RESULT:
- accumulated/ahistorical facts as starting point; discovered and 'ready-made'
- Language taken as *representation*; "morse-code" notion of teaching & learning
- expository language is impersonal; narrative language is missing

View #2: *Alternative view: HPS influenced epistemology and language*

*Science taken as "In Process" and Historically Dynamic

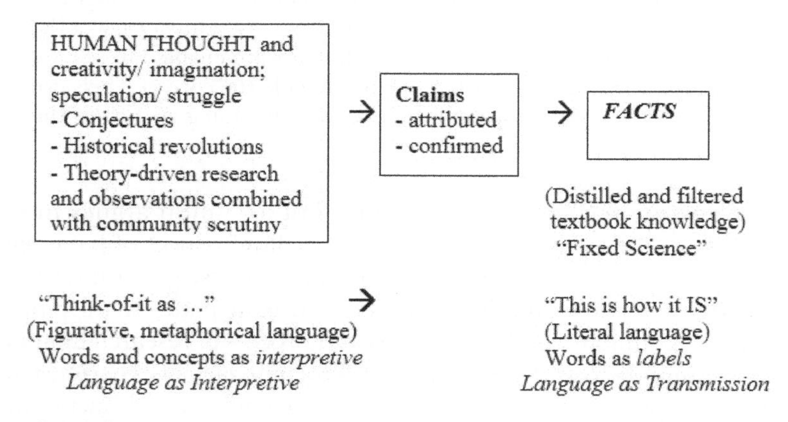

RESULT:
- Facts as the *end point* of a long historical development/struggle, and starting as tentative speculations/ideas in 'frontier science' *(Historicity* of concepts).
- Language is transformative; it serves different functions
- Narrative is partly personal/humanized ("science stories"; History of science)

Source: Based on Sutton's *Beliefs About Science and Beliefs About Language,* 1996.

Figure 5.2. Two views of science as influenced by language use.

In other words, a very restricted and (sadly) widespread notion of science, also termed "final form science" as stated (Duschl, 1990), is deliberately inaugurated and reinforced by how both textbooks and science disciplines *use language* to structure knowledge as curriculum—as a finished *product*. What results is a grave distortion of the NoS as a "rhetoric of conclusions" (Schwab) as said, as others have often pointed out.[37] But I echo Sutton that such consequences are closely correlated to how language is used and understood—invariably a deceptive epistemology of science becomes codified and reproduced in both textbook and teacher epistemologies. More to the point, such a consequence *must* come about according to a hermeneutic perspective, because language—ironically—is here shown to be expressly *not* functioning on two separate ontological planes. Dewey correctly identified the "final form" problem, as stated earlier, but pointed to the wrong conception of language as part of the solution.

Another answer regarding the role of interpretation and language in understanding science has been provided by Martin Eger (1992, 1993a, 1993b), as mentioned above, who presented a cogent and comprehensive analysis of Gadamerian hermeneutics and its relevance especially for science education. His work is complimentary to that of Sutton and I spotlight briefly some of his compelling insights, mentioning *three* important aspects of his analysis pertaining to: *interpreting* science, the problem of *misconceptions*, and the *contexts* of science and science education.

Eger accepts Gadamer's (1976/2008) universalist claim for hermeneutics, that it inheres to "the whole human experience of the world" (p. 15), especially in art, history and law, and that essentially any case of understanding requires interpretation, which necessarily presupposes the domineering nature and role of language. Yet he boldly extents this claim (alongside others) to an understanding of *natural science itself*, above all to *science education* in a novel move that Gadamer had not foreseen—though it can be rightfully taken to orbit in the region of Gadamer's broadly formulated original claim. Eger (1992) had initially distinguished between two different kinds of hermeneutic activities that are possible: one that we can characterize as the direct study of nature herself (science as *inquiry* or "frontier science") versus the study of science as a discipline (science as a *body of knowledge*, or "textbook science")[38]; the former comprises the probable hermeneutic activity of the researcher ("research situation") and the latter the hermeneutic activity of the science learner ("educational situation"). Both are also *immersed in language*: the first being traditionally associated with an assumed "language of nature" or "book of nature" (a term going back at least to Galileo), while the second is associated with a paradigm-bound "language of science," and now widely accepted even within different science education traditions as presenting an obstacle to learning for the novice (Lemke, 1990; Wellington & Osborne, 2001). It is to this second case, the "educational situation," that we briefly turn to, and avoid here completely the worthy and contentious conversation about the hermeneutic status of the first (Hiley et al., 1991).[39]

Eger equally fronts the problem of language as the central one in science study in correspondence with current concerns of language-focused research, yet provides distinctly alternative hermeneutic answers to this issue, including addressing the linked and well-researched *problem of student misconceptions or preconceptions*. By laying stress instead on several key Gadamerian notions such as surfacing *pre*judices or "fore-knowing" (equivalent to "pre-conceptions"), by having learners enter the "hermeneutic circle" of understanding in the correct way, and by providing opportunities for the student and the science "text" to "fuse their (interpretive) horizons" (the actual process in which *understanding* can be said to occur), he offers perceptive and practical solutions to some well-known problems of science learning.[40] An unusual but fruitful perspective is one where science study in general is looked upon "as the *interpretation* of the language of science, and upon the teacher as the chief *interpreter*" (Eger, 1992, p. 341), and in his papers he insightfully sketches out the ramifications of that view. The hermeneutic perspective may even prove beneficial when comparing where existing conceptual change theories of science learning are shown to be inadequate, rooted in either psychological-based constructivist models or PS analogies (Duschl & Hamilton, 1998).[41] This perspective has much to offer as potential for new directions in future research work, as some current researchers have discovered (Borda, 2007; Kalman, 2009, 2011).

Finally, a third aspect of Eger's analysis regards the *contexts* of science and science education. Traditionally only *two contexts* have been identified (originally formulated by positivist philosophers to define the proper field of study for philosophy of science), being the context of *justification* and of *discovery*. The segment of science education literature that has been critical of the epistemology of school science and focused on reforms linked to incorporating appropriate HPS themes into curricula and teacher education programs, has come to stress the value of the context of discovery ("how we know") along with the dominating context of justification ("what we know") in classrooms as essential for a more comprehensive understanding of science (Duschl, 1990; Matthews, 1994; Monk & Osborne, 1997). While I certainly agree with and support these reform efforts, I would nonetheless reinforce Eger's novel insight that a third context *of interpretation* (1993b, p. 321) has been fundamentally overlooked in the overall critical discussion within both philosophy of science and science education. And where not overlooked then certainly misunderstood and minimized, although its presence has always existed and been embedded within the historical and conceptual development of science (its theories, aims and methods) and within the "educational situation" especially.[42]

5.8. SUMMARY

This section has concentrated on the nature and role of language in science education with the specific focus of contrasting the different perspectives on language as found in two seminal 20th century thinkers, John Dewey and Hans-Georg Gadamer. It is argued that Dewey's linguistic assumptions behind his progressivist philosophy—especially as found in his many early and middle works—remain largely unexplored and problematic while they continue to resonate and detrimentally affect science learning. Although his legacy in science education was appraised, by critical comparison and analysis these assumptions were made explicit while providing exposure to an alternative understanding of language as offered by the philosophical hermeneutics of Gadamer. While this comprised the main body, an example was provided to illustrate how the "symbol and tool" conception of language in classrooms is dominated by a transmission model reinforced by a representational epistemology that hinders learning. It was recognized that Dewey was himself critical of such a view of language and epistemology taken alone, and especially in his last work (in 1949) had sought to overcome through his pragmatic "transactional inquiry" the dualism that characterized so much of Western epistemology (and educational philosophy, too). He moved towards a more "dialogic" view of *language-in-use*—very similar, or so it appears on the surface, to the later Wittgenstein—although his previous notions continue to be spread widely in science education. It was inferred instead that in typical science classroom discourse that language actually functions in ontological-hermeneutic mode and while this is generally unacknowledged in science education a shift towards a perspective that recognizes both the *dialogic* and *interpretive* dimensions would be beneficial. As well, explicit links were made to other research exploring hermeneutics in particular, and the language-based socio-cultural tradition in science education in general.

My own analysis goes hand in hand with these latter linguistic/hermeneutic approaches. But as was attempted in this section, things are not as simple as adding a linguistic approach here, or utilizing a hermeneutic style there, when it comes to proceeding toward a progressive science education that is truly *progressive*. It was shown that the problem runs much deeper, that a thorough treatment of language as something other than a symbol system must be established, and contextualized, in relation to what has formerly been considered "progressive" science education. This will be a linguistic treatment of science pedagogy that not only moves beyond the "two-world commitment," but gets past the verbalism/experience dualism that, as was asserted, has typified progressivist-inspired education since Dewey's time and the various iterations of his popularity. In a sense, then, what is being advocated is a science education, properly understood, that is "more Deweyan" than Dewey's. John Dewey indeed identified some of the central problems that would continue to plague science: science pedagogy is still too often preoccupied with a notion of language that assumes its primary role as a "pure" symbol system, as a tool either to be used in the transmission model or misused to reinforce an authoritarian image of science, of the textbook and the teacher's role while constraining or ignoring its vital socio-communal dimension and means for socio-educative ends. And while Dewey recognized that often language *appears* to function on two separate ontological levels (due more so, now widely recognized, as a fiction-category of our analytical thinking), contemporary classroom science education usually assumes a labeling system, a simple and direct correspondence between the representation and the object signified, itself revealing a simplistic objectivist epistemology underlying curricula and teacher talk. As a result the *nature of science* becomes distorted and science learning is stultified, but not because the character of language is as Dewey believed (although partly accepted as operating in dialogic mode later on), rather precisely because symbol, tool, and cognitive elements linger as a part of one world of *language as being*, as experienced and lived in its ontological and interpretive-hermeneutic dimensions.

NOTES

1. Lemke (1990) had scrutinized the discourse of science classrooms and teacher talk (especially questioning strategies) using a social semiotics perspective, including analyzing the semantic and thematic patterns of curriculum content and contrasting this to students' questions and answers that illustrated their own thematic discursive patterns. He also noted how the typical portrayal of science attitudes and myths of NoS (enforcing *scientism*) contributed to alienating students from science. He was among the first to argue that "mastery of a specialized subject like science is in large part mastery of its specialized ways of using language" (p. 21).
2. While it can be readily admitted that the "post-positivist" philosophy of science debate has been taken into consideration for some time in the research community, Brickhouse et al. (1993) pointed out it has tended to neglect the Continental version of the debate, including such names as Heidegger and Gadamer, especially with respect to the topic of fostering *practical reasoning*.
3. "Current attitudes toward foundationalism, as they have been since Descartes, are sharply divided. The minoritarian conviction (Chisholm, Apel, Habermas, Haack, Swinburne, and others) that some version of foundationalism is or is at least potentially viable is outweighed by the majoritarian belief that in the debate since Descartes, foundationalism has died a natural death and cannot be revived" (Rockmore, 2004, p. 56).
4. Rorty, of course, also surfaces representation, but he explicitly ties it to philosophy as a profession whose role as a foundational discipline (with its

"theory of knowledge" being essentially a "general theory of representation") was to adjudicate all cultural knowledge claims, eventually including scientific ones. His view is comparable to Taylor's: "To know is to represent accurately what is outside the mind; so to understand the possibility and the nature of knowledge is to understand the way in which the mind is able to construct such representation" (Rorty, 1979, p. 3).

5. Taylor (1987) links the success of "knowledge as correct representation" standpoint with two factors: its link with the rise of mechanistic science in the 17th century, whose mechanized worldview overthrew the Aristotelean one with its notion of "knowledge as participation" ("being informed by the same *eidos*, the mind participated in the being of the known object, rather than simply depicting it," p. 467); secondly, the influence of Cartesian philosophy that insisted a new reliable "method" was required that could guarantee certainty of the representation. Yet this method entailed, unlike in philosophical antiquity, the reflective and critical cast of individual *mind* performing a subjectivist inward turn. Rorty's view is similar (1979, p. 248).
6. He cites such authors as Rorty and Taylor (in philosophy), Gadamer (in language theory), and Kuhn and Hesse (in philosophy of science). Other philosophers of science endorsing hermeneutics are Heelan (1991) and Ihde (1998).
7. See especially Rorty (1979, Ch. 7) and Gadamer (1960/1975, p. 235). Gadamer comes at the epistemological issue from the problems created by the attempted objectivism of the historical school and the nature of *historical consciousness*, notably Dilthey: "The problem of epistemology acquires a new urgency through the historical sciences" (p. 216). He chides Dilthey for his scientistic attempt at making historical studies an "objective science," although he rightfully credits him with opposing the neo-Kantian school's return to the "epistemological subject" (p. 238). Instead of the restricted focus on the isolated cognitive subject (a philosopher's fiction), he comments how Husserl's idea of the *life-world* (extending Dilthey's "standpoint of life") replaces it: "The concept of *life-world* is the anti-thesis of objectivism" (p. 239), a concept he likewise foregrounds. Such a notion harbors its own *ontology* (as a structure embracing the worlds that man has experienced). "But this ontology of the world would still remain something quite different from what the natural sciences could even ideally achieve ... the *life-world* means something else, namely, the whole in which we live as historical creatures ... [it] is always at the same time a communal world that involves being with other people" (Gadamer, 1960/1975, p. 239). Later, in his critique of the Enlightenment's notion of objective, ahistorical reason (or "absolute reason"), he comes very close to Rorty: "The epistemological question must be asked here in a fundamentally different way.... The focus of subjectivity is a distorting mirror. The self-awareness of the individual is only a flickering in the closed circuits of historical life" (p. 278).
8. "Putnam now agrees with Goodman and Wittgenstein: to think of language as a picture of the world—a set of representations which philosophy needs to exhibit as standing in some sort of nonintentional relation to what they

represent—is *not* useful in explaining how language is learned or understood" (Rorty, 1979, p. 295; original italics).

9. This hermeneutic perspective on learning and understanding corresponds with the newer epistemological perspectives of the field: "increasingly, science education researchers are viewing meaning as public, interpreted by participants (and analysts) through interaction of people via discourse including signs, symbols, models, and ways of being" (Kelly et al., 2012, p. 288).
10. Hence his complaint that one can distinguish between "systematizers" (those engrossed in normal discourse) and "edifying" philosophers (anti-foundationalists like Dewey, and hermeneutic thinkers like Heidegger, Gadamer, who disrupt it) within the tradition—the latter whose status as "true" philosophers is often questioned by academic professionals.
11. Rockmore (2004) writes that Rorty maintains "a strict but wholly arbitrary distinction between epistemology and hermeneutics in order to equate the failure of foundationalism with a form of skepticism that cannot be alleviated through a hermeneutical turn" (p. 57). He accuses Rorty of still clinging to a standard of knowledge that he admits cannot be met. Rorty freely concludes that one can no longer hope to bring the mind in contact with the real and that *interpretation* must be the alternative, but just denies this will lead to knowledge in the conventional sense. Alternatively, Rockmore (2004) argues that "the main strategy for knowledge is, and always has been, interpretation" (p. 57), not to be taken as tantamount to skepticism.
12. This is not meant to imply this field of study is monolithic, and commentators commonly distinguish between "right wing" (Gadamer) and "left-wing" (Derrida) factions. Yet such a categorization is equally overly simplified. Those in educational studies—see Gallagher (1992)—distinguish four separate schools: conservative (Dilthey; Hirsch), moderate (Gadamer; Ricoeur), radical (Derrida; Foucault), and critical (Habermas; Apel).
13. Rockmore (2004) maintains that the shift leads to a *redefinition* of epistemology, from "knowing the way the mind-independent world is" to "the interpretation of experience," which is justified by the standards in use in a given cognitive domain. In this reformulation "then epistemology as hermeneutics presents itself as a viable successor to the traditional view of epistemology—indeed as the most likely approach at the start of the new century" (p. 11). Westphal (1999) criticizes Rorty for failing to distinguish between classical epistemology and hermeneutics seen as a generic epistemological task, hence, to differentiate the replacement of only one type (foundationalism): "*hermeneutics is epistemology,* generically construed…it belongs to the same genus precisely because like them it is a meta-theory about how we should understand the cognitive claims of common sense, of natural and social sciences, and even metaphysics and theology" (p. 416, original italics).
14. "To interpret … is to 'enter' the hermeneutic circle, to project, and to remain thus in motion between the text (or text-equivalent) and the projected fore-meanings. Therefore, a fundamental mode of human *being* is precisely *being in* this circle in which the mind is not with itself but is *drawn* by its own projections to that which it attempts to understand. The reason for

describing this as a state or mode of being, rather than simply as an activity among other(s) ... is precisely to identify this mode as 'primordial'—prior to all constructions—one without which the human *qua* human, cannot be understood" (Eger, 1993a, p. 12, original italics).

15. "Meaning *arises* in the interpretation itself. The subject/object cut does not lie between the interpreter on the one side and the text with its meaning on the other (objectivism). Neither does it lie between the text alone on the one side and the meaning *inside the interpreter's mind* on the other (subjectivism). As a fixed boundary, the cut is just not there; meaning is not disjoint either from the text or from the interpreter. Rather, a 'bare' text is to be thought of as an ontological core, around which *potential* meanings hover, so to speak, in a space of all possible meanings" (Eger, 1993a, p. 12, original italics).
16. Such a confining preoccupation (learning *of* science), nonetheless, has consistently failed to develop in students, regardless of their career aims, a proper scientific mind, and understanding *about* science (the development and nature of its concepts, theories, and social practices), as discussed in Chapter 4.
17. Dewey's preference was clear: "We may reject knowledge of the past as the *end* of education and thereby only emphasize its importance as a *means*" (1938, p. 23; original italics).
18. As discussed previously. References include Roberts (2007); Zeidler et al. (2005); Gaskell (2002); DeBoer (2000); Roberts and Oestman (1998); Shamos (1995); and AAAS (1993).
19. Which has helped shape, along with Schwab (1962), the popular "science as inquiry" approach in science education, and remains very active today (see the NSTA-supported book by Llewellyn, 2005).
20. Dewey writes, "...while language is not thought it is necessary for thinking" (1910/1991, p. 170).
21. See Stewart, 1995 and 1996.
22. It could be argued that Dewey Bentley in his last work (1949)—where he seeks to explicate his position at overcoming the dualism of the Cartesian subject-object epistemology that he saw behind Western thought in general—includes a view of language-in-use largely inconsistent with the "tool" notion and "two world commitment" as here so far described and analyzed. It is true that while Dewey takes "naming" and the "named" (or "designation") as a key inter-connected behavioral activity—one he calls "transaction" and which he states in effect is a "linguistic activity" (p. 144)—this does not exclude a two world commitment of understanding *per se* as to what language in essence is, although on the surface it may appear so. Indeed, linguistic activity might well connect naming and the named. Still, the fact that language acts as a tool during such connecting suggests nevertheless that language is of a different world than both the naming and the named. Language itself remains ephemeral while behavioral activity is of the world of naming and the named that Dewey attempts to connect. Dewey does object to the use of "name" as a tool, but this view is directly linked to his objection of such a notion as strictly related to an *objectivist* epistemology—where "word" is considered a "thing," in fact a "third thing" belonging to an indi-

vidual subjectivist mind (see p. 146). It does not, as far as I can judge, dismiss the idea of language as an *inter-subjective tool* (as here argued) brought about through a transaction of naming and language use in community. Hence, his entire emphasis on naming as a "behavioral process itself in action, with the understanding ...that many forms of behaviour ... operate as instrumental to other behavioral processes" (Dewey & Bentley 1949, p. 144).

23. "The designative tradition depicted language as a crucial instrument of knowledge, a very important representational tool, but nothing more than a tool. By contrast, the *expressive* tradition ... emphasized that language has more than instrumental value: it has *constitutive* value, for it constitutes who we are, how we think and how we live. On the Romantic view ... language, far from being a mere tool that we use, is part of who we are: it defines our humanity and sets the parameters of the life we lead" (Medina, 2005, p. 41, original italics).
24. For further analysis on the non-representational nature of language see Bingham (2002, 2005, 2009, pp. 41–63).
25. The study involved following an experienced physics teacher and his grade 12 physics students over a six-week period, and combined data from several sources (e.g., videotapes, observations, post-tests, interviews, and student notebooks).
26. These were a part of other instructional practices involving formula-driven lectures, traditional numerical problem-solving, computer simulations, and lab-based activities.
27. "The teacher in this study ... was typical of many teachers in his transmission view of learning and teaching.... He was very well-intentioned, but, bringing to this teaching technique an epistemological stance according to which the world was prestructured and knowledge matched this structure, overestimated what neophytes in physics could see in and learn from these demonstrations" (Roth et al., 1997, p. 526).
28. By "multi-modal" we defer to the authors' usage of the term to mean the various different ways to symbolically describe a vector "pointing up." Such a vector, as in this particular case with common academic science discourse, stands-in for (or *represents*) the physical phenomenon of angular momentum. The authors describe at least *four* such representations (three graphic, such as "L," and one physical "hand and thumb"; Roth et al., 1997, p. 527), although the typical (textbook) verbal or written definition should be included as a *fifth*.
29. In short, Dewey's (1938) main thrust, to be fair, would undoubtedly have been to reinforce his broader emphasis, that education as solely focused on pursuing "academic rationalism," as this case exemplifies, must fail, whereas his focus on "socialization" would be effective (using Eisner's, 1985 terms) since it would likely have framed the demonstration within some wider project of the *application* of momentum to life. The teacher's role requires exactly the successful bridging of this (difficult) task, through insight and ingenuity, to make the appropriate *linkage* of subject matter with the student's immediate life-world to eventually attain structured disciplinary knowledge. Needless to say, this does not seem to be possible with many key concepts

and theories of modern science (e.g., chemical bond or plate tectonics)—all of which are very far removed from students' language and life-world. One finds great problems also in going the other way—the Deweyan principle that starting out with life-world and using problem-solving inquiry would somehow lead both the educator and student to eventually arrive at such abstract, hidden, and theoretically complex schemes; that somehow students' questions and manipulations at the macroscopic level will open up formal knowledge at the super-microscopic or super-macroscopic (cosmic) "so that learners may gradually be led, through the extraction of facts and laws, to experience of a scientific order, sets one of his [teacher] main problems" (Dewey, 1938, p. 80).

30. Though, again, here someone may wish to object that it is the teacher and *not* the student that is being active, and so the learner is not directly involved and does not personally experience "the object" under discussion; hence for them, the physics symbols still remain detached from the object, the "word from the thing." This argument is a qualification of Dewey's view, and even if granted, it is the experience of my physics teaching that having had students practice on their own swivel chairs did not help them better understand the concept either. And this activity involved intense discussion on their part, hence making use of language as *intersubjective tool*. Such comments may be anecdotal and isolated, but they help point out that there is more to learning science that just skillfully linking the language of a scientific discipline ("the cart") to the personal experience and social environment of the learner via intersubjectivity ("the horse"; see his 1945 description).
31. Either because they erroneously assume an inductive epistemology lies behind science and undervalue the power of prior theory (Duschl, 1990, 1994) or assume that student inquiries and epistemologies are capable of constructing in some fashion established scientific ones very far removed from "common experience" (Hodson, 1996; Matthews, 2000; Osborne, 1996; also Kirschner, Sweller, & Clark, 2006).
32. See Leach and Scott (2003): "Following the process of internalisation, language provides the tools for individual thinking. Central to this view is the continuity between language and thought. It is not the case that language offers some 'neutral' means for communicating personally and internally generated thoughts; language provides the very tools through which those thoughts are first rehearsed on the intermental plane and then processed and used on the intramental plane" (p. 99).
33. Leach and Scott (2003) allude to limitations inherent to the Vygotskian notion of internalization, in that "there is no recognition of the different forms of intermental functioning which occur on the social plane" (p. 99), which must contribute to the process. Secondly, it is not clear how the so crucial internalization process happens, though as originally envisioned it was *not* one of simple transfer. Interesting for us and our hermeneutic perspective, is the view of some authors that it involves "personal interpretation," whereby "the individual comes to a personal understanding of the ideas encountered on the social plane" (p. 101).

34. Term employed by Yore and Treagust (2006): "A three-language (home language, instructional language, science language) problem exists for most science language learners that parallels English language learning and involves moving across discourse communities of their family, school, and science. No effective science education programme would be complete if it did not support students in acquiring the facility of oral science language and the ability to access, produce, and comprehend the full range of science text and representations" (p. 296).
35. We certainly recognize similarities in the hermeneutic language perspective to insights of researchers in the socio-cultural tradition who have been inspired by other comparable socio-linguistic-based views, such as, for example, Bakhtin (as cited in Mortimer & Scott, 2000): "Saying that different social languages 'are specific points of view on the world, forms for conceptualizing the world in words' [quoting Bakhtin] provides a warning against treating students' ideas as if they were solely individual constructs, independent of the language used to express them, and against treating language simply as a channel or conduit for communicating ideas" (p. 128). See also Kubli (2005).
36. The English word "interpretation" comes from the Latin *interpretatio*, which is itself a translation of the Greek verb *hermeneuein*, "which means to express aloud, to explain or interpret and to translate" (Schmidt, 2006, p. 6).
37. Students come to receive the impression that knowledge is static, ahistorical, and decontextualized; facts are to be inductively "read off nature"; scientific method is singular and straightforward; problems are to be solved algorithmically; and that essentially science is uncreative, impersonal, and value-free. In other words, "epistemologically flat," "ontologically exact," and "sociologically neutral."
38. These are terms that have been suitably borrowed from Bauer (1992).
39. We should mention, however, that Eger in his later papers (1993a, 1993b) presents fascinating arguments why also "research science"—and even conceptual developments *within* given scientific paradigms over time—exhibit typical hermeneutic dimensions and therefore science itself can be considered, to the contrary of many prominent views, as a true form of hermeneutic activity. (By providing evidence from the writings of some scientists, like the physicist Richard Feynman and the biologist Barbara McClintock, and akin to the views of some philosophers of science, like Heelan, 1991 and Hesse, 1980).
40. In Eger's view, what is required is a fundamental shift in the teacher's and student's notion of knowing and learning, away from an epistemological focus to an ontological experience of "being-in-the-world," that is, "bridging the horizon of physics and the horizon of the lifeworld implies an *extension* of language and of concepts" (Bevilacqua & Giannetto, 1995, p. 117).
41. The authors reference the research of Pea and the emphasis on the crucial role of *communication* (or *dialogue* in hermeneutics) in conceptual change: "crucial aspects of learning can occur within a context of conversations in which production and interpretation of communication drive the construction of meaning. Rather than being viewed as the messenger of knowledge

and meaning, communication is viewed as the location for the creation of knowledge and meaning" (Duschl & Hamilton, 1998, p. 1053). The role and nature of dialogue, as mentioned, is fundamental to Gadamer's ontological conception of language, meaning, and understanding.

42. The ramifications of this perspective have been discussed by some authors, pertaining to the nature of science (Hiley et al., 1991); the formalist versus interpretive debate in quantum mechanics (Cushing, 1998); and science education with respect to the historicity of science, of textbooks, and understanding (Bevilacqua & Giannetto, 1995).

CHAPTER 6

CONCLUSION

Philosophy and philosophy of education continue to remain outside the mainstream of thinking in science education. The chief purpose of this book has been to bring them closer into the fold. The value of philosophy *for* science education remains underappreciated at both pedagogical levels discussed, whether one examines the research field or classroom practice. The main concern has not only been to draw attention to these issues but to make the case for the academic and pragmatic establishment of "philosophy of science education."

Philosophy as a discipline of critical inquiry enables teachers to develop a thoughtful, critical capacity to reflect upon curricular, epistemological and popular media issues as they arise, whether during classroom discourse or professional policy deliberations. Philosophy is not far below the surface in any classroom, and in truth cannot be avoided. This holds especially when discussing common terms like "law," "theory," "model," and "proof," or for detecting curricular ideologies and conveyed textbook myths (e.g., academic rationalism, indoctrination into scientism, epistemological positivism, historically defined convergent realism, evolution versus intelligent design arguments, ambiguities and hazards of modern techno-science, cultural and personal bias, etc.). Its merit comes to the fore when teachers are required to justify content knowledge, or when analyzing national "standards" documents, or simply providing coherent educational rationales for their courses. Philosophy of education as a sub-discipline prepares a forum of informed analysis and discussion on a range of topics

Rethinking Science Education: Philosophical Perspectives, pp. 259–262

and issues that bear directly on science education as an educational project, which has deep roots in the historico-philosophical past.

A philosophy *of* science education (PSE) is understood as a *synthesis* of (at least) three academic fields of philosophy, philosophy of science (PS), and philosophy of education (PE), each of which have distinctive contributions to make in its development.

The worth of such a discipline-specific philosophy in essence lies in its power to perform critical inquiry at both levels, improving science education as a research field and science teacher education. In the first instance, to impel the discipline to clarify its identity and re-examine its goals (especially scientific literacy), foundations, "reform waves," research methods, and dependence on theories from outside disciplines, as well as determining the suitability of educational metatheories while scrutinizing the nature of scientific epistemology and practice germane to science education. In particular to deconstruct social-group motivated interests and ideologies. These are some key problems but by no means the only ones comprising the scope of its inquiry at the level of critique and research. But it implies that it should be inaugurated as a new *fourth* area of research inquiry (next to the common quantitative, qualitative, and emancipatory).

For the second case with respect to teacher education and transforming pedagogy, a PSE gives the ability to equip the classroom teacher with a "second order" reflective capacity when teaching or curriculum planning. As an integral component of practitioners' pedagogical content knowledge when decision-making, science teachers would recognize the need to problematize the common content knowledge for two reasons: firstly, to *transpose* the disciplinary-based science subject matter into a meaningful form accessible to the learner at age-appropriate grades as coordinated with educational philosophy; secondly, to *broaden* the knowledge base to reflect more authentic epistemological, historical, and sociological aspects of science and its development. (At minimum a critical-mindedness is sought to avoid perpetuating the "scientism" and pseudo-history of curricula plus helping dispel anti-science attitudes among pupils.) The former requires an understanding of the worth and function of *educational metatheory*, while the latter requires an understanding of *nature of science* discourse gained through "science studies," especially developments in the fields of history, philosophy, and sociology of science. The history and philosophy of science reform movement has made major contributions to support teachers in their duties.

Another further aspect of teachers' pedagogical content knowledge that a philosophy *of* science education can help develop is an improved understanding of the role and nature of language, whether found in active dialogue or how knowledge is framed in the historico-epistemological structuring of content in textbooks. Instructors need awareness and assistance to move beyond teaching only "final form science" of current paradigms, therefore the historicity of scientific concepts and theories must be recovered. Above all, the ubiquitous symbol and tool conception of language behind the "transmission" mode employed in classrooms must be amended for a hermeneutic influenced "interpretation" conception, with the perspective that "language as being" influences a student's life-world, hence how they learn and make meaning of texts.

To illustrate the tasks that philosophy of science education can perform when contributing to the advancement of the research field, *three* different philosophical analyses were undertaken, respective to studies in the fields of philosophy of education, philosophy of science, and philosophy of language. Respective to the latter two, detailed *case studies* examined the realism/instrumentalism debate in the *philosophy of science* and scrutinized Dewey's language views from a Gadamerian hermeneutic perspective as an example of a central dispute pertaining to *philosophy of language*.

When bringing to bear the worth of *philosophy of education*, two educational metatheories were compared and contrasted: the influential German-Norse *Bildung* tradition and Egan's cultural-linguistic conception. The merit of metatheories in general lies in creating curricular coherence, transposing subject matter for the age-appropriate learner, and defining and steering educational aims. It was argued that *Bildung* suffered disadvantages due both to its inherent conception and its implementation. Its defining characteristics harbor educational ideas in tension with each other while implementation efforts have resulted in opposing educational aims ensuing from diverse interpretations of these conflicting ideas. Egan sidesteps the common quandary of trying to balance three traditional educational ideas (knowledge, personal development, socialization) and reconceptualizes the educational endeavor to focus on developing students' cognitive-emotive tools of understanding at appropriate age-developmental stages. Such a conception of education and its goals could redefine the notion of *scientific literacy*, since current notions of the concept are also at odds—this being due, it was argued, to the definition suffering from the same conflicting three ideas that equally afflict *Bildung*. Moreover, Egan questions the worth of psychological metatheories for education, and seeks instead to gain autonomy for educational studies and theorizing. Educational development was contrasted with psychological development, and *narrative understanding* and *hermeneutic understanding* were chosen over against the more common conceptual change model for learning science. It was further argued that science education could be liberated from it dependence on learning theories in either psychology or philosophy of science analogs, as well as from socio-utilitarian ideologies, if it managed to accommodate educational metatheory and orientate itself primarily as a sub-discipline of educational studies. Because all "mature" sciences (natural or social) are marked by high-level theories that command their disciplines, it was advised that such an endorsement of educational metatheory could provide further grounds to establish its identity and progression as a research field, as a developing "mature" independent discipline.

The merit of philosophy in itself for science education will only be taken on board insofar as both researchers and practitioners will be convinced by the presentation of the arguments and the case studies. Philosophy does not come easy. It is demanding of time and considerable mental effort. Hopefully the benefits have been sufficiently elucidated to justify the exertion, including the merit of its contribution towards addressing and clarifying some fundamental issues and long-standing science educational problems—perhaps its real worth will only come to be fully recognized by the community when the "crisis" flag is raised once again, as has occurred too often in the past.

It is important to emphasize that the proposed case for a "philosophy of science education" at this point is an *initial* undertaking and a cross-disciplinary research project. The entire purpose of suggesting a search is to open up new ground for exploration and offer some solutions for science educational reform with the understanding that the issues raised and ideas presented will serve as a catalyst for further research and continued work—not only by the present author but along with other parties that have already expressed an interest in further collaboration. This book is the presentation of a new field of philosophy of science education.

REFERENCES

Abd-El-Khalick, F., & Akerson, V. L. (2007). On the role and use of "theory" in science education research: A response to Johnston, Southerland, and Sowell. *Science Education, 91*(1), 187–194.

Abd-El-Khalick, F., & Lederman, N. (2000). Improving science teacher's conceptions of the nature of science: A critical review of the literature. *International Journal of Science Education, 22*(7), 665–701.

Abell, S. K. (2007). Research on science teacher knowledge. In S. K. Abell & N. G. Lederman (Eds), *Handbook of research on science education* (pp. 1105–1145). Mahwah, NJ: Lawrence Erlbaum Associates.

Abell, S. K., & Lederman, N. G. (Eds.). (2007). *Handbook of research on science education*. Mahwah, NJ: Lawrence Erlbaum Associates.

Adler, M. J. (1983). *Paideia: Problems and possibilities*. New York, NY: Macmillan.

Adler, J. E. (2002). Knowledge, truth and learning. In R. R. Curren (Ed.), *A companion to the philosophy of education* (pp. 285–303). Oxford, England: Blackwell.

Aikenhead, G. (1996). Science education: Border crossing into the sub-culture of science. *Studies in Science Education, 27*, 1–52.

Aikenhead, G. (1997a). Towards a First Nations cross-cultural science and technology curriculum. *Science Education, 81*, 217–238.

Aikenhead, G. (1997b). STL and STS: Common ground or divergent scenarios? In E. Jenkins (Ed.), *Innovations in scientific and technological education, Vol. VI* (pp. 77–93). Paris: UNESCO.

Aikenhead, G. (1998). Border crossing: Culture, school science and the assimilation of students. In D. A. Roberts & L. Oestman (Eds.), *Problems of meaning in science curriculum* (pp. 86–100). New York, NY: Teachers College Press.

Aikenhead, G. (2002a). Whose scientific knowledge? The colonizer and the colonized. In W.-M. Roth & J. Désautels (Eds.), *Science education as/for sociopolitical action.* (pp. 151–166). New York, NY: Peter Lang.

Aikenhead, G. (2002b). The educo-politics of curriculum development: A response to Fensham's "Time to change drivers for scientific literacy." *Canadian Journal of Science, Mathematics and Technology Education, 2*(1), 49–57.

Aikenhead, G. (2006). *Science education for everyday life. Evidence-based practice.* New York, NY: Teachers College Press.

Aikenhead, G. (2007). Humanistic perspectives in the science curriculum. In S. K. Abell & N. G. Lederman (Eds.), *Handbook of research on science education* (pp. 880–930). Mahwah, NJ: Lawrence Erlbaum Associates.

Aldridge, J., Kuby, P., & Strevy, D. (1992). Developing a metatheory of education. *Psychological Reports, 70*, 683–687.

Allchin, D. (2000). How *not* to teach history in science. Retrieved from http://www1.umn.edu/ships/updates/hist-not.htm

Allchin, D. (2001). Values in science: An educational perspective. In F. Bevilacqua, E. Giannetto, & M. Matthews (Eds.), *Science education and culture: The contribution of history and philosophy* (pp. 185–196). Dordrecht: Kluwer.

Allchin, D. (2003). Scientific myth-conceptions. *Science Education, 87*, 329–351.

Allchin, D. (2004). Should the sociology of science be rated X? *Science Education, 88*, 934–946.

Allchin, D. (2012). The Minnesota case study collection: New historical inquiry case studies for nature of science education. *Science & Education, 21*, 1263–1282.

Allchin, D. (2013). *Teaching the nature of science. Perspectives and resources.* St. Paul, MN: SHiPS Education Press.

American Association for the Advancement of Science (AAAS). (1993). *Benchmarks for science literacy: Project 2061.* New York, NY: Oxford University Press.

Anderson, C. W. (2007). Perspectives on student learning. In S. K. Abell & N. G. Lederman (Eds.), *Handbook of research on science education* (pp. 3–30). Mahwah, NJ: Lawrence Erlbaum Associates.

Anderson, R. N. (1992). Perspectives on complexity: An essay on curricular reform. *Journal of Research in Science Teaching, 29*(8), 861–876.

Angell, C., Guttersrud, Ø., Hendriksen, E., & Isnes, A. (2004). Physics: Frightful, but fun. Pupils' and teachers' views of physics and physics teaching. *Science Education, 88*(5), 683–706.

Apple, M. (1990). *Ideology and curriculum* (2nd ed.) New York, NY: Routledge.

Apple, M. (1992). Educational reform and educational crisis. *Journal of Research in Science Teaching, 29*(8), 779–789.

Arabatzis, T. (2006). *Representing electrons. A biographical approach to theoretical entities.* Chicago, IL: University of Chicago Press.

Aristotle. (1962/1981). *The politics.* (Translated by T. A. Sinclair. Revised and re-presented by T. J. Saunders). London, England: Penguin Books.

Aristotle. (1998). *Metaphysics.* (Translated and introduction by H. Lawson-Tancred). London: Penguin books.

Aronson, J. L., Harré, R., & Way, E. C. (1995). *Realism rescued. How scientific knowledge is possible.* Chicago, IL: Open Court.

Atwater, M. M. (1996). Social constructivism: Infusion into the multicultural science education research agenda. *Journal of Research in Science Teaching, 33*(8), 821–837.

Ayala, F. J. & Arp, R. (Eds.) (2009). *Contemporary debates in philosophy of biology*. Malden, MA: John Wiley & Sons.

Bailey, R., Barrow, R., Carr, D., & McCarthy, C. (Eds.) (2010). *The SAGE handbook of the philosophy of education*. Thousand Oaks, CA: SAGE.

Bailin, S. (2002). Critical thinking and science education. *Science & Education, 11*, 361–375.

Bailin, S. & Battersby, M. (2010). *Reason in the balance: An inquiry approach to critical thinking*. Toronto: McGraw-Hill.

Bailin, S. & Siegel, H. (2003). Critical thinking. In N. Blake, P. Smeyers, R. Smith, & P. Standish (Eds.), *The Blackwell guide to the philosophy of education* (pp. 181–193). Malden, MA: Blackwell.

Bakhurst, D. (2005). Strong culturalism. In C. E. Erneling & D. M. Johnson (Eds.), *The mind as a scientific object. Between brain and culture* (pp. 413–431). Oxford, England: Oxford University Press.

Barnard, F. M. (2003). *Herder on nationality, humanity and history*. Montreal & Kingston: McGill-Queen's University Press.

Barrow, R. (2010). Schools of thought in philosophy of education. In R. Bailey, R. Barrow, D. Carr, & C. McCarthy (Eds.), *The SAGE handbook of the philosophy of education* (pp. 21–36). Thousand Oaks, CA: SAGE.

Barrow, R., & Woods, R. (2006). *An introduction to philosophy of education* (4th ed.). New York, NY: Routledge.

Barton, A. C., & Yang, K. (2000). The culture of power and science education: Learning from Miguel. *Journal of Research in Science Teaching, 37*, 871–889.

Bauer, H. H. (1992). *Scientific literacy and the myth of the scientific method*. Chicago, IL: University of Illinois Press.

Baynes, K., Bohman, J., & McCarthy, T. (Eds.). (1987). *After philosophy. End or transformation?* Cambridge, MA: MIT Press.

Beiser, F. (1998). A romantic education. The concept of *Bildung* in early German romanticism. In A. O. Rorty (Ed.), *Philosophers on education. Historical perspectives* (pp. 284–299). New York, NY: Routledge.

Bencze, L. (2001). Subverting corporatism in school science. *Canadian Journal of Science, Mathematics and Technology Education, 1*(3), 349–355.

Benner, D. (1990). Wissenschaft und Bildung. Überlegungen zu einem problematischen Verhältnis und zur Aufgabe einer bildenden Interpretation neuzeitlicher Wissenschaft [Science and education. Reflections on a problematic relationship and on the task of an educative interpretation of modern science]. *Zeitschrift für Pädagogik, 36*(4), 597–620.

Bernstein, R. J. (1983). *Beyond objectivism and relativism. Science, hermeneutics and praxis*. Philadelphia, PA: University of Pennsylvania Press.

Bevilacqua, F., & Giannetto, E. (1995). Hermeneutics and science education: The role of history of science. *Science & Education, 4*, 115–126.

Bevilacqua, F., Giannetto, E., & Matthews, M. (Eds.). (2001). *Science education and culture: The contribution of history and philosophy*. Dordrecht: Kluwer.

Beyer, L. E., & Apple, M. W. (Eds.). (1998). *The curriculum: Problems, politics and possibilities.* Albany, NY: Suny Press.

Bianchini J. A., & Kelly, G. J. (2003). Challenges of standards-based reform: The example of California's science content standards and textbook adoption process. *Science Education, 87*, 378–389.

Bingham, C. (2002). A dangerous benefit: Dialogue, discourse, and Michel Foucault's critique of representation. *Interchange, 33*(4), 51–69.

Bingham, C. (2005). The hermeneutics of educational questioning. *Educational Philosophy and Theory, 37*(4), 553–565.

Bingham, C. (2009). *Authority is relational: Rethinking educational empowerment*. Albany, NY: State University of New York Press.

Bird, A. (1998). *Philosophy of science*. Montreal: McGill-Queens University Press.

Bird, A. (2002). Kuhn's wrong turning. *Studies in History and Philosophy of Science, 33*, 443–463.

Blake, N.. & Masschelein, J. (2003). Critical theory and critical pedagogy. In N. Blake, P. Smeyers, R. Smith, & P. Standish (Eds.), *The Blackwell guide to the philosophy of education* (pp. 38–56). Malden, MA: Blackwell.

Blake, N., Smeyers, P., Smith, R., & Standish, P. (eds.) (2003). Introduction. In N. Blake, P. Smeyers, R. Smith, & P. Standish (Eds.), *The Blackwell guide to the philosophy of education* (pp. 1–17). Malden, MA: Blackwell.

Boersma, K., Goedhart, M., De Jong, O., & Eijkelhof, H. (Eds.). (2005). *Research and the quality of science education*. Dordrecht: Springer.

Bonjour, L., & Sosa, E. (2003). *Epistemic justification. Internalism vs. externalism, foundations vs. values.* Malden, MA: Blackwell.

Bontekoe, R. (1996). *Dimensions of the hermeneutic circle.* Atlantic Highlands, NJ: Humanities Press International.

Borda, E. J. (2007). Applying Gadamer's concept of dispositions to science and science education. *Science & Education, 16*(9), 1027–1041. doi:10.1007/s11191-007-9079-5

Boyd, R. (1983). On the current status of the issue of scientific realism. *Erkenntnis, 19*, 45–90.

Brackenridge, J. B. (1989). Education in science, history of science, and the textbook—necessary vs. sufficient conditions. *Interchange, 20*(2), 71–80.

Brickhouse, N. W., Stanley, W. B., & Whitson, J. A. (1993). Practical reasoning and science education: Implications for theory and practice. *Science & Education, 2*, 363–375.

Brighouse, H. (2009). Moral and political aims of education. In H. Siegel (Ed.), *The Oxford handbook of philosophy of education* (pp.35–51). Oxford, England: Oxford University Press.

Briscoe, C. (1990). John Dewey's philosophy of science: Its relationship to science education. In D. E. Herget (Ed.), *More history and philosophy of science in science teaching* (pp. 21–27). Tallahassee, FL: Florida State University.

Brockman, J. (Ed.). (2006). *Intelligent thought. Science versus the intelligent design movement.* New York, NY: Vantage Books.

Brouwer, W., Austen, D., & Martin, B. (1999). The unpopularity of physics lectures. *Physics in Canada, 55*(6), 361–370.

Brown, H. I. (2005). Incommensurability reconsidered. *Studies in the History and Philosophy of Science, 36*, 149–169.

Brown, J. R. (2001). *Who rules in science? An opinionated guide to the wars*. Cambridge, MA: Harvard University Press.

Brush, S. G. (2000). Thomas Kuhn as a historian of science. *Science and Education, 9*, 39–58.

Bruner, J. (2006a). Celebrating divergence. Piaget and Vygotsky. In J. S. Bruner *In search of pedagogy: The selected works of Jerome S. Bruner* (Vol. II, pp. 187–197). New York, NY: Routledge.

Bruner, J. (2006b). Folk pedagogy. In J. S. Bruner (Ed.), *In search of pedagogy. The selected works of Jerome S. Bruner* (Vol. II, pp. 160–174). New York, NY: Routledge.

Bruning, R. H., Schraw, G. J., & Ronning, R. R. (1995). *Cognitive psychology and instruction* (2nd ed). Englewood Cliffs, NJ: Merrill, Prentice Hall.

Burbules, N. C., & Linn, M. C. (1991). Science education and philosophy of science: Congruence or contradiction? *International Journal of Science Education, 13*, 227–241.

Bybee, R. (1998). National standards, deliberation, and design: The dynamics of developing meaning in science curriculum. In D. A, Roberts & L. Oestman (Eds.), *Problems of meaning in science curriculum* (pp. 150–165). New York, NY: Teachers College Press.

Bybee, R. W., Powell, J. C., & Trowbridge, L. W. (2008/1967). *Teaching secondary school science: Strategies for developing scientific literacy* (9th ed.). Upper Saddle River, NJ: Pearson Prentice Hall.

Bybee, R., & Ben-Zvi, N. (1998). Science curriculum: Transforming goals into practice. In B. J. Fraser & K. G. Tobin (Eds.), *International handbook of science education. Part I* (pp. 487–498). Dordrecht: Kluwer.

Bybee, R., & DeBoer, G. E. (1994). Research on goals for the science curriculum. In D. Gabel (Ed.), *Handbook of research on science teaching and learning* (pp. 357–387). New York, NY: Macmillan.

Carey, S. (1986). Cognitive psychology and science education. *American Psychologist, 41*, 1123–1130.

Carlsen, W. S. (2007). Language and science learning. In S. K. Abell & N. G. Lederman (Eds.), *Handbook of research on science education* (pp. 57–74). Mahwah, NJ: Lawrence Erlbaum Associates.

Carr, D. (Ed.). (1998). *Education, knowledge and truth: Beyond the postmodern impasse*. London, England: Routledge.

Carr, D. (2003). *Making sense of education*. London, England: Routledge Farmer.

Carr, D. (2009). Curriculum and the value of knowledge. In H. Siegel (Ed.), *The Oxford handbook of philosophy of education* (pp. 281–299). Oxford, England: Oxford University Press.

Carr, D. (2010). The philosophy of education and educational theory. In R. Bailey, R. Barrow, D. Carr, & C. McCarthy (Eds.), *The SAGE handbook of the philosophy of education* (pp. 37–54). Thousand Oaks, CA: SAGE.

Carson, R. N. (1998). Science and the ideals of liberal education. In B. J. Fraser & K. G. Tobin (Eds.), *International handbook of science education. Part II* (pp. 1001–1014). Dordrecht: Kluwer.

Carson, R. N. (2001). The epic narrative of intellectual culture as a framework for curricular coherence. In F. Bevilacqua, E. Giannetto, & M. Matthews (Eds.), *Science education and culture: The contribution of history and philosophy* (pp. 67–82). Dordrecht: Kluwer.

Cartwright, N. (1983). *How the laws of physics lie*. Oxford, England: Clarendon Press.

Cawthron, E. R., & Rowell, J. A. (1978). Epistemology and science education. *Studies in Science Education, 5,* 31–59.

Chambliss, J. J. (Ed.). (1996). *Philosophy of education: An encyclopedia*. New York, NY: Garland.

Claxton, G. (1997). Science of the times: A 2020 vision of education. In R. Levinson & J. Thomas (Eds.), *Science today. Problem or crisis?* (pp. 71–86). New York, NY: Routledge.

Clough, M., Berg, C., & Olson, N. (2009). Promoting effective science teacher education and science teaching: A framework for teacher decision-making. *International Journal of Science and Mathematics Education, 7,* 821–847.

Clough, M. & Olson, N. (2008). Teaching and assessing the nature of science: An introduction. *Science & Education, 17,* 143–145.

Cobern, W. W. (2000). The nature of science and the role of knowledge and belief. *Science & Education, 9*(3), 219–246.

Cobern, W. W. (2004). Apples and oranges. A rejoinder to Smith and Siegel. *Science & Education, 13,* 583–589.

Cobern, W., & Aikenhead, G. (1998). Cultural aspects of learning science. In B. J. Fraser & K. G. Tobin (Eds.), *International handbook of science education. Part I* (pp. 39–52). Dordrecht: Kluwer.

Cobern, W., & Loving, C. (2008). An essay for educators: Epistemological realism really is common sense. *Science & Education, 17*(4), 425–447.

Cole, M. (1996). *Cultural psychology: A once and future discipline*. Cambridge, MA: Harvard University Press.

Collins, H. (2007). The uses of sociology of science for scientists and educators. *Science & Education, 16,* 217–230.

Corrigan, D., Dillion, J., & Gunstone, D. (Eds.). (2007). *The re-emergence of values in the science curriculum.* Rotterdam, Holland: Sense.

Council of Ministers of Education, Canada. (1997). *Common framework of science learning outcomes K to 12: Pan-Canadian protocol for collaboration on school curriculum*. Toronto: Council of Ministers. Retrieved from http://www.cmec.ca/science/framework/index.htm

Curd, M., & Cover, J. A. (Eds.). (1998). *Philosophy of science. The central issues*. New York, NY: W.W. Norton.

Curren, R. R. (Ed.). (2003). *A companion to the philosophy of education*. Malden, MA: Blackwell.

Cushing, J. T. (1989). A tough act—history, philosophy and introductory physics (an American perspective). *Interchange, 20*(2), 54–59.

Cushing, J. T. (1998). *Philosophical concepts in physics. The historical relation between philosophy and scientific theories*. Cambridge, England: Cambridge University Press.

Cutnell, J. D., & Johnson, K. W. (1998). *Physics* (4th ed.). New York, NY: John Wiley.

Darling, J., & Nordenbo, S. E. (2003). Progressivism. In N. Blake, P. Smeyers, R. Smith, & P. Standish (Eds.), *The Blackwell guide to the philosophy of education* (pp. 288–308). Malden, MA: Blackwell.

Davson-Galle, P. (1994). Philosophy of science and school science. *Educational Philosophy and Theory, 26*(1), 34–53.

Davson-Galle, P. (1999). Constructivism: A curate's egg. *Educational Philosophy and Theory, 31*(2), 205–219.

Davson-Galle, P. (2004). Philosophy of science, critical thinking and science education. *Science & Education, 13*(6), 503–517.

Davson-Galle, P. (2008a). Why compulsory science education should *not* include philosophy of science. *Science & Education, 17*, 667–716.

Davson-Galle, P. (2008b). Against science education: The aims of science education and their connection to school science curricula. In T. Bertrand & L. Roux (Eds.), *Education Research Trends* (pp. 1–30). Hauppauge, NY: Nova.

DeBoer, G. E. (1991). *History of ideas in science education. Implications for practice*. New York, NY: Teachers College Press.

DeBoer, G. E. (2000). Scientific literacy: Another look at its historical and contemporary meanings and its relationship to science education reform. *Journal of Research in Science Teaching, 37*(6), 582–601.

Désautels, J. & Larochelle, M. (1998). The epistemology of students: The "thingified" nature of scientific knowledge. In B. J. Fraser & K. G. Tobin (Eds.), *International handbook of science education. Part I* (pp. 115–126). Dordrecht: Kluwer.

Devitt, M. (1997). *Realism and truth* (2nd ed.). Malden, MA: Blackwell.

de Vries, M. J. (2005). *Teaching about technology. An introduction to the philosophy of technology for non-philosophers*. Dordrecht: Springer.

Dewey, J. (1901). *The child and the curriculum*. Chicago, IL: University of Chicago Press.

Dewey, J. (1938). *Experience and education*. New York, NY: Touchstone.

Dewey, J. (1944). *Democracy and education. An introduction to the philosophy of education*. New York, NY: Free Press. (Original work published 1916)

Dewey, J. (1945). Method in science teaching. *Science Education, 29*, 119–123.

Dewey, J. (1981). Nature communication and meaning. In J. A. Boydston (Ed.), *The later works of John Dewey, Vol. 1, 1925–1953: Experience and Nature* (pp. 132–161). Carbondale, IL: Southern Illinois University Press.

Dewey, J. (1991). *How we think*. Amherst, NY: Prometheus Books. (Original work published 1901)

Dewey, J., & Bentley, A. F. (1949). *Knowing and the known*. Boston, MA: The Beacon Press.

DiSessa, A. A. (1982). Unlearning Aristotelian physics: A study of knowledge-based learning. *Cognitive Science, 6*, 37–75.

DiSessa, A. A. (1993). Toward an epistemology of physics. *Cognition and Instruction, 10*, 105–225.

Dolphin, G. (2009). Evolution of the theory of the earth: A contextualized approach for teaching the history of the theory of plate tectonics to ninth grade students. *Science & Education, 18*. doi:10.1007/slll91-007-9136-0

Donald, M. (1991). *Origins of the modern mind.* Cambridge, MA: Harvard University Press.

Donnelly, J. (2001). Instrumentality, hermeneutics and the place of science in the school curriculum. In F. Bevilacqua, E. Giannetto, & M. Matthews (Eds.), *Science education and culture: The contribution of history and philosophy* (pp. 109–127). Dordrecht: Kluwer.

Donnelly, J. (2004). Humanizing science education. *Science Education, 88*, 762–784.

Donnelly, J. (2006). The intellectual positioning of science in the curriculum, and its relationship to reform. *Journal of Curriculum Studies, 38*(6), 623–640.

Donnelly, J., & Jenkins, E. (2001). *Science education. Policy, professionalism and change.* London, England: Paul Chapman.

Driver, R. (1997). The application of science education theories: A response to Stephen P. Norris and Tone Kvernbekk. *Journal of Research in Science Teaching, 34*(10), 1007–1018.

Driver, R., & Erickson, G. (1983). Theories-in-action: Some theoretical and empirical issues in the study of students' conceptual frameworks in science. *Studies in Science Education, 10,* 37–60.

Driver, R., Leach, J., Millar, R., & Scott, P. (1996). *Young people's images of science.* Buckingham, England: Open University Press.

Drori, G. S. (2000). Science education and economic development. Trends, relationships, and research agenda. *Studies in Science Education, 35*, 27–58.

Duhem, P. (1998). Physical theory and experiment. In M. Curd & J. A. Cover (Eds.), *Philosophy of science. The central issues* (pp. 257–279). New York, NY: W. W. Norton. (Original work published 1954)

Duit, R. (2006). *Bibliography (new version): Students' and teachers' conceptions and science education.* Kiel, Germany: Institute for Science Education. Retrieved from http://www.ipn.uni-kiel.de/aktuell/stcse/bibint.html

Duit, R., Goldberg, F., & Niedderer, H. (Eds.). (1992). *Research in physics learning: Theoretical and empirical studies.* Kiel, Germany: Institute for Science Education.

Duit, R., Niedderer, H., & Schecker, H. (2007). Teaching physics. In S. K. Abell & N. G. Lederman (Eds.), *Handbook of research on science education* (pp. 599–630). Mahwah, NJ: Lawrence Erlbaum Associates.

Duit, R., & Treagust, D. (1998). Learning in science—from behaviourism towards social constructivism and beyond. In B. J. Fraser & K. G. Tobin (Eds.), *International handbook of science education. Part I* (pp. 3–26). Dordrecht: Kluwer.

Duit, R., & Treagust, D. (2003). Conceptual change—a powerful framework for improving science teaching and learning. *International Journal of Science Education, 25*(6), 671–688.

Dunne, J., & Pendlebury, S. (2003). Practical reason. In *The Blackwell guide to the philosophy of education* (pp. 194–211). Malden, MA: Blackwell.

Duschl, R. (1985). Science education and philosophy of science: Twenty-five years of mutually exclusive development. *School Science & Mathematics, 87*(7), 541–555.

Duschl, R. (1988). Abandoning the scientistic legacy of science education. *Science Education, 72*(1), 51–62.

Duschl, R. (1990). *Restructuring science education: The role of theories and their importance*. New York, NY: Teachers College Press.

Duschl, R. (1994). Research on the history and philosophy of science. In D. Gabel (Ed.), *Handbook of research on science teaching and learning* (pp. 443–465). New York, NY: Macmillan.

Duschl, R. (2008). Science education in three-part harmony: Balancing conceptual, epistemic, and social learning goals. *Review of Research in Education, 32*, 268–291.

Duschl, R., & Gitomer, D. (1991). Epistemological perspectives on conceptual change: Implications for educational practice. *Journal of Research in Science Teaching, 28*(9), 839–858.

Duschl, R., & Hamilton, R. J. (Eds.). (1992). *Philosophy of science, cognitive psychology, and educational theory and practice*. Albany, NY: State University of New York Press.

Duschl, R., & Hamilton, R. J. (1998). Conceptual change in science and in the learning of science. In B. J. Fraser & K. G. Tobin (Eds.), *International handbook of science education. Part II* (pp. 1047–1065). Dordrecht: Kluwer.

Duschl, R., Hamilton, R., & Grady, R. E. (1990). Psychology and epistemology: Match or mismatch when applied to science education? *International Journal of Science Education, 12*(3), 230–243.

Ebenezer, J. V., & Haggerty, S. M. (1998). *Becoming a secondary school science teacher*. Columbus, OH: Merrill.

Eflin, J. T., Glennan, S., & Reisch, G. (1999). The nature of science: A perspective from the philosophy of science. *Journal of Research in Science Teaching, 36*, 107–116.

Egan, K. (1983). *Education and psychology. Plato, Piaget, and scientific psychology*. New York, NY: Teachers College Press.

Egan, K. (1986). *Teaching as story telling*. London, England: Routledge.

Egan, K. (1997). *The educated mind. How cognitive tools shape our understanding*. Chicago, IL: University of Chicago Press.

Egan, K. (2002). *Getting it wrong from the beginning. Our progressivist inheritance from Herbert Spencer, John Dewey, and Jean Piaget*. New Haven, CT: Yale University Press.

Egan, K. (2005a). Students' development in theory and practice: The doubtful role of research. *Harvard Educational Review, 75*(1), 25–42.

Egan, K. (2005b). *An imaginative approach to teaching*. San Francisco, CA: Jossey-Bass.

Eger, M. (1992). Hermeneutics and science education: An introduction. *Science & Education, 1*, 337–348.

Eger, M. (1993a). Hermeneutics as an approach to science: Part I. *Science & Education, 2*, 1–29.

Eger, M. (1993b). Hermeneutics as an approach to science: Part II. *Science & Education, 2*, 303–328.

Einstein, A. (1949). Remarks concerning the essays brought together in this cooperative volume. In P. A. Schilpp (Ed.), *Albert Einstein: Philosopher-scientist. The Library of Living Philosophers* (Vol. 7, pp. 665–688). Evanston, IL.

Eisner, E. (1985). *The educational imagination: On the design and evaluation of school programs* (2nd ed.). New York, NY: Macmillan.

Eisner, E. (1992). Curriculum ideologies. In P. W. Jackson (Ed.), *Handbook of research on curriculum* (pp. 302–326). New York, NY: Macmillan.

Elby, A. & Hammer, D. (2001). On the substance of a sophisticated epistemology. *Science Education, 85*(5), 554–567.

Elkana, Y. (1970). Science, philosophy of science, and science teaching. *Educational Philosophy and Theory, 2*, 15–35.

Englund, T. (1998). Problematizing school subject content. In D. A. Roberts & L. Oestman (Eds.), *Problems of meaning in science curriculum* (pp. 13–24). New York, NY: Teachers College Press.

Erickson, G. (2000). Research programmes and the student science learning literature. In R. Millar, J. Leach, & J. Osborne (Eds.), *Improving science education. The contribution of research* (pp. 271–292). Buckingham, England: Open University Press.

Erickson, G. (2007). In the path of Linnaeus: Scientific literacy re-visioned with some thoughts on persistent problems and new directions for science education. In C. Linder, L. Östman, & P. E. Wickman (Eds.), *Promoting scientific literacy: Science education research in transaction. Proceedings of the Linnaeus Tercentenary Symposium, Uppsala, Sweden, May 28–29* (pp. 18–41). Uppsala, Sweden: Geotryckeriet.

Ernest, P. A. (1991). *Philosophy of mathematics education*. London, England: Routledge-Falmer.

Fensham, P. (1997). School science and its problems with scientific literacy. In R. Levinson & J. Thomas (Eds.), *Science today. Problem or crisis?* (pp. 119–136). New York, NY: Routledge.

Fensham, P. (1998). The politics of legitimating and marginalizing companion meanings: Three Australian case studies. In D. A. Roberts & L. Oestman (Eds.), *Problems of meaning in science curriculum* (pp. 178–192). New York, NY: Teachers College Press.

Fensham, P. (2000). Providing suitable content in the "science for all" curriculum. In R. Millar, J. Leach, & J. Osborne (Eds.), *Improving science education: The contribution of research* (pp. 147–164). Buckingham, England: Open University Press.

Fensham, P. (2002). Time to change drivers for scientific literacy. *Canadian Journal of Science, Mathematics and Technology Education, 2*(1), 9–24.

Fensham, P. (2004). *Defining an identity: The evolution of science education as a field of research*. Dordrecht: Kluwer.

Feyerabend, P. (1981). *Realism, rationalism and scientific method: Philosophical papers, Vol. 1*. Cambridge, England: Cambridge University Press.

Feyerabend, P. (1988). *Against method* (Revised ed.). London, Engalnd: Verso.

Feynman, R. (1965). *The character of physical law*. Cambridge, MA: MIT Press

Finkelstein, N. (2005). Learning physics in context: A study of student learning about electricity and magnetism. *International Journal of Science Education, 27*(10), 1187–1209.

Forster, M. R. (2000). Hard problems in the philosophy of science: Idealization and commensurability. In R. Nola & H. Sankey (Eds.), *After Popper, Kuhn and Feyerabend. Recent issues in theories of scientific method* (pp. 231–250). Dordrecht: Kluwer.

Foucault, M. (1980). *Power/knowledge. Selected interviews and other writings 1972–77.* New York, NY: Pantheon Books.

Foucault, M. (1989). *The archaeology of knowledge.* London, England: Routledge. (Original work published 1972)

Frank, P. (1957). *Philosophy of science: The link between science and philosophy.* Englewood Cliff, NJ: Prentice Hall.

Frankena, W. K. (1965). *Three historical philosophies of education. Aristotle, Kant, Dewey.* Chicago, IL: Scott Foresman.

Fraser, B. J., & Tobin, K. G. (Eds). (1998). *International handbook of science education. Two volumes.* Dordrecht: Kluwer.

Fraser, B. J., Tobin, K. G., & McRobbie, C. J. (Eds.). (2012). *Second international handbook of science education.* New York, NY: Springer.

Freire, P. (1970). *Pedagogy of the oppressed.* New York, NY: Continuum.

Gabel, D. (Ed). (1994). *Handbook of research on science teaching and learning.* New York, NY: Macmillan.

Gadamer, H. (1975). *Truth and method* (2nd rev. ed.). J. Weinsheimer & D. G. Marshall (Trans. and Ed.). New York, NY: Continuum. (Original work published 1960)

Gadamer, H. (2008). *Philosophical hermeneutics.* D. Linge (Trans. and Ed.). Berkeley, CA: University of California Press. (Original work published 1976)

Galili, I., & Hazan, A. (2001). "Experts" views on using history and philosophy of science in the practice of physics instruction. *Science & Education, 10,* 345–367.

Galison, P., & Stump, D. J. (Eds.). (1996). *The disunity of science: Boundaries, contexts, and power.* Stanford, CA: Stanford University Press.

Gallagher, J. J. (1991). Prospective and practicing secondary school science teachers' knowledge and beliefs about the philosophy of science. *Science Education, 75*(1), 121–133.

Gallagher, S. (1992). *Hermeneutics and education.* Albany, NY: State University Press.

Garrison, J. W., & Bentley, M. L. (1990). Science education, conceptual change and breaking with everyday experience. *Studies in Philosophy and Education, 10*(1), 19–36.

Gaskell, J. (2002). Of cabbages and kings: Opening the hard shell of science curriculum policy. *Canadian Journal of Science, Mathematics and Technology Education, 2*(1), 59–66.

Gardner, H. (1991). *The unschooled mind: How children think and how schools should teach.* New York, NY: Basic Books.

Gay, P. (1996). *The enlightenment: An interpretation. The science of freedom.* New York, NY: W.W. Norton. (Original work published 1969)

Geddis, A.N. (1993). Transforming subject-matter knowledge: The role of pedagogical content knowledge in learning to reflect on teaching. *International Journal of Science Education, 15,* 673–683.

Geelan, D. (1997). Epistemological anarchy and the many forms of constructivism. *Science & Education, 6*(1-2), 15–28.

Gibbs, W. W. & Fox, D. (1999). The false crisis in science education. *Scientific American, 281*(4), 87–92.

Giere, R. N. (1991). *Understanding scientific reasoning* (3rd ed.). Orlando, FL: Harcourt Brace Jovanovich.

Giere, R. N. (1999). *Science without laws.* Chicago, IL: University of Chicago Press.

Giere, R. N. (2004). How models are used to represent reality. *Philosophy of Science, 71*, 742–752.

Giere, R. N. (2005). Scientific realism: Old and new problems. *Erkenntnis, 63*, 149–165.

Gilbert, J. K. (2006). On the nature of "context" in chemical education. *International Journal of Science Education, 28*(9), 957–976.

Gilbert, J. K., & Boulter, C. J. (Eds.). (2000). *Developing models in science education.* Dordrecht: Kluwer.

Gilbert, J. K., Justi, R., Van Driel, J., De Jong, O., & Treagust, D. F. (2004). Securing a future for chemical education. *Chemistry Education: Research and Practice, 5*(1), 5–14.

Gilead, T. (2011). The role of education redefined: 18th century British and French educational thought and the rise of the Baconian conception of the study of nature. *Educational Philosophy and Theory, 43*(10), 1020–1034.

Girod, M. (2007). A conceptual overview of the role of beauty and aesthetics in science and science education. *Studies in Science Education, 43*, 38–61.

Good, R., Herron, J., Lawson, A., & Renner, J. (1985). The domain of science education. *Science Education, 69*, 139–141.

Good, R., & Shymansky, J. (2001). Nature-of-science literacy in *Benchmarks* and *Standards*: Post-modern/relativist or modern/realist? In F. Bevilacqua, E. Giannetto, & M. Matthews (Eds.), *Science education and culture: The contribution of history and philosophy* (pp. 53–66). Dordrecht: Kluwer.

Grandy, R. (2009). Constructivisms, scientific methods, and reflective judgment in science education. In H. Siegel (Ed.), *The Oxford handbook of philosophy of education* (pp. 358–380). Oxford: Oxford University Press.

Gredler, M. E. (1992). *Learning and instruction. Theory into practice* (2nd ed.). New York, NY: Macmillan.

Gregory, B. (1988). *Inventing reality: Physics as language*. New York, NY: Wiley.

Gross, P., & Levitt, N. (1994). *Higher superstitions. The academic left and its quarrels with science.* Baltimore, MD: John Hopkins Press.

Gross, P. R., Levitt, N., & Lewis, M. W. (Eds.). (1996). *The flight from science and reason*. New York, NY: New York Academy of Sciences.

Gundem, B. B., & Hopmann, S. (Eds.). (1998). *Didaktik and/or curriculum. An international dialogue*. New York, NY: Peter Lang.

Gunstone, R., & White, R. (2000). Goals, methods and achievements of research in science education. In R. Millar, J. Leach, & J. Osborne (Eds.), *Improving science education: The contribution of research* (pp. 293–307). Buckingham, England: Open University Press.

Haack, S. (2003). *Defending science—within reason*: *Between scientism and cynicism.* Amherst, NY: Prometheus Books.

Habermas, J. (1968). *Knowledge and human interests.* Oxford, England: Polity Press.

Habermas, J. (1987a). Philosophy as stand-in and interpreter. In K. Baynes, J. Bohman, & T. McCarthy (Eds.), *After philosophy. End or transformation* (pp. 296–315). Cambridge, MA: MIT Press.
Habermas, J. (1987b). *The philosophical discourse of modernity* (F. G. Lawrence, Trans.). Cambridge, MA: MIT Press.
Hadzigeorgiou, Y. (2006). Humanizing the teaching of physics through storytelling: The case of current electricity. *Physics Education, 41*(1), 42–45.
Hadzigeorgiou, Y. (2008). Rethinking science education as socio-political action. In M. Tomase (Ed.), *Science education in focus* (pp. 203–224). New York, NY: Nova Pubs.
Hadzigeorgiou, Y., Klassen, S., & Froese-Klassen, C. (2011). Encouraging a "romantic understanding" of science: The effect of the Nikola Tesla story. *Science & Education, 21*(8), 1111–1138.
Hake, R. R. (1998). Interactive engagement versus traditional methods: A six-thousand- student survey of mechanics test data for introductory physics courses. *American Journal of Physics, 66*(1), 64–74.
Halloun, I. (1996). Schematic modeling for meaningful learning of physics. *Journal of Research in Science Teaching, 33,* 1019–1041.
Halloun, I. (2004). *Modeling theory in science education.* Dordrecht: Springer.
Hansen, K.-H., & Olson J. (1996). How teachers construe curriculum integration: the science, technology, society (STS) movement as Bildung. *Journal of Curriculum Studies, 28*(6), 669–682.
Harris, D. & Taylor, M. (1983). Discovery learning in school science: The myth and the reality. *Journal of Curriculum Studies, 15,* 277–289.
Hebden, J. A. (1998). *Chemistry 11. A workbook for students.* Kamloops, BC, Canada: Hebden.
Hecht, E. (1994). *Physics.* Pacific Grove, CA: Brooks/Coley.
Heelan, P. A. (1991). Hermeneutical phenomenology and the philosophy of science. In H. J. Silverman (Ed.), *Continental philosophy IV. Gadamer and hermeneutics. science, culture, literature* (pp. 213–228). New York, NY: Routledge.
Heidegger, M. (1977). The end of philosophy and the task of thinking. In M. Heidegger & D. Krell (Ed.), *Basic writings* (pp. 373–392). San Francisco, CA: HarperCollins.
Heidegger, M. (2008). *Towards the definition of philosophy.* (Ted Sadler, Trans.). London, England: Continuum. (Original work published 1919)
Hempel, C. (1966). *Philosophy of natural science.* Englewood, NJ: Prentice Hall.
Henry, N. B. (Ed.). (1955). *Modern philosophies and education: The fifty-fourth yearbook of the national society for the study of education.* Chicago, IL: University of Chicago Press.
Hesse, M. (1980). *Revolutions and reconstructions in the philosophy of science.* Brighton, Sussex: Harvester Press.
Hestenes, D. (1998). Who needs physics education research? *American Journal of Physics, 66,* 465–467.
Hestenes, D., Wells, M., & Swackhammer, G. (1992). Force concept inventory. *The Physics Teacher, 30,* 141–158.
Hickman, L. (2001). *Philosophical tools for a technological culture: Putting pragmatism to work.* Bloomington, IN: Indiana University Press.

Hiley, D. R., Bohman, J. F., & Shusterman, R. (Eds.). (1991). *The interpretative turn. Philosophy, science, culture.* Ithaca, NY: Cornell University Press.

Hirst, P. (1966). Educational theory. In J. W. Tibble (Ed.), *The study of education* (pp. 29–58). New York, NY: Routledge and Kegan Paul.

Hirst, P. (1974). *Knowledge and the curriculum.* London, England: Routledge.

Hirst, P. (2003). Foreword. In N. Blake, P. Smeyers, R. Smith & P. Standish (Eds.), *The Blackwell guide to the philosophy of education* (pp. xv–xvi). Oxford, England: Blackwell.

Hirst, P. (2008a). In pursuit of reason. In L. J. Waks (Ed.), *Leaders in philosophy of education. Intellectual self portraits* (pp. 113–124). Rotterdam: Sense.

Hirst, P. (2008b). Philosophy of education in the UK. In L. J. Waks (Ed.), *Leaders in philosophy of education. Intellectual self portraits* (pp. 305–310). Rotterdam: Sense.

Hodson, D. (1985). Philosophy of science, science and science education. *Studies in Science Education, 12,* 25–57.

Hodson, D. (1988). Toward a philosophically more valid science curriculum. *Science Education, 72*(1), 19–40.

Hodson, D. (1991). Philosophy of science and science education. In M. R. Matthews (Ed.), *History, philosophy and science teaching* (pp. 19–32). Toronto: OISE Press.

Hodson, D. (1993a). In search of a rationale for multicultural science education. *Science Education, 77*(6), 685–711.

Hodson, D. (1993b). Philosophic stance of secondary school science teachers, curriculum experiences, and children's understanding of science: Some preliminary findings. *Interchange, 24*(1&2), 41–52.

Hodson, D. (1994). Seeking directions for change: The personalisation and politicisation of science education. *Curriculum Studies, 2*(1), 71–98.

Hodson, D. (1996). Laboratory work as scientific method: Three decades of confusion and distortion. *Journal of Curriculum Studies, 28*(2), 115–135.

Hodson, D. (1998). Science fiction: The continuing misrepresentation of science in the school curriculum. *Curriculum Studies, 6*(2), 191–216.

Hodson, D. (2003). Time for action: Science education for an alternative future. *International Journal of Science Education, 25*(6), 645–670.

Hodson, D. (2008). *Towards scientific literacy. A teacher's guide to the history, philosophy and sociology of science.* Rotterdam: Sense.

Hodson, D. (2009). *Teaching and learning about science. Language, theories, methods, history, traditions and values.* Rotterdam: Sense.

Hofer, B. K., & Pintrich, P. R. (1997). The development of epistemological theories: Beliefs about knowledge and knowing and their relation to learning. *Review of Educational Research, 67*(1), 88–140.

Holton, G. (2003). What historians of science and science educators can do for one another. *Science & Education, 12,* 603–616.

Holton, G., & Brush, S. G. (2001). *Physics, the human adventure. From Copernicus to Einstein and beyond.* New Brunswick, NJ: Rutgers University Press. (Original work published 1972)

Höttecke, D., Henke, A., & Riess, F. (2012). Implementing history and philosophy in science teaching: Strategies, methods, results and experiences from the European HIPST project. *Science & Education, 21*, 1233–1261.

Höttecke, D. & Silva, C. C. (2011). Why implementing history and philosophy in school science education is a challenge. An analysis of the obstacles. *Science & Education, 20*, 293–316.

Hurd, P. D. (1994). New minds for a new age: Prologue to modernizing the science curriculum. *Science Education, 78*(1), 103–116.

Hurd, P. D. (2000). Science education for the 21st century. *School Science and Mathematics, 100*(6), 282–288.

Hyslop-Margison, E. J. & Naseem M. A. (2007). Philosophy of education and the contested nature of empirical research: A rejoinder to D. C. Phillips. *Philosophy of Education, Dec.*, 310–318.

Ihde, D. (1998). *Expanding hermeneutics. Visualism in science.* Evanston, IL: Northwestern University Press.

International Association for the Evaluation of Educational Achievement (IEA). (2003). *Third International Science and Mathematics Study 2003* (TIMSS). Retrieved from http://nces.ed.gov/TIMSS/publication.asp

Irzik, G. & Nola, R. (2009). Worldviews and their relation to science. In M. R. Matthews (Ed.), *Science, worldviews and education* (pp. 81–98). Dordrecht: Springer

Irzik, G., & Nola, R. (2011). A family resemblance approach to the nature of science for science education. *Science & Education, 20*, 591–607.

Izquierdo-Aymerich, M., & Aduriz-Bravo, A. (2003). Epistemological foundations of school science. *Science & Education, 12*, 47–43.

Jammer, M. (1961). *Concepts of mass in classical and modern physics.* Cambridge, MA: Harvard University Press.

Jenkins, E. (1992). School science education: Towards a reconstruction. *Journal of Curriculum Studies, 24*(3), 229–246.

Jenkins, E. (1994). Public understanding of science and science education for action. *Journal of Curriculum Studies, 26*(6): 601–611.

Jenkins, E. (1997). Scientific and technological literacy: Meanings and rationales. In E. Jenkins (Ed.), *Innovations in scientific and technological education* (Vol. VI, pp. 11–39). Paris: UNESCO.

Jenkins, E. (2000). "Science for all": time for a paradigm shift? In R. Millar, J. Leach, & J. Osborne (Eds), *Improving science education. The contribution of research* (pp. 207–226). Buckingham, England: Open University Press.

Jenkins, E. (2001). Science education as a field of research. *Canadian Journal of Science, Mathematics and Technology Education, 1*(1), 9–21.

Jenkins, E. (Ed.). (2003). *Innovations in scientific and technological education.* (Vol. VIII). Paris: UNESCO.

Jenkins, E. (2007). School science: A questionable construct? *Journal of Curriculum Studies, 39*(3), 265–282.

Jenkins, E. (2009). Reforming school science education: A commentary on selected reports and policy documents. *Studies in Science Education, 45*(1), 65–92.

Jung, W. (2012). Philosophy of science and education. *Science & Education, 21*(8).

Kalman, C. (2002). Developing critical thinking in undergraduate courses: A philosophical approach. *Science & Education, 11*, 83–94.

Kalman, C. (2008). *Successful science and engineering teaching. Theoretical and learning perspectives*. Dordrecht: Springer Verlag.

Kalman, C. (2009). The need to emphasize epistemology in teaching and research. *Science & Education, 18*, 325–347.

Kalman, C. (2010). Enabling students to develop a scientific mindset. *Science & Education, 19*(2), 147–164.

Kalman, C. (2011). Enhancing student's conceptual understanding by engaging science text with reflective writing as a hermeneutical circle. *Science & Education, 20*(2), 159–172.

Kant, I. (1974). "Beantwortung der Frage: Was ist Aufklärung?" In E. Bahr (Ed.), *Was ist Aufklärung? Thesen und Definitionen*. Stuttgart: Reklam. (Original work published 1784)

Keeves, J. P. & Aikenhead, G. S. (1995). Science curricula in a changing world. In B. J. Fraser & H. J. Walberg (Eds.), *Improving science education* (pp. 13–45). Chicago, IL: University of Chicago Press.

Kelly, G. J. (1997). Research traditions in comparative context: A philosophical challenge to radical constructivism. *Science Education, 81*, 355–375.

Kelly, G, (2007). Discourse in science classrooms. In S. K. Abell & N. G. Lederman (Eds.), *Handbook of research on science education* (pp. 443–469). Mahwah, NJ: Lawrence Erlbaum Associates.

Kelly, G. J., Carlsen, W. S., & Cunningham, C. M. (1993). Science education in sociocultural context: Perspectives from the sociology of science. *Science Education, 77*, 207–220.

Kelly, G. J., McDonald, S., & Wickman, P. O. (2012). Science learning and epistemology. In K. Tobin, B. Fraser, & C. McRobbie (Eds.), *Second international handbook of science education* (pp. 281–291). Dordrecht: Springer.

Kind, V. (2009). Pedagogical content knowledge in science education: Perspectives and potential for progress. *Studies in Science Education, 45*(2), 169–204.

Kirschner, P. A., Sweller, J., & Clark, R. E. (2006). Why minimal guidance during instruction does not work: An analysis of the failure of constructivist, discovery, problem-based, experiential, and inquiry-based teaching. *Educational Psychologist, 41*(2), 75–86.

Kitchener, R. (1992). Piaget's genetic epistemology: Epistemological implications for science education. In R. A. Duschl & R. J. Hamilton (Eds.), *Philosophy of science, cognitive psychology, and educational theory and practice* (pp. 116–145). Albany, NY: State University of New York Press.

Kitcher, P. (1998). 1953 and all that: A tale of two sciences. In M. Curd & J. A. Cover (Eds.), *Philosophy of science. The central issues* (pp. 971–1003). New York, NY: W.W. Norton.

Klaassen, C. W. J. M., & Lijnse, P. L. (1996). Interpreting students' and teachers' discourse in science classes: An underestimated problem? *Journal of Research in Science Teaching, 33*, 115–134.

Klassen, S. (2002). *A theoretical framework for the incorporation of history in science education* (Unpublished doctoral dissertation). University of Manitoba, Winnipeg.

Klassen, S. (2006). A theoretical frame for the contextual teaching of science. *Interchange, 37*(1/2), 31–62.

Klafki, W. (1995). Didactic analysis as the core of the preparation for instruction. *Journal of Curriculum Studies, 27*(1), 13–30.

Klopfer, L. E. & Champagne, A. B. (1990). Ghosts of crisis past. *Science Education, 74*(2), 133–153.

Knorr Cetina, K. (1999). *Epistemic cultures. How the sciences make knowledge*. Cambridge, MA: Harvard University Press.

Kokkotas, P. V., Malamitsa, K. S., & Rizaki, A. A. (Eds.). (2011). *Adapting historical knowledge production to the classroom.* Rotterdam: Sense.

Kozulin, A. (1998). *Psychological tools. A sociocultural approach to education.* Cambridge, MA: Harvard University Press.

Kruckeberg, R. (2006). A Deweyan perspective on science education: Constructivism, experience, and why we learn science. *Science & Education, 15*, 1–30.

Kubli, F. (2005). Science teaching as a dialogue—Bakhtin, Vygotsky and some applications in the classroom. *Science & Education, 14*, 501–534.

Kuhn, D. (1993). Science as argument: Implications for teaching and learning scientific thinking. *Science Education, 77*(3), 319–337.

Kuhn, T. (1970). *The structure of scientific revolutions* (2nd ed.). Chicago, IL: University of Chicago Press.

Kuhn, T. (1977). *The essential tension. Selected studies in scientific tradition and change.* Chicago, IL: University of Chicago Press.

Kuhn, T. (1985). *The Copernican revolution. Planetary astronomy in the development of western thought.* Cambridge, MA: Harvard University Press. (Original work published 1957)

Kuhn, T. (2000). *The road since structure: Philosophical essays, 1970–1993 with an autobiographical interview.* Edited by J. Conant & J. Haugeland. Chicago, IL: University of Chicago Press.

Kwan, Y., & Lawson, A. E. (2000). Linking brain growth with the development of scientific reasoning ability and conceptual change during adolescence. *Journal of Research in Science Teaching, 37*(1), 44–62.

Kyle, W. C. Jr., Abell, S. K., Roth, W.-M., & Gallagher, J. J. (1992). Toward a mature discipline of science education. *Journal of Research of Science Teaching, 29*, 1015–1018.

Lacey, H. (2008). Science, respect for nature, and human well-being. *Scientiae Studia, 8*(3), 297–327.

Ladyman, J. (2002). *Understanding philosophy of science*. London, England: Routledge.

Lakatos, I. (1970). Falsification and the methodology of scientific research programmes. In I. Lakatos & A. Musgrave (Eds.), *Criticism and the growth of knowledge* (pp. 91–196). Cambridge, England: Cambridge University Press.

Lakatos, I. & Musgrave, A. (Eds.). (1970). *Criticism and the growth of knowledge.* Cambridge: Cambridge University Press.

Lange, M. (2002). *An introduction to the philosophy of physics*. Malden, MA: Blackwell.

Langer, J. (1988). A note on the comparative psychology of mental development. In S. Strauss (Ed.), *Ontogeny, phylogeny and historical development* (pp. 68–85). Norwood, NJ: Ablex.

Laudan, L. (1990). *Science and relativism. Some key controversies in the philosophy of science*. Chicago, IL: University of Chicago Press.

Laudan, L. (1998a). A confutation of convergent realism. In M. Curd & J. A. Cover (Eds.), *Philosophy of science. The central issues* (pp. 1114–1135). New York, NY: W.W. Norton.

Laudan, L. (1998b). Dissecting the holist picture of scientific change. In M. Curd & J. A. Cover (Eds.), *Philosophy of science. The central issues* (pp. 139–169). New York, NY: W.W. Norton.

Laudan, L., Donovan, A., Laudan, R., Barker, P., Brown, H., Leplin, J., Thagard, P., & Wykstra, S. (1986). Scientific change: Philosophical models and historical research. *Synthese, 69*, 141–223.

Laugksch, R. (2000). Scientific literacy: A conceptual overview. *Science Education, 84*, 71–94.

Layton, D. (1993). *Technology's challenge to science education*. Buckingham, England: Open University Press.

Leach, J., & Scott, P. (2003). Individual and socio-cultural views of learning in science education. *Science & Education, 12*, 91–113.

Leahey, T. H. (2005). Mind as a scientific object: A historical-philosophical exploration. In C. E. Erneling & D. M. Johnson (Eds.), *The mind as a scientific object. Between brain and culture* (pp. 35–78). Oxford, England: Oxford University Press.

Lederman, N. G. (1998). The state of science education: Subject matter without context. *Electronic Journal of Science Education, 3(2)*. Retrieved from http://unr.edu/homepage/jcannon/ejse/lederman.html

Lederman, N. G. (2007). Nature of science: Past, present and future. In S. K. Abell & N. G. Lederman (Eds.), *Handbook of research on science education* (pp. 831–879). Mahwah, NJ: Lawrence Erlbaum Associates.

Lederman, N. G., & Abd-El-Khalick, F. (1998). Avoiding de-natured science: Activities that promote understandings of the nature of science. In W. F. McComas (Ed.), *The nature of science in science education: Rationales and strategies*. (pp. 83–126). Dordrecht: Kluwer.

Lee, M., Wu, Y., & Tsai, C. (2009). Research trends in science education from 2003 to 2007. A content analysis of publications in selected journals. *International Journal of Science Education, 31*(15), 1999–2020.

Lehrman, R. L. (1982). Physics texts: An evaluative review. *Physics Teacher, 20*, 508–523.

Lemke, J. (1990). *Talking science. Language, learning, values*. Norwood, NJ: Ablex.

Leonard, C. S. (2003). Response: Ideal or real: What is the nature of science? *Philosophy of Education, January*, 293–295.

Levinson, R. (2010). Science education and democratic participation: An uneasy congruence? *Studies in Science Education, 46*(1), 69–119.

Llewellyn, D. (2005). *Teaching high school science through inquiry. A case study approach*. Thousand Oaks, CA: Corwin-NSTA Press.

Litt, T. (1963). *Naturwissenschaft und Menschenbildung* (3rd ed.). (Science and education) Heidelberg: Quelle und Meyer.

Lijnse, P. (2000). Didactics of science: The forgotten dimension in science education research? In R. Millar, J. Leach, & J. Osborne (Eds.), *Improving science education: The contribution of research.* Buckingham: Open University Press.

Locke, J. (1964). Some thoughts concerning education. In P. Gay (Ed.), *John Locke on education* (pp. 19–176). New York, NY: Teachers College Press. (Original work published 1693)

Loving, C. C. (1997). From the summit of truth to its slippery slopes: Science education's journey through positivist-postmodern territory. *American Educational Research Journal, 34*(3), 421–452.

Lucas, C. (1972). *Our western educational heritage*. New York, NY: Macmillan.

Lyle, S. (2000). Narrative understanding: Developing a theoretical context for understanding how children make meaning in classroom settings. *Journal of Curriculum Studies, 32*(1), 45–63.

Lyotard, J.-F. (1984). *The postmodern condition. A report on knowledge*. G. Bennington & B. Massumi (Trans.). Minneapolis, MN: University of Minnesota Press. (Original work published 1979)

Martin, B. E., Kass, H., & Brouwer, W. (1990). Authentic science: A diversity of meanings. *Science Education, 74*(5), 541–554.

Mas, C. J. F., Perez, J. H., & Harris, H. H. (1987). Parallels between adolescents' conceptions of gases and the history of chemistry. *Journal of Chemical Education, 64*(7), 616–618.

Mason, C. L., & Gilbert, S. W. (2004). A science education research organization's perspective on reform in teaching undergraduate science. In D. W. Sunal, E. L. Wright, & J. B. Day (Eds.), *Reform in undergraduate science teaching for the 21st century* (pp. 301–316). Greenwich, CT: Information Age.

Mason, R. (2003). *Understanding understanding*. Albany, NY: State University of New York Press.

Matthews, M. (1989). History, philosophy and science teaching: A brief review. *Synthese, 80*, 1–7.

Matthews, M. (1991). Ernst Mach and contemporary science education reforms. In M. R. Matthews (Ed.), *History, philosophy and science teaching: Selected readings* (pp. 9–18). Toronto: OISE Press.

Matthews, M. (1994a). *Science teaching. The role of history and philosophy of science*. New York, NY: Routledge.

Matthews, M. R. (1994b). Philosophy of science and science education. In T. Husen & T. N. Postlethwaite (Eds.), *The international encyclopedia of education* (2nd ed., pp. 4461–4464). London: Pergamon Press.

Matthews, M. (1997). Scheffler revisited on the role of history and philosophy of science in science teacher education. In H. Siegel (Ed.), *Reason and education: Essays in honor of Israel Scheffler* (pp. 159–173). Dordrecht: Kluwer.

Matthews, M. (1998a). The nature of science and science teaching. In B. J. Fraser & K. G. Tobin (Eds.), *International handbook of science education. Part II* (pp. 981–999). Dordrecht: Kluwer.

Matthews, M. (1998b). In defense of modest goals when teaching about the nature of science. *Journal of Research in Science Teaching, 35*, 161–174.

Matthews, M. (2000). Appraising constructivism in science and mathematics education. In D. C. Phillips (Ed.), *Constructivism in education. Opinions and second opinions on controversial issues* (pp. 161–192). Chicago, IL: University of Chicago Press.

Matthews, M. (2003a). Thomas Kuhn's impact on science education: What can be learned? *Science Education, 88*(2), 90–118.

Matthews, M. (2003b). Data, phenomena, and theory: How clarifying the concepts can illuminate the nature of science. *Philosophy of Education, January*, 283–292.

Matthews, M. (2007). Models in science and in science education: An introduction. *Science & Education*, Special Issue. doi:10.1007/s11191-007-9089-3

Matthews, M. (Ed.). (2009a). *Science, worldviews and education*. Dordrecht: Springer.

Matthews, M. (2009b). Book notes: Review of Fensham's (2004) *Defining an identity*. Newsletter of the International History, Philosophy and Science Teaching Group, May 2009, pp. 21–39. Retrieved from http://ihpst.net/newsletters/may2009.pdf

Matthews, M. (2014). *Science teaching. The role of history and philosophy of science.* (20th anniversary and expanded edition). New York, NY: Routledge.

Matthews, M. R. (1988). A role for history and philosophy in science teaching. *Educational Philosophy and Theory, 20*(2), 67–75.

Matthews, M. R. (2002). Teaching science. In R. R. Curren (Ed.), *A companion to the philosophy of education* (pp. 342–353). Oxford, Englalnd: Blackwell.

Maurines, L., & Beaufils, D. (2013). Teaching the nature of science in physics courses: The contribution of classroom historical inquiries. *Science & Education, 22*, 1443–1466.

McCarty, L. P. (1990). Philosophy, Dewey and education: The Deweyan perspective. In D. E. Herget (Ed.), *More history and philosophy of science in science teaching* (pp. 13–20). Tallahassee, FL: Florida State University.

McCloskey, M. (1983). Intuitive physics. *Scientific American, 248*, 122–130.

McComas, W. F. (1998). The principal elements of the nature of science: Dispelling the myths. In W. F. McComas (Ed.), *The nature of science in science education: Rationales and strategies* (pp. 53–69). Dordrecht: Kluwer.

McComas, W. F., Almazroa, H., & Clough, M. P. (1998). The nature of science in science education: An introduction. *Science & Education, 7*, 511–532.

McComas, W. F., & Olson, J. K. (1998). The nature of science in international science education standards documents. In W. F. McComas (Ed.), *The nature of science in science education: Rationales and strategies* (pp. 41–52). Dordrecht: Kluwer.

McDermott, L. C. (1984). Research on conceptual understanding in mechanics. *Physics Today, 37*, 24–32.

McDermott, L. C. (1991). Millikan lecture 1990: What we teach and what is learned—Closing the gap. *American Journal of Physics, 59*, 301–315.

McDermott, L. C., Heron, P. R. L., Shaffer, P. S., & MacKenzie, R. S. (2006). Improving the preparation of K–12 teachers through physics education research. *American Journal of Physics, 74*(9), 763–767.

McDermott, L. C. & Reddish, E. F. (1999). Resource letter: PER-1: physics education research. *American Journal of Physics, 67*(9), 755–767.

McDonnell, F. (2005). Editorial—Why so few choose physics: An alternative explanation for the leaky pipeline. *American Journal of Physics, 73*(3), 583–586.

McEneaney, E. (2003). Worldwide cachet of scientific literacy. *Comparative Education Review, 47*(2), 217–237.
McFarland, E., & Hirsch, A. (1992). Physics teaching in Canada. *The Physics Teacher, 30,* 226–234.
McIntyre, L. (2007). The philosophy of chemistry: Ten years later. *Synthese, 155,* 291–292.
McKee, J. (2005). Editorial-physics, science and the changing university. *Physics in Canada, 68,* 1–6.
McMullin, E. (1998). Rationality and paradigm change in science. In M. Curd & J. A. Cover (Eds.), *Philosophy of science. The central issues* (pp. 119–138). New York, NY: W.W. Norton.
Medina, J. (2005). *Language. Key concepts in philosophy.* New York, NY: Continuum.
Meichstry, Y. (1993). The impact of science curricula on students' views about the nature of science. *Journal of Research in Science Teaching, 39*(5), 429–443.
Mercan, F. C. (2012). Epistemic beliefs about justification employed by physics students and faculty in two different problem contexts. *International Journal of Science Education, 34*(8), 1411–1441.
Merzyn, G. (1987). The language of school science. *International Journal of Science Education, 9*(4), 483–469.
Mestre, J. P. (1991). Learning and instruction in pre-college physical science. *Physics Today, 44*(9), 56–62.
Metz, D., Klassen, S., McMillan, B., Clough, M., & Olson, N. (2007). Building a foundation for use of historical narratives. *Science & Education, 16,* 313–334.
Millar, R. (1997). Science education for democracy: What can the school curriculum achieve? In R. Levinson & J. Thomas (Eds.), *Science today. Problem or crisis?* (pp. 87–101). New York, NY: Routledge.
Millar, R., Leach, J., & Osborne, J. (Eds.). (2000). *Improving science education. The contribution of research.* Buckingham, England: Open University Press.
Millar, R., & Osborne, J. (Rds.) (1998). *Beyond 2000: Science educationfor the future.* London, England: King's College School of Education Press.
Miller, J. D. (1996). Scientific literacy for effective citizenship. In R. E. Yager (Ed.), *Science/technology/society as reform in science education* (pp. 185–204). Albany, NY: State University Press.
Miller, J. D. (1998). The measurement of civic scientific literacy. *Public Understanding of Science, 7*(3), 203–223.
Mitchell, K. (2001). Education for democratic citizenship: Transnationalism, multiculturalism, and the limits of liberalism. *Harvard Educational Review, 71*(1), 51–78.
Moll, L. C. (Ed.). (1990). *Vygotsky and education. Instructional implications and applications of sociohistorical psychology.* Cambridge, England: Cambridge University Press.
Monk, M., & Osborne, J. (1997). Placing the history and philosophy of science on the curriculum: A model for the development of pedagogy. *Science Education, 81*(4), 405–424.
Morris, R. (1997). *Achilles in the quantum universe. The definitive history of infinity.* New York, NY: Henry Holt.

Mortimer, E. F., & Scott, P. H. (2000). Analysing discourse in the science classroom. In R. Millar, J. Leach, & J. Osborne (Eds.), *Improving science education: The contribution of research* (pp. 126–142). Buckingham, Engalnd: Open University Press.

Munby, A. H. (1976). Some implications for language in science education. *Science Education, 60*, 115–124.

Nadeau, R., & Désautels, J. (1984). *Epistemology and the teaching of science*. Ottawa: Science Council of Canada.

Nakhleh, M. B. (1992). Why some students don't learn chemistry. *Journal of Chemical Education, 69*(3), 191–196.

Nakhleh, M. B. & Mitchell, R. C. (1993). Concept learning versus problem-solving: There is a difference. *Journal of Chemical Education, 70*(3), 190–192.

Nashon, M. N., & Nielsen, W. (2007). Participation rates in physics 12 in BC: Science teachers' and students' views. *Canadian Journal of Science, Mathematics and Technology Education,* 7(2/3), 93–106.

Nashon, S., Nielsen, W., & Petrina, S. (2008). Whatever happened to STS? Pre-service physics teachers and the history of quantum mechanics. *Science & Education, 17*, 387–401.

National Research Council (NRC). (1996). *National science education standards*. Washington, DC: National Academy Press.

National Science Foundation (NSF). (1996). *Shaping the future. New expectations for undergraduate education in science, mathematics, engineering and technology*. (U.S. NSF 96-139). Retrieved from http://www.edr.nsf.gov/HER/DUE/documents/reviews/96139/threea.htm

Nersessian, N. (1989). Conceptual change in science and in science education. *Synthese, 80*(1), 163–184.

Nersessian, N. (1992). Constructing and instructing: The role of "abstraction techniques" in creating and learning physics. In R. A. Duschl & R. J. Hamilton (Eds.), *Philosophy of science, cognitive psychology, and educational theory and practice* (pp. 48–67). Albany, NY: State University of New York Press.

Nersessian, N. (1995). Should physicists preach what they practice? Constructive modeling in doing and learning physics. *Science & Education, 4*, 203–226.

Nersessian, N. (1998). Conceptual change. In W. Bechtel & G. Graham (Eds.), *A companion to cognitive science* (pp. 155–166). Malden, MA: Blackwell.

Nersessian, N. (2003). Kuhn, conceptual change, and cognitive science. In T. Nickles (Ed.), *Thomas Kuhn* (pp. 178–211). Cambridge, England: Cambridge University Press.

Niaz, M. (1993). "Progressive problem shifts" between different research programs in science education: A Lakatosian perspective. *Journal of Research in Science Teaching, 30*(7), 757–765.

Niaz, M. (2000). The oil-drop experiment: A rational reconstruction of the Millikan Ehrenhaft controversy and its implications for chemistry textbooks. *Journal of Research in Science Teaching, 37*(5), 480–508.

Niaz, M. (2002). How in spite of the rhetoric, history of chemistry has been ignored in presenting atomic structure in textbooks. *Science & Education, 11*, 423–441.

Niaz, M. (2005). Do general chemistry textbooks facilitate conceptual understanding? *Quimica Nova, 28*(2), 335–336.

Niaz, M. (2009). Progressive transitions in chemistry teachers' understanding of nature of science based on historical controversies. *Science & Education, 18,* 43–65.

Niaz, M. (2010). Science curriculum and teacher education: The role of presuppositions, contradictions, controversies and speculations vs. Kuhn's "normal science." *Teaching and Teacher Education, 26,* 891–899.

Niaz, M., Abd-El-Khalick, F., Benarroch, A., Cardellini, L., Laburu, C. E., & Marin, N. (2003). Constructivism: Defense or a continual critical appraisal—A response to Gil-Perez et al. *Science & Education, 12,* 787–797.

Niaz, M. & Rodriguez, M. A. (2001). Do we have to introduce history and philosophy of science or is it already "inside" chemistry? *Chemistry education: Research and practice in Europe, 2*(2), 159–164.

Nielsen, H., & Thomsen, P. V. (1990). History and philosophy of science in physics education. *International Journal of Science Education, 12*(3), 308–316.

Noddings, N. (2011). *Philosophy of education* (3rd ed.). Boulder, CO: Westview Press.

Nola, R. (1980). Fixing the reference of theoretical terms. *Philosophy of Science, 47,* 505–531.

Nola, R. (Ed.). (1988). *Relativism and realism in science. Australasian studies in the history and philosophy of science* (Vol. 6). Dordrecht: Kluwer.

Nola, R. (1994). Post-modernism, a French cultural Chemobyl: Foucault on power/knowledge. *Inquiry, 37,* 3–43.

Nola, R. (1997). Constructivism in science and science education: A philosophical critique. *Science & Education, 6*(1/2), 55–83.

Nola, R. (2001). Saving Kuhn from the sociologists of science. In F. Bevilacqua, E. Giannetto, & M. Matthews (Eds.), *Science education and culture: The contribution of history and philosophy* (pp. 213–226). Dordrecht: Kluwer.

Nola, R. & Irzik, G. (2005). *Philosophy, science, education and culture*. Dordrecht: Springer.

Nola, R., & Sankey, H. (Eds.). (2000). *After Popper, Kuhn and Feyerabend. Recent issues in theories of scientific method*. Dordrecht: Kluwer.

Norris, C. (1997). *Against relativism. Philosophy of science, deconstruction and critical theory.* Oxford, England: Blackwell.

Norris, S. P. (1984). Cynicism, dogmatism, relativism, and scepticism: Can all these be avoided? *School Science and Mathematics, 84,* 484–495.

Norris, S. P. (1995). Learning to live with scientific expertise: Towards a theory of intellectual communalism for guiding science teaching. *Science Education, 79,* 201–217.

Norris, S. P. (1997). Intellectual independence for non-scientists and other content-transcendent goals of science education. *Science Education, 81,* 239–258.

Norris, S. P., Guilbert, S. M., Smith, M. L., Hakimelahi, S., & Phillips, L. M. (2005). A theoretical framework for narrative explanation in science. *Science Education, 89,* 535–563.

Norris, S. & Kvernbekk, T. (1997). The application of science education theories. *Journal of Research in Science Teaching, 34*(10), 977–1005.

Novak, J. (1977). *A theory of education.* Ithaca, NY: Cornell University Press.

Novak, J. (1978). An alternative to Piagetian psychology for science and mathematics education. *Science Education, 61*, 453–477.

Nyberg, D. & Egan, K. (1981). *The erosion of education. Socialization and the schools*. New York, NY: Teachers College Press.

Ogborn, J. (1995). Recovering reality. *Studies in Science Education, 25*, 3–38.

Olesko, K. M. (2006). Science pedagogy as a category of historical analysis: Past, present and future. *Science & Education, 15*, 863–880.

O'Meara, J. M. (2006). Motivations and influences in choice of undergraduate science major at the University of Guelph. *Physics in Canada*, 19–24.

Organization for Economic Co-operation and Development (OECD). (2004). PISA 2003. Retrieved from http://www.oecd.org/pisa

Orpwood, G. (1998). The logic of advice and deliberation: Making sense of science curriculum talk. In D. A. Roberts & L. Oestman (Eds.), *Problems of meaning in science curriculum* (pp. 133–149). New York, NY: Teachers College Press

Osborne, J. (1996). Beyond constructivism. *Science Education, 80*(1), 53–82.

Osborne, J., Collins, S., Millar, R., & Duschl, R. (2003). What "ideas-about-science" should be taught in school science? A Delphi study of the expert community. *Journal of Research in Science Teaching, 40*(7), 692–720.

Osborne, J. & Dillon, J. (2008). *Science education in Europe: Critical reflections. A report to the Nuffield Foundation*. London, England: Kings College.

Osborne, J., Erduran, S., & Simon, S. (2004). Enhancing the quality of argumentation in school science. *Journal of Research in Science Teaching, 41*(10), 994–1020.

Osborne, R., & Freyberg, P. (1985). *Learning in science. The implications of children's science*. Hong Kong: Heinemann.

Palmer, D. (1997). The effect of context on students' reasoning about forces. *International Journal of Science Education, 19*(6), 681–696.

Pedretti, E. G., Bencze, L., Hewitt, J., Romkey, L., & Jivraj, A. (2008). Promoting issues-based STSE perspectives in science teacher education: Problems of identity and ideology. *Science & Education, 17*, 941–960.

Pedretti, E. G., & Nazir, J. (2011). Currents in STSE education: Mapping a complex field, 40 years on. *Science Education, 95*, 601–626.

Peters, R. S. (1966). The philosophy of education. In J. W. Tibble (Ed.), *The study of education* (pp. 59–89). New York, NY: Routledge and Kegan Paul.

Phillips, D. C. (1997). Coming to terms with radical social constructivisms. *Science & Education, 6*(112), 85–104.

Phillips, D. C. (Ed.). (2000). *Constructivism in education. Opinions and second opinions on controversial issues*. Chicago, IL: NSSE, University of Chicago Press.

Phillips, D. C. (2002). Theories of teaching and learning. In R. R. Curren (Ed.), *A companion to the philosophy of education* (pp. 233–237). Malden, MA: Blackwell.

Phillips, D. C. (2005). The contested nature of empirical research. *Journal of Philosophy of Education, 39*(4), 577–597.

Phillips, D. C. (2007). Getting it wrong from the beginning, but maybe (just maybe) it's a start. *Philosophy of Education*, 319–322.

Phillips, D. C. (2008). Philosophy of education. *Stanford encyclopedia of philosophy*. Retrieved from http://plato.stanford.edu/entries/education-philosophy/

Phillips, D. C. (2009). Empirical educational research: Charting philosophical disagreements in an undisciplined field. In H. Siegel (Ed.), *The Oxford handbook of philosophy of education* (pp. 381–408). Oxford & New York: Oxford University Press.

Phillips, D. C. (2010). What is philosophy of education? In R. Bailey, R. Barrow, D. Carr, & C. McCarthy (Eds.), *The SAGE handbook of the philosophy of education* (pp. 3–19). Thousand Oaks, CA: SAGE.

Plato. (1970). *The Laws*. (T. Saunders, Trans.). New York, NY: Penguin Classics.

Plato. (1974). *The Republic* (2nd ed.). (D. Lee, Trans.). New York, NY: Penguin Classics.

Plato. (1975). *Protagoras and Meno*. (W. K. C. Guthrie, Trans.). Penguin Classics.

Polito, T. (2005). Educational theory as theory of culture: A Vichian perspective on the educational theories of John Dewey and Kieran Egan. *Educational Philosophy and Theory, 37*(4), 475–494.

Popper, K. (2002). *Conjectures and refutations. The growth of scientific knowledge*. New York, NY: Routledge. (Original work published 1963)

Posner, G. J. (1998). Models of curriculum planning. In L. E. Beyer & M. W. Apple (Eds.), The curriculum: Problems, politics and possibilities (2nd ed., pp. 79–100). Albany, NY: Suny Press.

Posner, G., Strike, K., Hewson, P., & Gertzog, W. (1982). Accommodation of a scientific conception: Toward a theory of conceptual change. *Science Education, 66*(2), 211–227.

Pring, R. (2010). The philosophy of education and educational practice. In R. Bailey, R. Barrow, D. Carr, & C. McCarthy (Eds.), *The SAGE handbook of the philosophy of education* (pp. 56–66). Thousand Oaks, CA: SAGE.

Psillos, S. (1999). *Scientific realism. How science tracks truth*. London, England: Routledge.

Psillos, S. (2000). The present state of the scientific realism debate. *British Journal for the Philosophy of Science, 51,* 705–728.

Psillos, D., Kariotoglou, P., Tselfes, V., Hatzikraniotis, E., Fassoulopoulos, G., & Kallery, M. (Eds.). (2003). *Science education research in the knowledge-based society*. Dordrecht: Kluwer.

Pyle, A. (2000). The rationality of the chemical revolution. In R. Nola & H. Sankey (Eds.), *After Popper, Kuhn and Feyerabend. Recent issues in theories of scientific method* (pp. 99–124). Dordrecht: Kluwer.

Redish, E. A. (1999). Millikan lecture 1998: Building a science of teaching physics. *American Journal of Physics, 67*(7), 562–573.

Reid, W. A. (1998). Systems and structures or myths and fables? A cross-cultural perspective on curriculum content. In B. B. Gundem & S. Hopmann (Eds.), *Didaktik and/or curriculum. An international dialogue* (pp. 11–27). New York, NY: Peter Lang.

Reiss, F. (2000). Problems with German science education. *Science & Education, 9,* 327–331.

Rigden, J., & Tobias, S. (1991). Tune in, turn off, drop out. Why do so many college students abandon science after introductory courses? *The Sciences, January,* 16–20.

Roberts, D. A. (1982). Developing the concept of "curriculum emphases" in science education. *Science Education, 66*, 243–260.

Roberts, D. A. (1988). What counts as science education? In P. Fensham (Ed.), *Development and dilemmas in science education* (pp. 27–54). Philadelphia, PA: Falmer.

Roberts, D. A. (2007). Scientific literacy/science literacy. In S. K. Abell & N. G. Lederman (Eds.), *Handbook of research on science education* (pp. 729–780). Mahwah, NJ: Lawrence Erlbaum Associates.

Roberts, D. A., & Oestman, L. (Eds.). (1998). *Problems of meaning in science curriculum*. New York, NY: Teachers College Press.

Roberts, D. A., & Russell, T. L. (1975). An alternative approach to science education: Drawing from philosophical analysis to examine practice. *Curriculum Theory Network, 5*(2), 107–125.

Robertson, E. (2009). The epistemic aims of education. In H. Siegel (Ed.), The Oxford handbook of philosophy of education (pp. 11–34). Oxford & New York, NY: Oxford University Press.

Robinson, J. T. (1969). Philosophy of science: Implications for teacher education. *Journal of Research in Science Teaching, 6*, 99–104.

Rockmore, T. (2004). *On foundationalism. A strategy for metaphysical realism*. Oxford: Rowman & Littlefield.

Rodriguez, M. A., & Niaz, M. (2002). How in spite of the rhetoric, history of chemistry has been ignored in presenting atomic structure in textbooks. *Science & Education, 11*, 423–441.

Rogers, P. J. (1982). Epistemology and history in the teaching of school science. *European Journal of Science Education, 4*(1), 1–10.

Rohrlich, F. (1988a). Four philosophical issues essential for good science teaching. *Educational Philosophy and Theory, 20*(2), 1–6.

Rohrlich, F. (1988b). Pluralistic ontology and theory reduction in the physical sciences. *British Journal of the Philosophy of Science, 39*, 295–312.

Rorty, R. (1984). Habermas and Lyotard on postmodernity. In I. Hoesterey (Ed.), *Zeitgeist in babel. The postmodernist controversies* (pp. 84–97). Indiana Press.

Rorty, A. O. (1998). The ruling history of education. In A. O. Rorty (Ed.), *Philosophers on education. Historical perspectives* (pp. 1–13). London, England: Routledge.

Rorty, R. (1979). *Philosophy and the mirror of nature*. Princeton, NJ: Princeton University Press.

Rorty, R. (1982). Pragmatism and philosophy. In R. Rorty, *Consequences of pragmatism* (xii–xlvii). Minneapolis, MN: University of Minneapolis Press.

Roscoe, K., & Mrazek, R. (2005). *Scientific literacy for Canadian students. Curriculum, instruction, and assessment*. Calgary, Alberta: Detselig.

Roth, W.-M.. & Barton, A. (2004). *Rethinking scientific literacy*. New York, NY: Routledge-Falmer.

Roth, W.-M., & Désautels, J. (Eds.). (2002). *Science education as/for sociopolitical action*. New York, NY: Peter Lang.

Roth, W.-M., & Lucas, K. (1997). From "truth" to "invented reality": A discourse analysis of high school physics students' talk about scientific knowledge. *Journal of Research in Science Teaching, 34*(2), 145–179.

Roth, W.-M., & McGinn, M. (1998). Knowing, researching and reporting science education: Lessons from science and technology studies. *Journal of Research in Science Teaching, 35*(2), 213–235.

Roth, W.-M., McRobbie, C. J., Lucas, K. B., & Boutonne, S. (1997). Why may students fail to learn from demonstrations? A social practice perspective on learning in physics. *Journal of Research in Science Teaching, 34*(5), 509–533.

Roth, W.-M., & Roychoudhury, A. (1994). Physics students' epistemologies and views about knowing and learning. *Journal of Research in Science Teaching, 31*(1), 5–30.

Rousseau, J.-J. (1979). *Emile* (A. Bloom, Trans.). New York, NY: Basic Books. (Original work published 1762)

Rowlands, S. (2000). Turning Vygotsky on his head: Vygotsky's "scientifically based method" and the socioculturalist's "social other." *Science & Education, 9,* 537–575.

Rowlands, S., Graham, T., & Berry, J. (2011). Problems with fallibilism as a philosophy of mathematics education. *Science & Education, 20*(7/8), 625–686.

Rudolph, J. L. (2000). Reconsidering the "nature of science" as a curriculum component. *Journal of Curriculum Studies, 32*(3), 403–419.

Rudolph, J. L. (2002). Portraying epistemology: School science in historical context. *Science Education, 87*, 64–79.

Ruse, M. (1990). Making use of creationism. A case study for the philosophy of science classroom. *Studies in Philosophy of Education, 10*(1), 81–92.

Ryan, A. G., & Aikenhead, G. S. (1992). Students' preconceptions about the epistemology of science. *Science Education, 76*, 559–580.

Säther, J. (2003). The concept of ideology in analysis of fundamental questions in science education. *Science & Education, 12*, 237–260.

Scerri, E. (2001). The new philosophy of chemistry and its relevance to chemical education. *Chemistry Education: Research and Practice in Europe, 2*(2), 165–170.

Scerri, E. (2003). Philosophical confusion in chemical education research. *Journal of Chemical Education, 80*(5), 468–474.

Scerri, E. (2007). *The periodic table. Its story and its significance*. Oxford: Oxford University Press.

Scerri, E., & McIntyre, L. (1997). The case for the philosophy of chemistry. *Synthese, 111*, 213–232.

Scharff, R. C. (Ed.). (2002). *Philosophy of technology: The technological tradition. An anthology.* Malden, MA: Blackwell.

Scheffler, I. (1992). Philosophy and the curriculum. Reprinted in *Science & Education, 1*(1), 71–76. (Original work published 1970)

Sherman, W. (2004). Science studies, situatedness, and instructional design in science education. A summery and critique of the promise. *Canadian Journal of Science, Mathematics and Technology Education, 4*, 443–465.

Sherman, W. (2005). A reply to Roth. *Canadian Journal of Science, Mathematics and Technology Education, 5*(2), 199–207.

Schiller, F. (1993). Letters on the aesthetic education of man. In F. Schiller Essays (Edited by W. Hinderer and D. 0. Dahlstrom; pp. 86–178). New York, NY: Continuum. (Original work published 1795)

Schmidt, L. K. (2006). *Understanding hermeneutics*. Stocksfield, UK: Acumen.

Schulz, R. M. (2007). Lyotard, postmodernism and science education. A rejoinder to Zembylas. *Educational Philosophy and Theory, 39*(6), 633–656.

Schulz, R. M. (2009a). Reforming science education: Part I. The search for a *philosophy* of science education. *Science & Education, 18*, 225–249.

Schulz, R. M. (2009b). Reforming science education: Part II. Utilizing Kieran Egan's educational metatheory. *Science & Education, 18*, 251–273.

Schulz, R. M. (2011). Developing a philosophy of science education. In F. Seroglou, V. Koulountzos, & A. Siatras (Eds.), *Science and culture: Promise, challenge and demand—Proceedings of the 11th IHPST & 6th Greek History, Philosophy & Science Teaching Joint Conference, July 2011, Thessaloniki, Greece* (pp. 672–677). Thessaloniki: Epikentro.

Schulz, R. M., & Sivia, A. (2007, May). *Developing "theoretic thinking": using history, model-based reasoning and epistemology to reform science education*. Paper presented at the UNESCO sponsored 2nd Advanced International Colloquium on "Building the Scientific Mind." Retrieved from http://www.learndev.org/dl/BtSM2007/RolandSchulz-AwneetSivia.pdf

Schwab, J. J. (1962). The teaching of science as enquiry. In J. J. Schwab & P. F. Brandwein, *The teaching of science* (pp. 3–103). Cambridge, MA: Harvard University Press.

Schwab, J. J. (1978). *Science, culture and liberal education. Selected essays*. Chicago, IL: Chicago University Press.

Schwartz, R. S., Lederman, N. G., & Crawford, B. A. (2004). Developing views of nature of science in an authentic context: An explicit approach to bridging the gap between nature of science and scientific inquiry. *Science Education, 88*(4), 610–645.

Scott, D. (2008). *Critical essays on major curriculum theorists.* New York, NY: Routledge.

Scott, P., Asoko, H., & Leach, J. (2007). Student conceptions and conceptual learning in science. In S. K. Abell & N. G. Lederman (Eds.), *Handbook of research on science education* (pp. 31–56). Mahwah, NJ; Lawrence Erlbaum Associates.

Selley, N. J. (1989). The philosophy of school science. *Interchange, 20*(2), 24–32.

Shamos, M. H. (1995). *The myth of scientific literacy*. New Brunswick, NJ: Rutgers University Press.

Shulman, L. S. (1986). Paradigms and research programs in the study of teaching: A contemporary perspective. In M. C. Wittrock (Ed.), *Handbook of research in teaching* (3rd ed. pp. 3–36). New York, NY: Macmillan.

Shulman, L. S. (1987). Knowledge and teaching: Foundations of the new reform. *Harvard Educational Review, 57*(1), 1–22.

Shymansky, J. & Kyle, W. C., Jr. (1992). Establishing a research agenda: Critical issues of science curriculum reform. *Journal of Research in Science Teaching, 29*(8), 749–778.

Siegel, H. (1978). Kuhn and Schwab on science texts and the goals of science education. *Educational Theory, 28*(4), 302–309.

Siegel, H. (1979). On the distortion of the history of science in science education. *Science Education, 63*(1), 111–118.

Siegel, H. (1987a). Farewell to Feyerabend. *Inquiry, 32,* 343–369.

Siegel, H. (1987b). *Relativism refuted*. Dordrecht: Reidel.

Siegel, H. (1988). *Educating reason: Rationality, critical thinking and education.* London, England: Routledge.

Siegel, H. (1989). The rationality of science, critical thinking, and science education. *Synthese, 80*(1), 9–32.

Siegel, H. (1992). Two perspectives on reason as an educational aim: The rationality of reasonableness. *Philosophy of Education*, 225–233.

Siegel, H. (Ed.). (1997). *Reason and education: Essays in honor of Israel Scheffler.* Dordrecht: Kluwer Academic Publishers.

Siegel, H. (2001). Incommensurability, rationality, and relativism: In science, culture and science education. In P. Hoyningen-Huene & H. Sankey (Eds.), *Incommensurability and related matters* (pp. 207–224). Dordrecht: Kluwer.

Siegel, H. (2003). Cultivating reason. In R. R. Curren (Ed.), *A companion to the philosophy of education* (pp. 305–319). Oxford, England: Blackwell.

Siegel, H. (2007). Philosophy of education. In *Encyclopædia Britannica*. Retrieved from http://www.britannica.com/EBchecked/topic/179491/philosophy-of-education

Siegel, H. (Ed.). (2009). *The Oxford handbook of philosophy of education*. Oxford, England: Oxford University Press.

Siegel, H. (2010). Knowledge and truth. In R. Bailey, R. Barrow, D. Carr, & C. McCarthy (Eds.), *The SAGE handbook of the philosophy of education* (pp. 283–295). Thousand Oaks, CA: SAGE.

Siemsen, H., & Siemsen, K. H. (2009). Resettling the thoughts of Ernst Mach and the Vienna Circle in Europe: The cases of Finland and Germany. *Science & Education, 18*, 299–323.

Simon, J. (Ed.). (2013). Thematic special issue: Cross-national and comparative history of science education. *Science & Education, 22*(4).

Sjøberg, S. (2003). Science and technology education in Europe: Current challenges and possible solutions. In E. Jenkins (Ed.), *Innovations in scientific and technological education* (Vol. VIII, pp. 201–228). Paris: UNESCO.

Sjöström, J. (2013). Towards *Bildung*-oriented chemistry education. *Science & Education, 22*(7), 1873–1890.

Slezak, P. (1994a). Sociology of scientific knowledge and scientific education: Part I. *Science & Education, 3*(3), 265–294.

Slezak, P. (1994b). Sociology of scientific knowledge and science education. Part II: laboratory life under the microscope. *Science & Education, 3*(4), 329–355.

Slezak, P. (2007). Is cognitive science relevant to teaching? *Journal of Cognitive Science, 8*, 171–208.

Slezak, P., & Good, R. (Eds.). (2011). Thematic special issue: Pseudoscience in society and schools. *Science & Education, 10*(5/6).

Smeyers, P. (1994). Philosophy of education: Western European perspectives. In T. Husen & T. Postlethwaite (Eds.), *The international encyclopedia of education* (2nd ed., Vol. 8, pp. 4456–4461). London, England: Pergamon Press.

Smith, M., & Siegel, H. (2004). Knowing, believing and understanding: What goals for science education? *Science and Education, 13*(6), 553–582.

Smolicz, J. J., & Nunan, E. E. (1975). The philosophical and sociological foundations of science education: The demythologizing of school science. *Studies in Science Education, 2*, 101–143.

Smolin, L. (2006). *The trouble with physics. The rise of string theory, the fall of a science and what comes next.* New York, NY: Mariner Books.

Snively, G., & MacKinnon, A. (Eds.). (1995). *Thinking globally about mathematics and science education.* Vancouver, BC: University of British Columbia.

Snook, I. A. (Ed.). (1972). *Concepts of indoctrination*. London, England: Routledge & Kegan Paul.

Sokal, A. (1996a). Transgressing the boundaries: Toward a transformative hermeneutics of quantum gravity. *Social Text, 4617*(14.1–2), 217–252.

Sokal, A. (1996b). A physicist experiments with cultural studies. *Lingua Franca. (July/August)*, 62–64.

Sokal, A., & Bricmont, J. (1998). *Fashionable nonsense. Postmodern intellectuals' abuse of science.* New York, NY: Picador.

Solomon, J. (1999). Meta-scientific criticisms, curriculum innovation and the propagation of scientific culture. *Journal of Curriculum Studies, 31*(1), 1–15.

Solomon, J., & Aikenhead, G. (Eds.). (1994). *STS education. International perspectives on reform.* New York, NY: Teachers College Press.

Southerland, S. A., Sinatra, G. M., & Matthews, M. R. (2001). Belief, knowledge, and science education. *Educational Psychology Review, 13*(4), 325–351.

Stenhouse, D. (1986). Conceptual change in science education: Paradigms and language games. *Science Education, 70*(4), 413–425.

Stewart, J. (1995). *Language as articulate contact: Toward a post-semiotic philosophy of communication*. Albany, NY: State University of New York Press.

Stewart, J. (Ed.). (1996). *Beyond the symbol model: Reflections on the representational nature of language.* Albany, NY: State University of New York Press.

Stinner, A. (1989). Science, humanities and society—the Snow-Leavis controversy. *Interchange, 20(2),* 16–23.

Stinner, A. (1992). Science textbooks and science teaching: From logic to evidence. *Science Education, 76*, 1–16.

Stinner, A. (1995a). Contextual settings, science stories, and large context problems: Towards a more humanistic science education. *Science Education, 79*(5), 555–581.

Stinner, A. (1995b). Science textbooks: Their present role and future form. In S. M. Glynn & R. Duit (Eds.), *Learning science in the schools: Research reforming practice* (pp. 275–296). Mahwah, NJ: Lawrence Erlbaum Associates

Stinner, A. (2001). Linking "the book of nature" and "the book of science": Using circular motion as an *exemplar* beyond the textbook. *Science & Education, 10*, 323–344.

Stinner, A., Mcmillan, B., Metz, D., Jilek, J. M., & Klassen, S. (2003). The renewal of case studies in science education. *Science & Education, 12*, 617–643.

Strike, K. A., & Posner, G. J. (1992). A revisionist theory of conceptual change. In R. A. Duschl & R. J. Hamilton (Eds.), *Philosophy of science, cognitive psychology, and educational theory and practice* (pp. 147–176). Albany, NY: State University of New York Press.

Suchting, W. (1992). Constructivism deconstructed. *Science & Education, 1*(3), 223–254.

Suchting, W. A. (1995). Much ado about nothing: Science and hermeneutics. *Science & Education, 4*(2), 161–171.

Sula, T. (2009). Unified view of science and technology for education: Technoscience and technoscience education. *Science & Education, 18*, 275–298.

Sullenger, K., Turner, S., Caplan, H., Crummey, J., Cuming, R., Charron, C., & Corey, B. (2000). Culture wars in the classroom: Prospective teachers question science. *Journal of Research in Science Teaching, 37*(9), 895–915.

Summers, M. K. (1982). Philosophy of science in the science teacher education curriculum. *International Journal of Science Education, 4*(1), 19–27.

Sunal, D. W., Wright, E. L., & Day, J. (Eds.). (2004). *Reform in undergraduate science teaching for the 21st century.* Greenwich, CT: Information Age.

Suppe, P. (1977). *The structure of scientific theories*. Chicago, IL: University of Chicago Press.

Sutton, C. (1996). Beliefs about science and beliefs about language. *International Journal of Science Education, 18*(1), 1–18.

Sutton, C. (1998). New perspectives on language in science. In B. J. Fraser & K. G. Tobin (Eds.), *International handbook of science education, Part I* (pp. 27–38). Dordrecht: Kluwer.

Tasker, R. & Osborne, R. (1985). Facing the mismatches in the classroom. In R. Osborne & P. Freyberg (Eds), *Learning in science. The implications of children's science* (pp. 66–80). Hong Kong: Heinemann.

Taylor, C. (1987). Overcoming epistemology. In K. Baynes, J. Bohman, & T. McCarthy (Eds.), *After philosophy. End or transformation?* (pp. 464–485). London, England: MIT Press.

Taylor, C. (1991). *The malaise of modernity*. Toronto: Anansi Press.

Thomas, G. (1997). What's the use of theory? *Harvard Educational Review, 67*, 75–104.

Thomsen, P. V. (1998). The historical-philosophical dimensions in physics teaching: Danish experiences. *Science & Education, 7*, 493–503.

Tibble, J. W. (Ed.). (1966). *The study of education.* New York, NY: Routledge and Kegan Paul.

Tobin, K., McRobbie, C., & Anderson, D. (1997). Dialectical constraints to the discursive practices of a high school physics community. *Journal of Research in Science Teaching, 34*(5), 491–507.

Toulmin, S. (Ed.). (1970). *Physical reality. Philosophical essays on twentieth-century physics*. New York, NY: Harper & Row.

Toulmin, S. E. (1972). Human understanding. Princeton, NJ: Princeton University Press.

Tsaparlis, G. (2001). Editorial: theories in science education at the threshold of the third millennium. *Chemical Education Research and Practice in Europe, 2*, 1–4.

Turner, S. & Sullenger, K. (1999). Kuhn in the classroom, Lakatos in the lab: Science educators confront the nature-of-science debate. *Science, Technology and Human Values, 24*(1), 5–30.

Ungerleider, C. (2003). *Failing our kids. How we are ruining our public schools*. Toronto: McClelland & Stewart.

Van den Akker, J. (1998). The science curriculum: Between ideals and outcomes. In B. J. Fraser & K. G. Tobin (Eds.), *International handbook of science education: Part I* (pp. 421–448). Dordrecht: Kluwer.

Van Driel, J. H. & Abell, S. (2010). Science teacher education. In D. Peterson, E. Baker, & B. McGraw (Eds.), *International encyclopedia of education* (pp. 712–718). New York, NY: Elsevier.

Van Driel, J. H., Verloop, N., & de Vos, W. (1998). Developing science teachers' pedagogical content knowledge. *Journal of Research in Science Teaching, 35*(6), 673–695.

Van Fraassen, B. (1998). Arguments concerning scientific realism. In M. Curd & J. A. Cover (Eds.), *Philosophy of science. The central issues* (pp. 1064–1087). New York, NY: W.W. Norton.

Vásquez-Levy, D. (2002). Essay review. *Bildung*-centred Didaktik: a framework for examining the educational potential of subject matter. *Journal of Curriculum Studies, 34*(1), 117–128.

Von Glasersfeld, E. (1989). Cognition, construction of knowledge and teaching. *Synthese, 80*(1), 121–140.

Vygotsky, L. S. (1978). *Mind in society. The development of higher psychological processes.* Cambridge, MA: Harvard University Press.

Vygotsky, L. S. (1986). *Thought and language.* London: MIT Press.

Waks, L. J. (Ed.). (2008). *Leaders in philosophy of education. Intellectual self-portraits.* Rotterdam: Sense Publishers.

Walker, D. F. (2003). *Fundamentals of curriculum. Passion and professionalism* (2nd ed.). Mahwah, NJ: Lawrence Erlbaum and Associates.

Walker, D. F., & Soltis, J. F. (1997). *Curriculum and aims* (3rd ed.). New York, NY: Teachers College Press. (Original work published 1986)

Wallace, J., & Louden, W. (1998). Curriculum change in science: Riding the waves of reform. In B. J. Fraser & K. G. Tobin (Eds.), *International handbook of science education: Part I* (pp. 471–486). Dordrecht: Kluwer.

Wandersee, J. H., Mintzes, J. L., & Novak, J. D. (1994). Research on alternative conceptions in science. In D. Gabel (Ed.), *Handbook of research on science teaching and learning* (pp. 177–210). New York, NY: Macmillan.

Weinberg, S. (1992). *Dreams of a final theory.* New York, NY: Pantheon Books.

Weaver, A., Morris, M., & Appelbaum, P. (Eds.). (2001). *(Post)modern science (education). Propositions and alterative paths.* Albany, NY: State University of New York Press.

Welch, W. W., Klopfer, L., Aikenhead, G., & Robinson, J. (1981). The role of inquiry in science education: Analysis and recommendations. *Science Education, 65*, 33–50.

Wellington, J., & Osborne, J. (2001). *Language and literacy in science education.* Buckingham: Open University Press.

Wertsch, J. V. (1985). *Vygotsky and the social formation of mind.* Cambridge, MA: Harvard University Press.

Westfall, R. S. (1971). *The construction of modern science. Mechanism and mechanics.* Cambridge, England: Cambridge University Press.

Westphal, M. (1999). Hermeneutics as epistemology. In J. Greco & E. Sosa (Eds.), *The Blackwell guide to epistemology* (pp. 415–435). Malden, MA: Blackwell.

Whitehead, A. N. (1957). *The aims of education and other essays.* New York, NY: Free Press. (Original work published 1929)

Winch, C. & Gingell, J. (1999). *Key concepts in the philosophy of education.* New York, NY: Routledge.

Wittgenstein, L. (1958). *Philosophical investigations* (3rd ed.). (G. E. M. Anscomb, Trans.). Englewood Cliffs, NJ.: Prentice Hall.

Witz, K. (1996). Science with values and values for science education. *Journal of Curriculum Studies, 28*(5), 597–612.

Witz, K. (2000). OP-ED: The "academic problem." *Journal of Curriculum Studies, 32*(1), 9–23.

Witz, K., & Lee, H. (2009). Science as an ideal: Teachers' orientations to science and science education reform. *Journal of Curriculum Studies, 41*(3), 409–431.

Wolpert, L. (1992). *The unnatural nature of science.* London, England: Faber & Faber.

Wong, D., Pugh, K., & the Dewey Ideas Group. (2001). Learning science: The Deweyan perspective. *Journal of Research in Science Teaching, 38*(3), 317–336.

Yager, R. E. (1984). Defining the discipline of science education. *Science Education, 68*(1), 35–37.

Yager, R. E. (Ed.). (1996). *Science/technology/society as reform in science education.* Albany, NY: State of New York University Press.

Yore, L. D., & Treagust, D. F. (2006). Current realities and future possibilities: Language and science literacy—empowering research and informing instruction. *International Journal of Science Education, 28*(2/3), 291–314.

Zeidler, D.L. and Sadler, T.D. (2008). Social and ethical issues in science education: A prelude to action. *Science & Education, 77*(8-9), 799–803.

Zeidler, D. L., Sadler, T. D., Simmons, M. L., & Howes, E. V. (2005). Beyond STS: A research-based framework for socioscientific issues education. *Science Education, 89,* 357–377.

Zembylas, M. (2000). Something "paralogical" under the sun: Lyotard's postrnodem condition and science education. *Educational Philosophy and Theory, 32,* 159–184.

Zembylas, M. (ed.) (2006). Special issue: Philosophy of science education. *Educational Philosophy and Theory, 38*(5).

Zoller, U. (1990). Students' misunderstandings and misconceptions in college freshman chemistry. *Journal of Research in Science Teaching, 27*(10), 1053–1065.

ABOUT THE AUTHOR

Roland M. Schulz is a researcher with the *Imaginative Education Research Group* (IERG) at Simon Fraser University in Vancouver, and a sessional instructor in science teacher education. He holds a PhD in curriculum theory focusing on philosophy of education, philosophy of science and science education. His undergraduate degrees are in physics and in physical science education. He has been active as a secondary science teacher, having taught physics and chemistry for many years in Canada, and physics in Istanbul, Turkey. His research interests include science education reform, philosophy of education, teacher education, hermeneutics and language theory, and the use of narrative and models to incorporate the history and philosophy of science for improving curriculum and instruction. Previous publications have appeared in the international journals *Science & Education* and *Educational Philosophy and Theory*.

CPSIA information can be obtained at www.ICGtesting.com
Printed in the USA
LVOW04*2240240914

405757LV00004B/12/P

9 781623 967154